TA 169 .K43 1991 v.2
Kececioglu, Dimitri.
Reliability engineering
 handbook

S0-ARI-199

Reliability Engineering Handbook

Volume 2

Books by Dimitri Kececioglu, Ph.D., P.E.

Reliability Engineering Handbook Volume 1
Reliability Engineering Handbook Volume 2

Reliability Engineering Handbook

Volume 2

Dimitri Kececioglu, Ph.D., P.E.

Department of Aerospace and Mechanical Engineering
The University of Arizona

PTR Prentice Hall, Englewood Cliffs, New Jersey 07632

Library of Congress Cataloging-in-Publication Data

D, HIDEN RAMSEY LIBRARY
U.N.C. AT ASHEVILLE
ASHEVILLE, N. C. 28804

Kececioglu, Dimitri.
    Reliability engineering handbook.

    Includes bibliographical references and indexes.
    1. Reliability (Engineering)--Handbooks, manuals,
etc. I. Title.
TA169.K43 1991   620'.00452   90-14277
ISBN 0-13-772294-X (v. 1)
ISBN 0-13-772302-4 (v. 2)

Editorial/production supervision: *Harriet Tellem*
Cover design: *Lundgren Graphics, Ltd.*
Manufacturing buyer: *Kelly Behr/Susan Brunke*
Acquisitions editor: *Mike Hays*

© 1991 by PTR Prentice-Hall, Inc.
A Simon & Schuster Company
Englewood Cliffs, New Jersey 07632

The publisher offers discounts on this book when ordered
in bulk quantities. For more information, write:

    Special Sales/Professional Marketing
    Prentice-Hall, Inc.
    Professional & Technical Reference Division
    Englewood Cliffs, New Jersey 07632

All rights reserved. No part of this book may be
reproduced, in any form or by any means,
without permission in writing from the publisher.

Printed in the United States of America
10 9 8 7 6 5 4 3

ISBN 0-13-772302-4

Prentice-Hall International (UK) Limited, *London*
Prentice-Hall of Australia Pty. Limited, *Sydney*
Prentice-Hall Canada Inc., *Toronto*
Prentice-Hall Hispanoamericana, S.A., *Mexico*
Prentice-Hall of India Private Limited, *New Delhi*
Prentice-Hall of Japan, Inc., *Tokyo*
Simon & Schuster Asia Ptc. Ltd., *Singapore*
Editora Prentice-Hall do Brasil, Ltda., *Rio de Janeiro*

To my wonderful wife Lorene,
daughter Zoe,
and son John.

CONTENTS

PREFACE

NEED FOR RELIABILITY ENGINEERING

Reliability engineering is a very fast growing and very important field in consumer and capital goods industries, in space and defense industries, and in NASA and DOD agencies. Reliability engineering provides the theoretical and practical tools whereby the probability and capability of parts, components, products, and systems to perform their required functions in specified environments for the desired period without failure can be specified, designed-in, predicted, tested, and demonstrated, and the results fed back to engineering, manufacturing, quality control, inspection, testing, packaging, shipping, purchasing, receiving, sales, and service for improvements and necessary corrective actions.

This book is intended to be used as a first text in reliability engineering by upper-level undergraduate and first-year graduate students, and as a working book by the practicing reliability engineer or the initiate. Great emphasis has been placed on clarity of presentation and practicality of subject matter. Some knowledge of the mathematics of probability and statistics would be helpful but is not absolutely necessary, as the needed mathematical background and tools are presented in the text whenever deemed necessary.

It is hoped that, in the not too distant future, deans of colleges of engineering will require at least one course in probability and statistics of all their students, so that students taking reliability engineering courses would already have the necessary mathematical background. Furthermore, most engineering courses today recognize the fact that nature and everything in it is probabilistic and statistical in nature; consequently, stochastic approaches have to be incorporated in them. It is similarly hoped that engineering students will be required to take at least one course in reliability engineering based on the contents of this book, because what good is it to design and build a product which meets the requirements of performance specifications during the relatively short period of the input-output test and of the efficiency of performance test, only to fail shortly thereafter or before the required period of function or mission is consummated. To optimally minimize such product failures, most engineers should be knowledgeable of reliability engineering. This book is intended to fulfill this objective.

It is hoped that efforts will be extended to attract students from all fields of engineering, as well as from operations research, mathematics, statistics, probability, and physics, into reliability engineering, because this field is interdisciplinary, the demand for qualified reliability engineers is great, and their supply is meager.

For any country's industry and technology to survive in today's highly competitive world marketplace, it is essential for all countries to know the reliability of their products and be able to control it, and produce products at their optimum reliability level that yields the minimum life-cycle cost to their users.

HOW TO USE THIS BOOK

The material in the two volumes of this book is intended to be used by reliability, maintainability and product assurance engineers and practitioners and as a college textbook, preferably covered in two regular semesters. Every major section within each chapter has a statement of its objectives and what the reader is expected to learn. Theoretical development is followed by illustrative examples. Each chapter has at least five problems to be worked on by the practitioner or the student. All tables necessary to solve the problems are given in the appropriate chapter or in the Appendixes at the end of this book for the convenience of the user, so that he or she will not have to hunt for them elsewhere. Each chapter is completely referenced, including the specific pages involved, for further in-depth study by the reader.

Volume 1 should be covered first, in a one semester course in reliability engineering, and is for practitioners, seniors and graduate students. Chapter 1 provides the important objectives of reliability engineering, and it may be covered in two sessions. Chapter 2 covers the history of reliability engineering and may be left up to the reader to pursue. Chapter 3 defines reliability in great detail. Chapter 4 quantifies the concepts of times-to-failure distributions, reliability, conditional reliability, failure rate, and mean life and provides the necessary background in statistics and probability which may not be covered by the instructor but may be made required reading for the students. Chapters 5 through 11 cover the most frequently used distributions in reliability engineering in great detail: exponential, Weibull, normal, lognormal, extreme value, Rayleigh, and uniform. Chapter 12 covers early, chance (useful life), and wear-out reliability, as experienced in the three life periods of components, equipment and systems. Chapter

13 combines the three types of life characteristics into a unified approach of quantifying the full-life behavior of components, equipment and systems. Finally, Chapter 14 provides five unique models for quantifying the reliability bath-tub curve which enables the determination of the full life behavior of components, equipment and systems; their burn-in, breaking-in and de-bugging period; their spare parts provisioning; their preventive maintenance schedules; the effects of different stress levels of use; the effects of corrective and preventive maintenance policies; the effects of quality control and preventive maintenance efficiency and concistency; and the effects of different field use strategies on the reliability and life characteristics of all types of units.

Volume 2 should be used as a follow-up to Volume 1. Chapters 1 through 4 cover the prediction of equipment and system reliability for the series, parallel, standby, and conditional function configuration cases and should preferably be covered in the first semester course in reliability engineering, in conjunction with Volume 1. The remaining chapters in Volume 2 are for practitioners and graduate students who have been exposed to Volume 1 material already. Chapter 5 covers the prediction of the reliability of complex components, equipment, and systems; Chapter 6, multimode function and logic; Chapter 7, multistress level of function; Chapter 8, load sharing function mode; Chapter 9, static switches; Chapter 10, cyclic switches; and Chapter 11, fault tree analysis. Chapter 12 is unique in that it covers five practical and very comprehensive case histories of predicting equipment and system reliabilities and comparing them with their reliability goals. Chapter 13 covers Drenick's theorem of complex systems times-to-failure distributions. Chapter 14 covers the reliability of components with a policy of replacing those that fail by a prescribed operating time. Chapter 15 covers methods of allocation or apportionment of an equipment's or system's reliability goal to its subsystems, all the way down to its components. Chapter 16 covers reliability growth and test-analyze-and-fix models to quantify when the MTBF and reliability goals of products under development will be attained, and finally Chapter 17 covers failure modes, effects, and criticality analysis (FAMECA) methods to identify design improvement areas.

It is recommended that instructors teaching this course assign three to five problems every week for homework, the specific number depending on the degree of difficulty of the problems assigned, to be handed in within a week. After the homework is corrected and returned it should be discussed in class and all subtleties in the solutions brought out.

ACKNOWLEDGMENTS

The author would like to thank all his colleagues and friends for making this book possible. Special thanks are due to the Allis-Chalmers Manufacturing Company for starting him in the field of reliability and maintainability engineering; Mr. Igor Bazovsky for his personal support of the author's activities in these fields, for helping make the *Annual Reliability Engineering and Management Institutes* a success, and for helping lay the foundation for his book; Dr. Austin Bonis, Dr. Leslie W. Ball, Dr. Myron Lipow, and Dr. David K. Lloyd for their inspiration to write this book; and to Dr. Harvey D. Christensen, former Head, Aerospace and Mechanical Engineering Department, for starting the instruction of reliability and maintainability engineering courses, Dr. Walter J. Fahey, former Dean, College of Engineering, Dr. Lawrence B. Scott, Jr. and Dr. J.T. Chen, past Heads, Aerospace and Mechanical Engineering Department, Dr. Ernest T. Smerdon, present Dean, College of Engineering and Mines, Dr. Steve Crow, present Head of Aerospace and Mechanical Engineering Department, and Dr. Pitu Mirchandani, present Head, Systems and Industrial Engineering Department, all of The University of Arizona, for staunchly supporting the Reliability Engineering Program at this university; the many companies and government agencies he consulted for who enabled him to gather the practical material included in this book and for inspiring him to arrive at the effective format of his book; the many reviewers for their suggestions and help; and the many outstanding graduate students who worked under the author and helped work out many of the problems.

The author is deeply indebted to his wife Lorene June Kececioglu, his daughter Zoe Diana Kececioglu/Draelos, M.D. in Dermatology, and his son Dr. John Dimitri Kececioglu, Ph.D. in Computer Science.

The author is indebted greatly to the numerous undergraduate students who took his reliability engineering courses at The University of Arizona and to over 90 graduate students who got their Master's Degree in reliability engineering and their Ph.D. Degree with a reliability engineering minor under him, and who made many suggestions to improve coverage and correct errors, and in particular to Dr. Dingjun Li, Dr. Siyuan Jiang, and Mr. Phuong Hung Nguyen, his outstanding graduate students, who contributed substantially to several chapters in this book, to the theory therein and to the problems whose solutions they helped work out, and who keyed in the manuscript in the LATEX language and made it camera ready, including all figures and tables.

The author is much indebted to the many secretaries who tirelessly typed the original manuscript, and in particular to Florence Conant, Dorothy A. Long and Dolores L. Meyer. Mrs. Long contributed the lion's share with her superspeed typing and accurate work.

The Prentice Hall staff were very cooperative and contributed much to the imaginative format and to the excellence of this book.

Dr. Dimitri Kececioglu
Tucson, Arizona

Chapter 1

RELIABILITY OF SERIES SYSTEMS

1.1 N UNIT RELIABILITYWISE SERIES SYSTEM

A system is said to have units which are reliabilitywise in series when the failure of any one or more units before the mission is completed results in the failure of that system. Conversely, for a series system to succeed for the duration of the intended mission, all of its units have to succeed. The reliability block diagram of a series system is as shown in Fig. 1.1. Probabilistically, the reliability of this system is the probability that Unit 1 succeeds, or does not fail, during the mission, and Unit 2 succeeds, or does not fail, during the mission, ..., and that Unit n succeeds, or does not fail, during the mission. Mathematically, the reliability of this series system, R_{ss}, is given by

$$R_{ss} = R_1 \times R_2 \times R_3 \times \cdots \times R_n = \prod_{i=1}^{n} R_i, \qquad (1.1)$$

where $R_1, R_2, \ldots, R_n$ are the reliabilities of Unit 1, Unit 2, ..., Unit n, respectively. This expression assumes that the $R_i's$ are independent. If they are dependent, then the parameters in the pdf of each unit should be adjusted to reflect the effects, if any, of the application and operation stresses imposed thereupon by all units in the series system. Then Eq. (1.1) will apply.

Fig. 1.1 – The reliability block diagram of a reliabilitywise
series system.

1.2 EXPONENTIAL UNITS

If the distribution of the times to failure of such units is the exponential,
then each has a constant failure rate, λ_i, and the reliability, R_i, of each
unit is given by

$$R_i(t) = e^{-\int_0^t \lambda_i(t)dt} = e^{-\lambda_i \int_0^t dt} = e^{-\lambda_i t},$$

where λ_i are the respective failure rates of these units and t is the
mission time.

The reliability of the series system for a mission of t duration,
$R_{ss}(t)$, is then given by

$$R_{ss}(t) = e^{-\lambda_1 t}e^{-\lambda_2 t} \dots e^{-\lambda_n t} = e^{-(\sum\limits_{i=1}^{n} \lambda_i)t}, \tag{1.2}$$

where $\sum\limits_{i=1}^{n} \lambda_i$ is the failure rate of the series system, λ_{ss}, or

$$\lambda_{ss} = \sum_{i=1}^{n} \lambda_i. \tag{1.3}$$

The mean time between failures of the series system is given by the
first moment of the system's failure distribution function, or by

$$\text{MTBF}_{ss} = \int_0^\infty T f_{ss}(T)\, dT,$$

where $f_{ss}(T)$ is the series system's failure probability density function,
or by

$$\text{MTBF}_{ss} = \int_0^\infty R_{ss}(t)\, dt. \tag{1.4}$$

Using the latter approach, we get

$$\text{MTBF}_{ss} = \int_0^\infty e^{-(\sum_{i=1}^{n} \lambda_i)t} \, dt,$$

$$\text{MTBF}_{ss} = \int_0^\infty e^{-\lambda_{ss}t} \, dt,$$

or

$$\text{MTBF}_{ss} = \frac{1}{\lambda_{ss}} = \frac{1}{\sum_{i=1}^{n} \lambda_i}. \tag{1.5}$$

The failure probability density function of the reliabilitywise series system, $f_{ss}(T)$, is given by

$$f_{ss}(T) = -\frac{d[R_{ss}(T)]}{dT} = -\frac{d[e^{-\lambda_{SS}T}]}{dT},$$

or

$$f_{ss}(T) = \lambda_{ss}e^{-\lambda_{SS}T}. \tag{1.6}$$

1.3 WEIBULLIAN UNITS

If the times to failure of the units in the system are Weibull distributed, then the series system's reliability is given by

$$R_{ss}(T) = \prod_{i=1}^{n} e^{-(\frac{T-\gamma_i}{\eta_i})^{\beta_i}} = e^{-\sum_{i=1}^{n}(\frac{T-\gamma_i}{\eta_i})^{\beta_i}}. \tag{1.7}$$

The *pdf* of times to failure of the series system is given by

$$f_{ss}(T) = -\frac{d[R_{ss}(T)]}{dT},$$

or

$$f_{ss}(T) = \left[\sum_{i=1}^{n} \frac{\beta_i}{\eta_i}(\frac{T-\gamma_i}{\eta_i})^{\beta_i-1} \right] e^{-\sum_{i=1}^{n}(\frac{T-\gamma_i}{\eta_i})^{\beta_i}}. \tag{1.8}$$

The failure rate function of the series system is given by

$$\lambda_{ss}(T) = \frac{f_{ss}(T)}{R_{ss}(T)},$$

or

$$\lambda_{ss}(T) = \sum_{i=1}^{n} \frac{\beta_i}{\eta_i} \left(\frac{T - \gamma_i}{\eta_i} \right)^{\beta_i - 1}. \tag{1.9}$$

The mean time to failure of the series system is then given by

$$\text{MTTF}_{ss} = \int_{\gamma}^{\infty} R_{ss}(T) \, dT.$$

For a system of two different Weibullian units in series, the mean time to the first failure, MTTF, is given in Case 4 of Table 1.1, and for n Weibullian units in series, with identical shape parameter, β, and location parameter, γ, the MTTF is given in Case 6 of Table 1.1. Additional cases are given in Table 1.1. The derivations of the results for Cases 2, 4, 6, and 7 are given in Appendix 1A.

EXAMPLE 1-1

A system consists of three units which function reliabilitywise in series. The failure rates of these units are as follows:

λ_1 = 3.5 failures per million hours,

or

λ_1 = 3.5 fr/10^6 hr,

which is also equal to

λ_1 = 0.35 percent failures per thousand hours,

or

λ_1 = 0.35% fr/1,000 hr, or λ_1 = 0.35%/1,000 hr.

These are some of the various ways failure rates may be found to be given in the literature. See Table 4.8 in Vol. 1 of this handbook for more ways. Also,

λ_2 = 16 fr/10^6 hr and λ_3 = 1.2 fr/10^6 hr.

Determine the following parameters of this system:

1. Failure rate, $\lambda_{ss}(T)$.

2. Reliability, $R_{ss}(t)$, for a 100-hr mission.

3. Mean time between failures, MTBF_{ss}.

4. Failure probability density function, $f_{ss}(T)$.

TABLE 1.1 – Equations for the calculation of the MTBF of various systems with units that are reliabilitywise in series.

Case number	System structure	MTTF	Comment
1	$\lambda(T) = \lambda$	$\dfrac{1}{\lambda}$	Single, constant failure rate unit.
2	$\lambda(T) = kT$	$\left(\dfrac{\pi}{2k}\right)^{\frac{1}{2}}$	Single, linearly increasing failure rate unit. In the Rayleigh pdf $k = \dfrac{2}{\eta^2}$. In the Weibull pdf $\beta = 2$, $k = \dfrac{2}{\eta^2}$, and $\gamma = 0$.
3	$\lambda(T) = \dfrac{\beta}{\eta}\left(\dfrac{T-\gamma}{\eta}\right)^{\beta-1}$	$\gamma + \eta\Gamma\left(\dfrac{1}{\beta} + 1\right)$ $\Gamma(n) = \displaystyle\int_{0}^{\infty} e^{-x}\, x^{n-1}\, dx$ is the gamma function.	Single, Weibull failure rate unit.

5

TABLE 1.1 – (Continued)

Case number	System structure	MTTF	Comment
4	$\lambda_1(T) = \dfrac{\beta_1}{\eta_1}\left(\dfrac{T-\gamma_1}{\eta_1}\right)^{\beta_1-1}$, $\lambda_2(T) = \dfrac{\beta_2}{\eta_2}\left(\dfrac{T-\gamma_2}{\eta_2}\right)^{\beta_2-1}$, $\gamma_2 < \gamma_1$. 	$\gamma_2 + \dfrac{\eta_2}{\beta_2} G\left[\dfrac{1}{\beta_2}, \left(\dfrac{\gamma_1-\gamma_2}{\eta_2}\right)^{\beta_2}\right] +$ $\displaystyle\int_{\gamma_1}^{\infty} e^{-\left[\left(\frac{T-\gamma_1}{\eta_1}\right)^{\beta_1}+\left(\frac{T-\gamma_2}{\eta_2}\right)^{\beta_2}\right]}\, dT$, where $G(n, z) = \displaystyle\int_0^z e^{-x}\, x^{n-1}\, dx$ is the incomplete gamma function. If $\gamma_1 = \gamma_2 = \gamma$ and $\beta_1 = \beta_2 = \beta$, then the MTTF becomes $\gamma + \Gamma\left(\dfrac{1}{\beta}+1\right) / \left[\left(\dfrac{1}{\eta_1}\right)^{\beta}+\left(\dfrac{1}{\eta_2}\right)^{\beta}\right]^{\frac{1}{\beta}}$, where $\Gamma(n) =$ gamma function.	Two series, Weibull failure rate units.
5	$\lambda_1(T)=\lambda_1$, $\lambda_2(T)=\lambda_2, \cdots$ $\lambda_n(T) = \lambda_n$. 	$\dfrac{1}{\displaystyle\sum_{i=1}^{n} \lambda_i}$	n series, constant failure rate units.

TABLE 1.1 – (Continued)

Case number	System structure	MTTF	Comment
6	$\lambda_1(T) = \dfrac{\beta}{\eta_1}\left(\dfrac{T-\gamma}{\eta_1}\right)^{\beta-1}$, $\lambda_2(T) = \dfrac{\beta}{\eta_2}\left(\dfrac{T-\gamma}{\eta_2}\right)^{\beta-1}$, $\cdots$ $\lambda_n(T) = \dfrac{\beta}{\eta_n}\left(\dfrac{T-\gamma}{\eta_n}\right)^{\beta-1}$. $\rightarrow\boxed{1}\boxed{2}\cdots\boxed{n}\rightarrow$	$\gamma + \Gamma\left(\dfrac{1}{\beta}+1\right)\Big/\left[\displaystyle\sum_{i=1}^{n}\left(\dfrac{1}{\eta_i}\right)^{\beta}\right]^{\frac{1}{\beta}}$, where $\Gamma(n)$ = gamma function.	n series Weibull failure rate units, all with the same shape parameter β and location parameter γ.
7	$\lambda_1(T)=\lambda_1$, $\lambda_2(T)=\lambda_2$, $\cdots$, $\lambda_n(T)=\lambda_n$, $\lambda_{n+1}(T)=k_1 T$, $\lambda_{n+2}(T)=k_2 T$, $\cdots$, $\lambda_{n+r}(T)=k_r T$, $\rightarrow\boxed{1}\boxed{2}\cdots\boxed{n}$ $\boxed{n+1}$ $\boxed{n+2}\cdots\boxed{n+r}\rightarrow$	$\left(\dfrac{\pi}{k}\right)^{\frac{1}{2}} e^{\frac{\lambda^2}{k}}\left\{1-\Phi\left[\lambda\left(\dfrac{2}{k}\right)^{\frac{1}{2}}\right]\right\}$, where $\lambda = \dfrac{1}{2}\displaystyle\sum_{i=1}^{n}\lambda_i$, $k = \dfrac{1}{2}\displaystyle\sum_{i=1}^{r}k_i$, and $\Phi(z) = \dfrac{1}{(2\pi)^{\frac{1}{2}}}\displaystyle\int_{-\infty}^{z} e^{-\frac{x^2}{2}}\, dx$.	n+r series units, n with a constant failure rate and r units with a linearly increasing failure rate.

SOLUTIONS TO EXAMPLE 1–1

1. The series system's failure rate is

$$\lambda_{ss}(T) \;=\; \lambda_{ss} = \sum_{i=1}^{3} \lambda_i = \lambda_1 + \lambda_2 + \lambda_3,$$

$$= \; 3.5 \times 10^{-6} + 16 \times 10^{-6} + 1.2 \times 10^{-6},$$

$$= \; 20.7 \times 10^{-6} \ \text{fr/hr},$$

or

$$\lambda_{ss} \;=\; 20.7 \ \text{fr}/10^6 \ \text{hr}.$$

2. The series system's reliability is given by

$$R_{ss}(t) = e^{-\lambda_{ss}t},$$

where $\lambda_{ss} = 20.7 \ \text{fr}/10^6 \ \text{hr}$, and $t = 100$ hr; hence

$$R_{ss}(t = 100\text{hr}) \;=\; e^{20.7 \times 10^{-6} \times 100},$$

or

$$R_{ss} = e^{-0.00207} \;=\; 0.9979.$$

It may be pointed out that $e^{-\lambda t} \simeq 1 - \lambda t$ to four decimal place accuracy when $0.0045 \le \lambda t \le 0.01$, $e^{-\lambda t} \simeq 1 - \lambda t$ to five decimal place accuracy when $0.0014 \le \lambda t < 0.0045$, and $e^{-\lambda t} \simeq 1 - \lambda t$ to six decimal place accuracy when $\lambda t < 0.0014$. Also make sure that the exponent λt in

$$R = e^{-\lambda t} \tag{1.10}$$

is in units of failures only. Thus, if t is in hours, convert λ to failures per hour before substituting in Eq. (1.10).

3. The series system's MTBF is

$$\text{MTBF}_{ss} = \frac{1}{\lambda_{ss}},$$

or

$$\text{MTBF}_{ss} = \frac{1}{20.7 \times 10^{-6}} = 48,309 \ \text{hr}.$$

This means that on the average such identical systems, operating under identical application and operation stresses, would fail after 48,400 hr of operation.

4. The series system's *pdf* is given by

$$f_{ss}(T) = \lambda_{ss} e^{-\lambda_{ss}T},$$

or

$$f_{ss}(T) = 20.7 \times 10^{-6} \ e^{-20.7 \times 10^{-6}T}.$$

This *pdf* may be used to determine the probability of failure of the system from

$$Q_{ss}(t) = \int_0^t f_{ss}(T)\, dT$$

or the reliability of the system from

$$R_{ss}(t) = 1 - Q_{ss}(t) = 1 - \int_0^t f_{ss}(T)\, dT,$$

or

$$R_{ss}(t) = \int_{T=t}^{\infty} f_{ss}(T)\, dT.$$

The number of such series systems failing, N_f, and surviving, N_s, while operating for a mission of t duration is given by

$$N_f = N Q_{ss}(t), \tag{1.11}$$

where N is the number of systems operating for a mission of t duration. Similarly,

$$N_s = N R_{ss}(t). \tag{1.12}$$

It must be pointed out that t and T have been used in the equations of this chapter. $\lambda(T)$ and $f(T)$ are functions of age, T. The reliability is a function of mission duration, t. However, for the first mission $t = T$. For subsequent missions the conditional reliability should be used; i.e.,

$$R(T,t) = \frac{R(T+t)}{R(T)}.$$

Furthermore, *during useful life*, or when $\lambda(T)$ is constant, the reliability is only a fuction of mission duration, t, and is independent of age, T. This is the reason t is used in Eqs. (1.2) and (1.4).

In Table 1.1 the term MTTF is used even though the title says MTBF. The more exact term to use is MTTFF, or the Mean Time To First Failure, because most time-to-failure *pdf* parameters are determined by testing each unit until it fails the first time. If the failled unit fails and gets repaired, there is no guarantee that the repair action has not altered the failure rate, or the time-to-failure *pdf* of this unit. Then the only correct term to use is MTTFF, or MTTF for short. If the failure rate of the unit is constant and repairs after failures do not alter this failure rate, then MTBF is the correct term to use because the MTBF will also be constant and the same from repair to repair.

EXAMPLE 1–2

Two exponential subsystems make up a system. Subsystem 1 has a reliability of 99.95% for a 100-hr mission, and Subsystem 2 has a reliability of 99.95% for a 10-hr mission. What is this system's reliability for a 100-hr mission when the two subsystems are reliabilitywise in series?

SOLUTIONS TO EXAMPLE 1–2

SOLUTION 1

For Subsystem 1 it is known that

$$R_1(t = 100 \text{ hr}) = 99.95\% = 0.9995,$$

and for Subsystem 2 it is known that

$$R_2(t = 10 \text{ hr}) = 99.95\% = 0.9995.$$

For exponential units it is known that

$$R(t) = e^{-\lambda t},$$

and

$$\lambda = -\frac{\ln[R(t)]}{t}.$$

Therefore,

$$\lambda_1 = -\frac{\ln(0.9995)}{100} = 0.0000050013,$$

and

$$\lambda_2 = -\frac{\ln(0.9995)}{10} = 0.000050013.$$

It is also known that

$$R_{ss}(t = 100 \text{ hr}) = [R_1(t = 100 \text{ hr}) \times R_2(t = 100 \text{ hr})],$$
$$= e^{-\lambda_1(100)} e^{-\lambda_2(100)},$$

or

$$R_{ss}(t = 100 \text{ hr}) = e^{-(\lambda_1+\lambda_2)(100)}.$$

Therefore,

$$R_{ss}(t = 100 \text{ hr}) = e^{-(0.0000050013+0.000050013)(100)} = e^{-0.0055014},$$

or

$$R_{ss}(t = 100 \text{ hr}) = 0.9945 = 99.45\%.$$

SOLUTION 2

Since $R_1(t = 100 \text{ hr})$ is known, the system's reliability is also given by

$$R_{ss}(t = 100 \text{ hr}) = [R_1(t = 100 \text{ hr})]e^{-\lambda_2(100)}$$
$$= 0.9995e^{-(0.000050013)(100)},$$

or

$$R_{ss}(t = 100 \text{ hr}) \quad = \quad 0.9945 = 99.45\%.$$

PROBLEMS

1-1. Two subsystems make up a system. Subsystem 1 has a reliability of 99.9% for 100 hr, and Subsystem 2 has a reliability of 98.9% for 10 hr. What is the system's reliability for a 200-hr mission when the two subsystems are reliabilitywise in series and each has a constant failure rate?

1-2. A target flight reliability of 0.999 is specified for a four-engine airplane. The flight consists of two legs. The first leg is 5 hr long and the second is 7 hr long. The flight is not allowed to continue on its second leg unless all four engines are functioning satisfactorily and *no delay for repairs is permitted.* Two out of the four engines are required for successful flight. Determine the failure rate that should be designed into each engine to meet the target flight reliability. Assume all four engines are equal and have a constant failure rate.

1-3. In Problem 1-2, if each engine has a Weibullian pdf with $\beta = 2.5$, $\gamma = 0$ hr, and $\eta = 1,000$ hr, find the reliability of each flight, assuming that the Weibull *pdf* parameters do not change when one or two engines fail.

1-4. A small plug-in type of transistor circuit has been tested to a reliability of 0.9998 for a 2-hr operating period. A total of 1,000 of these plug-in circuits is used in a special–purpose airborne digital computer. Find the following:

   (1) The MTBF of each plug-in circuit.

   (2) The failure rate, in failures per million hours, of each plug-in circuit.

   (3) The probability of successful operation of the digital computer during a 3-hr flight.

1-5. An amplifier-computer unit consists of the following parts:

Part	Quantity	Failure rate, fr/10^6 hr per unit
Relay	12	30
Saturable reactor	25	7
Diode	239	10
Resistor	418	4
Capacitor	54	2
Miscellaneous	–	1,052 (total)

(1) Determine the failure rate of this amplifier-computer unit.

(2) If a mission reliability of 0.9942 is required, what maximum length of mission time is allowable?

APPENDIX 1A

DERIVATION OF THE EQUATIONS FOR
CASES 2, 4, 6, AND 7 OF TABLE 1.1

Case 2 – If

$$\lambda(T) = kT,$$

then

$$R(T) = e^{-k\frac{T^2}{2}},$$

and

$$\text{MTTF} = \int_0^\infty R(T)\, dT,$$

or

$$\text{MTTF} = \int_0^\infty e^{-k\frac{T^2}{2}}\, dT. \tag{A1.1}$$

Let

$$x = k\frac{T^2}{2};$$

then

$$T = \left(\frac{2x}{k}\right)^{\frac{1}{2}} \quad \text{and} \quad dT = \left(\frac{1}{2k}\right)^{\frac{1}{2}} x^{\frac{1}{2}-1} dx,$$

and Eq. (A1.1) becomes

$$\begin{aligned}
\text{MTTF} &= \int_0^\infty \left(\frac{1}{2k}\right)^{\frac{1}{2}} x^{\frac{1}{2}-1} e^{-x} dx, \\
&= \left(\frac{1}{2k}\right)^{\frac{1}{2}} \int_0^\infty x^{\frac{1}{2}-1} e^{-x} dx, \\
&= \left(\frac{1}{2k}\right)^{\frac{1}{2}} \Gamma\left(\frac{1}{2}\right), \\
&= \left(\frac{1}{2k}\right)^{\frac{1}{2}} (\pi)^{\frac{1}{2}},
\end{aligned}$$

or

$$\text{MTTF} = \left(\frac{\pi}{2k}\right)^{\frac{1}{2}}. \tag{A1.2}$$

Case 4 – In this case the failure rate functions are

$$\lambda_1(T) = \begin{cases} \frac{\beta_1}{\eta_1}\left(\frac{T-\gamma_1}{\eta_1}\right)^{\beta_1-1}, & T > \gamma_1, \\ 0, & T \le \gamma_1 \end{cases}$$

and

$$\lambda_2(T) = \begin{cases} \frac{\beta_2}{\eta_2}\left(\frac{T-\gamma_2}{\eta_2}\right)^{\beta_2-1}, & T > \gamma_2, \\ 0, & T \le \gamma_2. \end{cases} \qquad (A1.3)$$

The reliability functions are

$$R_1(T) = \begin{cases} e^{-\left(\frac{T-\gamma_1}{\eta_1}\right)^{\beta_1}}, & T > \gamma_1, \\ 1, & T \le \gamma_1, \end{cases}$$

$$R_2(T) = \begin{cases} e^{-\left(\frac{T-\gamma_2}{\eta_2}\right)^{\beta_2}}, & T > \gamma_2, \\ 1, & T \le \gamma_2. \end{cases} \qquad (A1.4)$$

The MTTF is given by

$$\text{MTTF} = \int_0^\infty R_{ss}(T)\, dT,$$

or

$$\text{MTTF} = \int_0^\infty R_1(T)R_2(T)\, dT. \qquad (A1.5)$$

Substituting Eq. (A1.4) for each time interval into Eq. (A1.5), assuming that $\gamma_2 < \gamma_1$, yields

$$\begin{aligned} \text{MTTF} &= \int_0^{\gamma_2} dT + \int_{\gamma_2}^{\gamma_1} e^{-\left(\frac{T-\gamma_2}{\eta_2}\right)^{\beta_2}} dT \\ &\quad + \int_{\gamma_1}^\infty e^{-\left[\left(\frac{T-\gamma_1}{\eta_1}\right)^{\beta_1}+\left(\frac{T-\gamma_2}{\eta_2}\right)^{\beta_2}\right]} dT, \\ &= \text{I+II+III}, \end{aligned}$$

where

$$\text{I} = \int_0^{\gamma_2} dT = \gamma_2,$$

$$\text{II} = \int_{\gamma_2}^{\gamma_1} e^{-\left(\frac{T-\gamma_2}{\eta_2}\right)^{\beta_2}} dT,$$

and

$$\text{III} = \int_{\gamma_1}^\infty e^{-\left[\left(\frac{T-\gamma_1}{\eta_1}\right)^{\beta_1}+\left(\frac{T-\gamma_2}{\eta_2}\right)^{\beta_2}\right]} dT$$

Let

$$(\frac{T - \gamma_2}{\eta_2})^{\beta_2} = x;$$

then

$$T = \eta_2 x^{\frac{1}{\beta_2}} + \gamma_2,$$

and

$$dT = \frac{\eta_2}{\beta_2} x^{\frac{1}{\beta_2} - 1} dx, \quad x \to 0 \text{ when } T \to \gamma_2.$$

Now the second integral becomes

$$\text{II} = \frac{\eta_2}{\beta_2} \int_0^{(\frac{\gamma_1 - \gamma_2}{\eta_2})^{\beta_2}} e^{-x} x^{\frac{1}{\beta_2} - 1} dx,$$

or

$$\text{II} = \frac{\eta_2}{\beta_2} G\left[\frac{1}{\beta_2}, (\frac{\gamma_1 - \gamma_2}{\eta_2})^{\beta_2}\right], \tag{A1.6}$$

where $G(n, z)$ is the incomplete gamma function and is defined by

$$G(n, z) = \int_0^z e^{-x} x^{n-1} dx, \quad |z| < \infty, \quad n > 0. \tag{A1.7}$$

To calculate the third integral, numerical methods of integration should be used. Now the MTTF can be written as

$$\text{MTTF} = \gamma_2 + \frac{\eta_2}{\beta_2} G\left[\frac{1}{\beta_2}, (\frac{\gamma_1 - \gamma_2}{\eta_2})^{\beta_2}\right] + \int_{\gamma_1}^{\infty} e^{-\left[(\frac{T-\gamma_1}{\eta_1})^{\beta_1} + (\frac{T-\gamma_2}{\eta_2})^{\beta_2}\right]} dT. \tag{A1.8}$$

where $\gamma_1 < \gamma_2$ and $G\left[\frac{1}{\beta_2}, (\frac{\gamma_1 - \gamma_2}{\eta_2})^{\beta_2}\right]$ is defined by Eq. (A1.7).

In the special case where $\gamma_1 = \gamma_2 = \gamma$ and $\beta_1 = \beta_2 = \beta$, Eq. (A1.8) becomes

$$\text{MTTF} = \gamma + \frac{\eta_2}{\beta} G\left[\frac{1}{\beta}, 0\right] + \int_{\gamma}^{\infty} e^{-\left[(\frac{T-\gamma}{\eta_1})^{\beta} + (\frac{T-\gamma}{\eta_2})^{\beta}\right]} dT,$$

or

$$\text{MTTF} = \gamma + 0 + \int_{\gamma}^{\infty} e^{-\left[(\frac{1}{\eta_1})^{\beta} + (\frac{1}{\eta_2})^{\beta}\right](T-\gamma)^{\beta}} dT. \tag{A1.9}$$

Let

$$\left[(\frac{1}{\eta_1})^{\beta} + (\frac{1}{\eta_2})^{\beta}\right](T - \gamma)^{\beta} = x;$$

then

$$T = \left[\frac{x}{(\frac{1}{\eta_1})^{\beta} + (\frac{1}{\eta_2})^{\beta}}\right]^{\frac{1}{\beta}} + \gamma,$$

and

$$dT = \frac{x^{\frac{1}{\beta}-1}}{\beta\left[(\frac{1}{\eta_1})^\beta + (\frac{1}{\eta_2})^\beta\right]^{\frac{1}{\beta}}} dx.$$

When $T \to \gamma$, $x \to 0$, and when $T \to \infty$, $x \to \infty$. Equation (A1.9) now becomes

$$MTTF = \gamma + \frac{1}{\beta\left[(\frac{1}{\eta_1})^\beta + (\frac{1}{\eta_2})^\beta\right]^{\frac{1}{\beta}}} \int_0^\infty e^{-x} x^{\frac{1}{\beta}-1} dx,$$

$$= \gamma + \frac{1}{\beta\left[(\frac{1}{\eta_1})^\beta + (\frac{1}{\eta_2})^\beta\right]^{\frac{1}{\beta}}} \Gamma(\frac{1}{\beta}),$$

or

$$MTTF = \gamma + \frac{1}{\left[(\frac{1}{\eta_1})^\beta + (\frac{1}{\eta_2})^\beta\right]^{\frac{1}{\beta}}} \Gamma(\frac{1}{\beta} + 1). \qquad (A1.10)$$

Case 6 – In this case, where n Weibullian units are reliabilitywise in series and each unit has the same shape parameter β and the same location parameter γ, the MTTF is given by

$$MTTF = \int_0^\infty R_{ss}(T)\, dT,$$

$$= \int_0^\infty R_1(T)R_2(T)\ldots R_n(T)\, dT,$$

$$= \int_0^\gamma dT + \int_\gamma^\infty e^{-\sum_{i=1}^n (\frac{T-\gamma}{\eta_i})^\beta}\, dT,$$

or

$$MTTF = \gamma + \int_\gamma^\infty e^{-(T-\gamma)^\beta \sum_{i=1}^n (\frac{1}{\eta_i})^\beta}\, dT. \qquad (A1.11)$$

Let

$$x = \left[\sum_{i=1}^n (\frac{1}{\eta_i})^\beta\right](T-\gamma)^\beta;$$

then

$$T = \left[\frac{x}{\sum_{i=1}^n (\frac{1}{\eta_i})^\beta}\right]^{\frac{1}{\beta}} + \gamma,$$

and

$$dT = \frac{1}{\beta} \frac{x^{\frac{1}{\beta}-1}}{\left[\sum\limits_{i=1}^{n} (\frac{1}{\eta_i})^{\beta} \right]^{\frac{1}{\beta}}} dx.$$

$x \to 0$ when $T \to \gamma$, and $x \to \infty$ when $T \to \infty$; then Eq. (A1.11) becomes

$$\text{MTTF} = \gamma + \frac{1}{\beta \left[\sum\limits_{i=1}^{n} (\frac{1}{\eta_i})^{\beta} \right]^{\frac{1}{\beta}}} \int_0^{\infty} e^{-x} x^{\frac{1}{\beta}-1} dx,$$

$$= \gamma + \frac{1}{\beta \left[\sum\limits_{i=1}^{n} (\frac{1}{\eta_i})^{\beta} \right]^{\frac{1}{\beta}}} \Gamma(\frac{1}{\beta}),$$

or

$$\text{MTTF} = \gamma + \frac{1}{\left[\sum\limits_{i=1}^{n} (\frac{1}{\eta_i})^{\beta} \right]^{\frac{1}{\beta}}} \Gamma(\frac{1}{\beta} + 1). \tag{A1.12}$$

Case 7 – In this case the MTTF is given by

$$\text{MTTF} = \int_0^{\infty} R_{ss}(T)\, dT,$$

or

$$\text{MTTF} = \int_0^{\infty} \exp(-\sum_{i=1}^{n} \lambda_i T) \exp(-\sum_{i=1}^{r} k_i \frac{T^2}{2})\, dT.$$

Let

$$\lambda = \frac{1}{2} \sum_{i=1}^{n} \lambda_i \quad \text{and} \quad k = \frac{1}{2} \sum_{i=1}^{r} k_i;$$

then

$$\text{MTTF} = \int_0^{\infty} e^{-(2\lambda T + kT^2)} dT,$$

$$= \int_0^{\infty} e^{-(kT^2 + 2\lambda T + \frac{\lambda^2}{k} - \frac{\lambda^2}{k})} dT,$$

or

$$\text{MTTF} = e^{\frac{\lambda^2}{k}} \int_0^{\infty} e^{-\frac{1}{k}(kT + \lambda)^2} dT. \tag{A1.13}$$

Let

$$\frac{x^2}{2} = \frac{1}{k}(kT + \lambda)^2;$$

then

$$x = (\frac{2}{k})^{\frac{1}{2}}(kT + \lambda), \quad \text{and} \quad dx = (2k)^{\frac{1}{2}}dT.$$

$$x \to (\frac{2}{k})^{\frac{1}{2}}\lambda \quad \text{when} \quad T \to 0, \quad \text{and} \quad x \to \infty \quad \text{when} \quad T \to \infty.$$

Then Eq. (A1.13) becomes

$$\text{MTTF} \;=\; (\frac{\pi}{k})^{\frac{1}{2}} e^{\frac{\lambda^2}{k}} \int_{[(\frac{2}{k})^{\frac{1}{2}}\lambda]}^{\infty} \frac{1}{(2\pi)^{\frac{1}{2}}} e^{-\frac{x^2}{2}} dx,$$

or

$$\text{MTTF} \;=\; (\frac{\pi}{k})^{\frac{1}{2}} e^{\frac{\lambda^2}{k}} \left[1 - \Phi\left[(\frac{2}{k})^{\frac{1}{2}}\lambda \right] \right], \qquad (A1.14)$$

where

$$\lambda = \frac{1}{2}\sum_{i=1}^{n}\lambda_i, \quad k = \frac{1}{2}\sum_{i=1}^{r}k_i,$$

and

$$\Phi(z) = \int_{-\infty}^{z} \frac{1}{n(2\pi)^{\frac{1}{2}}} e^{-\frac{x^2}{2}} dx.$$

Chapter 2

RELIABILITY OF PARALLEL SYSTEMS

2.1 N UNIT RELIABILITYWISE PARALLEL SYSTEM

A system is said to have n units which are reliabilitywise in parallel when only the failure of all n units in the system results in system failure. Conversely, for a parallel system to succeed at least one of the n parallel units in the system needs to succeed, or operate without failure, for the duration on the intended mission

The reliability block diagram of a parallel system is shown in Fig. 2.1. Probabilistically, the unreliability of this system, Q_{sp}, is the probability that all units fail, or Unit 1 fails, and Unit 2 fails, ..., and Unit n fails, and the reliability is obtained from $R_{sp} = 1 - Q_{sp}$.

Mathematically, the unreliability of this parallel system is then given by

$$Q_{sp} = Q_1 Q_2 ... Q_n,$$

and the reliability by

$$R_{sp} = 1 - Q_{sp} = 1 - (Q_1 Q_2 \ ... \ Q_n),$$

or

$$R_{sp} = 1 - \prod_{i=1}^{n} Q_i = 1 - \prod_{i=1}^{n} (1 - R_i).$$

In other words, the reliability of a system in parallel is one minus the

19

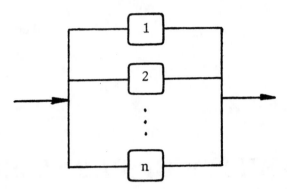

Fig. 2.1 – The reliability block diagram of a reliabilitywise parallel system.

product of the unreliabilities of all parallel units in the system.

2.2 EXPONENTIAL PARALLEL UNITS

If the units have constant failure rates, then the parallel system's reliability is given by

$$R_{sp}(T) = 1 - \prod_{i=1}^{n}(1 - e^{-\lambda_i T}).$$

The system failure rate is given by

$$\lambda_{sp}(T) = \frac{f_{sp}(T)}{R_{sp}(T)},$$

which requires the knowledge of $f_{sp}(T)$. This is given by

$$f_{sp}(T) = -\frac{d[R_{sp}(T)]}{dT}.$$

$f_{sp}(T)$ for n units is very complex, hence we will write it for two units only, to illustrate the procedure. Then

$$R_{sp}(T) = 1 - (1 - e^{-\lambda_1 T})(1 - e^{-\lambda_2 T}),$$
$$R_{sp}(T) = e^{-\lambda_1 T} + e^{-\lambda_2 T} - e^{-(\lambda_1 + \lambda_2)T},$$
$$= R_1(T) + R_2(T) - R_1(T)R_2(T),$$

and for equal units

$$R_{sp}(T) \quad = 2R(T) - R^2(T) = 2e^{-\lambda T} - e^{-2\lambda T},$$

$$R_{sp}(T,t) = \frac{R_{sp}(T+t)}{R_{sp}(T)},$$

or

$$R_{sp}(T,t) = \frac{e^{-\lambda_1(T+t)} + e^{-\lambda_2(T+t)} - e^{-(\lambda_1+\lambda_2)(T+t)}}{e^{-\lambda_1 T} + e^{-\lambda_2 T} - e^{-(\lambda_1+\lambda_2)T}}.$$

This is age and mission dependent even through the units are exponential!

If the units are equal then

$$R_{sp}(T,t) = \frac{2e^{-\lambda(T+t)} - e^{-2\lambda(T+t)}}{2e^{-\lambda T} - e^{-2\lambda t}},$$

or

$$R_{sp}(T,t) = \frac{2R(T+t) - R^2(T+t)}{2R(T) - R^2(t)}.$$

Also,

$$f_{sp}(T) = -\frac{d[e^{-\lambda_1 T} + e^{-\lambda_2 T} - e^{-(\lambda_1+\lambda_2)T}]}{dT},$$

and

$$f_{sp}(T) = \lambda_1 e^{-\lambda_1 T} + \lambda_2 e^{-\lambda_2 T} - (\lambda_1 + \lambda_2)e^{-(\lambda_1+\lambda_2)T}$$

for two different units in parallel.

The failure rate of this parallel system is given by

$$\lambda_{sp}(T) = \frac{f_{sp}(T)}{R_{sp}(T)} = \frac{\lambda_1 e^{-\lambda_1 T} + \lambda_2 e^{-\lambda_2 T} - (\lambda_1 + \lambda_2)e^{-(\lambda_1+\lambda_2)T}}{e^{-\lambda_1 T} + e^{-\lambda_2 T} - e^{-(\lambda_1+\lambda_2)T}},$$

and if the units are equal, then

$$\lambda_{sp}(T) = \frac{2\lambda e^{-\lambda T} - 2\lambda e^{-2\lambda T}}{2e^{-\lambda T} - e^{-2\lambda T}}.$$

It must be observed that $\lambda_{sp}(T)$ is not constant but a function of age although each unit has a constant λ.

The mean time between failures of the system is given by

$$\text{MTBF}_{sp} = \int_0^\infty R_{sp}(T)\, dT,$$

or for two different units in parallel

$$\text{MTBF}_{sp} = \int_0^\infty \left[e^{-\lambda_1 T} + e^{-\lambda_2 T} - e^{-(\lambda_1 + \lambda_2)T} \right] dT,$$

or

$$\text{MTBF}_{sp} = \frac{1}{\lambda_1} + \frac{1}{\lambda_2} - \frac{1}{\lambda_1 + \lambda_2}.$$

For equal units

$$\text{MTBF}_{sp} = \frac{1.5}{\lambda},$$

or

$$\text{MTBF}_{sp} = 1.5m,$$

where m is the MTBF of each unit. It must be observed that

$$\lambda_{sp}(T) \neq \frac{1}{\text{MTBF}_{sp}}$$

for parallel systems even with units that have a constant failure rate.

2.3   WEIBULLIAN UNITS

If the times to failure of the units in the parallel system are Weibull distributed, then the system's reliability is given by

$$R_{sp}(T) = 1 - \prod_{i=1}^{n} \left\{ 1 - e^{-[(T-\gamma_i)/\eta_i]^{\beta_i}} \right\}.$$

For a two-unit parallel system the system's reliability is given by

$$R_{sp}(T) = 1 - \left\{ 1 - e^{-[(T-\gamma_1)/\eta_1]^{\beta_1}} \right\} \left\{ 1 - e^{-[(T-\gamma_2)/\eta_2]^{\beta_2}} \right\},$$

or

$$R_{sp}(T) = e^{-[(T-\gamma_1)/\eta_1]^{\beta_1}} + e^{-[(T-\gamma_2)/\eta_2]^{\beta_2}} - e^{-\{[(T-\gamma_1)/\eta_1]^{\beta_1} + [(T-\gamma_2)/\eta_2]^{\beta_2}\}}.$$

The *pdf* of the times to failure for the two-unit parallel system is given by

$$
\begin{aligned}
f_{sp}(T) &= -\frac{d[R_{sp}(T)]}{dT}, \\
&= \frac{\beta_1}{\eta_1}\left(\frac{T-\gamma_1}{\eta_1}\right)^{\beta_1-1} e^{-[(T-\gamma_1)/\eta_1]^{\beta_1}} + \frac{\beta_2}{\eta_2}\left(\frac{T-\gamma_2}{\eta_2}\right)^{\beta_2-1} e^{-[(T-\gamma_2)/\eta_2]^{\beta_2}} \\
&\quad - \left[\frac{\beta_1}{\eta_1}\left(\frac{T-\gamma_1}{\eta_1}\right)^{\beta_1-1} + \frac{\beta_2}{\eta_2}\left(\frac{T-\gamma_2}{\eta_2}\right)^{\beta_2-1}\right] \\
&\quad \cdot e^{-\{[(T-\gamma_1)/\eta_1]^{\beta_1} + [(T-\gamma_2)/\eta_2]^{\beta_2}\}},
\end{aligned}
$$

or

$$f_{sp}(T) = \frac{\beta_1}{\eta_1}(\frac{T-\gamma_1}{\eta_1})^{\beta_1-1}e^{-[(T-\gamma_1)/\eta_1]^{\beta_1}}\{1 - e^{-[(T-\gamma_2)/\eta_2]^{\beta_2}}\}$$
$$+ \frac{\beta_2}{\eta_2}(\frac{T-\gamma_2}{\eta_2})^{\beta_2-1}e^{-[(T-\gamma_2)/\eta_2]^{\beta_2}}\{1 - e^{-[(T-\gamma_1)/\eta_1]^{\beta_1}}\}.$$

The two-unit parallel system's failure rate is given by

$$\lambda_{sp}(T) = \frac{f_{sp}(T)}{R_{sp}(T)},$$

or

$$\lambda_{sp}(T) = \left\{\frac{\beta_1}{\eta_1}(\frac{T-\gamma_1}{\eta_1})^{\beta_1-1}e^{-[(T-\gamma_1)/\eta_1]^{\beta_1}}\{1 - e^{-[(T-\gamma_2)/\eta_2]^{\beta_2}}\}\right.$$
$$\left.+ \frac{\beta_2}{\eta_2}(\frac{T-\gamma_2}{\eta_2})^{\beta_2-1}e^{-[(T-\gamma_2)/\eta_2]^{\beta_2}}\{1 - e^{-[(T-\gamma_1)/\eta_1]^{\beta_1}}\}\right\}\Bigg/$$
$$\left\{e^{-[(T-\gamma_1)/\eta_1]^{\beta_1}} + e^{-[(T-\gamma_2)/\eta_2]^{\beta_2}}\right.$$
$$\left.- e^{-\{[(T-\gamma_1)/\eta_1]^{\beta_1} + [(T-\gamma_1)/\eta_1]^{\beta_1}\}}\right\}.$$

The mean time to failure of the system is given by

$$\text{MTBF}_{sp} = \int_0^\infty R_{sp}(T)\,dT.$$

From the derivation given in Appendix 2A, Case 2, the mean time to failure for the system with two Weibullian units, reliabilitywise in parallel, is

$$\text{MTBF}_{sp} = \gamma_1 + \eta_1\Gamma(\frac{1}{\beta_1}+1) + \eta_2\Gamma(\frac{1}{\beta_2}+1)$$
$$- \frac{\eta_2}{\beta_2}G[\frac{1}{\beta_2}, (\frac{\gamma_1-\gamma_2}{\eta_2})^{\beta_2}]$$
$$- \int_{\gamma_1}^\infty e^{-\{[(T-\gamma_1)/\eta_1]^{\beta_1}+[(T-\gamma_2)/\eta_2]^{\beta_2}\}}\,dT,$$

where

$\Gamma(n)$ = gamma function,

and

$G(n, z) = \int_0^z e^{-x}x^{n-1}\,dx$, is the incomplete gamma function.

TABLE 2.1 – Equations for the calculation of the MTTF of various systems with units that are reliabilitywise in parallel.

Case number	System structure	MTTF	Comment
1	$\lambda_1(T) = \lambda_1,$ $\lambda_2(T) = \lambda_2.$ 	$\dfrac{1}{\lambda_1} + \dfrac{1}{\lambda_2} - \dfrac{1}{\lambda_1+\lambda_2}.$	Two parallel, constant failure rate units.
2	$\lambda_1(T) = \dfrac{\beta_1}{\eta_1}\left(\dfrac{T-\gamma_1}{\eta_1}\right)^{\beta_1-1},$ $\lambda_2(T) = \dfrac{\beta_2}{\eta_2}\left(\dfrac{T-\gamma_2}{\eta_2}\right)^{\beta_2-1},$ $\gamma_2 < \gamma_1.$ 	$\gamma_1 + \eta_1\,\Gamma\!\left(\dfrac{1}{\beta_1}+1\right) + n_2\,\Gamma\!\left(\dfrac{1}{\beta_2}+1\right)$ $- \dfrac{n_2}{\beta_2}\,G\!\left[\dfrac{1}{\beta_2},\left(\dfrac{\gamma_1-\gamma_2}{n_2}\right)^{\beta_2}\right]$ $- \displaystyle\int_{\gamma_1}^{\infty} e^{-\left[\left(\frac{T-\gamma_1}{n_1}\right)^{\beta_1} + \left(\frac{T-\gamma_2}{n_2}\right)^{\beta_2}\right]}\, dT,$ where $\Gamma(n)$ = gamma function, and $G(n,z) = \displaystyle\int_0^z e^{-x}\, x^{\,n-1}\, dx,$ is the incomplete gamma function. If $\gamma_1=\gamma_2=\gamma$ and $\beta_1=\beta_2=\beta$, then the MTTF becomes $\gamma + \Gamma\!\left(\dfrac{1}{\beta}+1\right)\left\{ \eta_1+\eta_2 - \dfrac{1}{\left[\left(\frac{1}{\eta_1}\right)^{\beta}+\left(\frac{1}{\eta_2}\right)^{\beta}\right]^{\frac{1}{\beta}}} \right\}.$	Two parallel, Weibull failure rate units.

TABLE 2.1 – (Continued)

Case number	System structure	MTTF	Comment
3	$\lambda_i(T) = \lambda_i$, $i = 1,\ldots,n$.	$\left(\frac{1}{\lambda_1} + \frac{1}{\lambda_2} + \ldots + \frac{1}{\lambda_n}\right) - \left(\frac{1}{\lambda_1+\lambda_2} + \frac{1}{\lambda_1+\lambda_3}\right)$ $+ \ldots + \frac{1}{\lambda_{n-1}+\lambda_n} + \ldots + (-1)^{n+1}\frac{1}{\sum\limits_{i=1}^{n}\lambda_i}.$ If $\lambda_i = \lambda$, $i=1,\ldots,n$, then $\sum\limits_{k=1}^{n}\frac{1}{k\lambda}$	n parallel, constant failure rate units.
4	$\lambda_i(T) = \frac{\beta}{n_i}\left(\frac{T-\gamma}{n_i}\right)^{\beta-1}$, $i = 1,\ldots,n$	$\gamma + \Gamma(\tfrac{1}{\beta}+1)\left\{\sum\limits_{i=1}^{n} n_i - \frac{1}{\left[\left(\frac{1}{n_1}\right)^\beta + \left(\frac{1}{n_2}\right)^\beta\right]^{\frac{1}{\beta}}} + \right.$ $\frac{1}{\left[\left(\frac{1}{n_1}\right)^\beta + \left(\frac{1}{n_3}\right)^\beta\right]^{\frac{1}{\beta}}} + \ldots + \frac{1}{\left[\left(\frac{1}{n_{n-1}}\right)^\beta + \left(\frac{1}{n_n}\right)^\beta\right]^{\frac{1}{\beta}}}$ $\left. + \ldots + (-1)^{n+1}\frac{1}{\left[\sum\limits_{i=1}^{n}\left(\frac{1}{n_i}\right)^\beta\right]^{\frac{1}{\beta}}}\right\},$ where $\Gamma(n)$ = gamma function.	n parallel, Weibull units having the same shape parameter β and location parameter γ.

For solutions with more units reliabilitywise in parallel, see Table 2.1, and for the derivations of the equations for Cases 2 and 4 in Table 2.1, see Appendix 2A.

EXAMPLE 2-1

A system consists of the same units as in Example 1-1; however, these units are now reliabilitywise in parallel. Find the following for this system:

1. The reliability for a 100-hr mission starting the mission at age zero.

2. The MTBF.

SOLUTIONS TO EXAMPLE 2-1

1. The reliability of this system for t = 100 hr is

$$R_{sp}(t) = 1 - (1 - e^{-\lambda_1 t})(1 - e^{-\lambda_2 t})(1 - e^{-\lambda_3 t}),$$

$$R_{sp}(100 \text{ hr}) = 1 - (1 - e^{-(3.5)(10^{-6})(100)}) \cdot (1 - e^{-(16.0)(10^{-6})(100)})$$

$$\cdot (1 - e^{-(1.2)(10^{-6})(100)}),$$

$$= 1 - (1 - e^{-0.00035})(1 - e^{-0.00160})(1 - e^{-0.00012}),$$

$$= 1 - (0.00035)(0.00160)(0.00012),$$

or

$$R_{sp}(100 \text{ hr}) = 0.9_{10}328.$$

This is a high reliability and compares with $R_{ss}(100 \text{ hr}) = 0.9979$ when all three units were reliabilitywise in series.

2. The MTBF of this system is

$$\text{MTBF}_{sp} = \int_0^\infty R_{sp}(t)\,dt,$$

$$= \frac{1}{\lambda_1} + \frac{1}{\lambda_2} + \frac{1}{\lambda_3} - \frac{1}{\lambda_1 + \lambda_2} - \frac{1}{\lambda_2 + \lambda_3}$$

$$- \frac{1}{\lambda_3 + \lambda_1} + \frac{1}{\lambda_1 + \lambda_2 + \lambda_3},$$

$$\text{MTBF}_{sp} = \frac{1}{(3.5)(10^{-6})} + \frac{1}{(16)(10^{-6})} + \frac{1}{(1.2)(10^{-6})}$$

$$- \frac{1}{(3.5 + 16)(10^{-6})} - \frac{1}{(16 + 1.2)(10^{-6})}$$

$$- \frac{1}{(1.2 + 3.5)(10^{-6})} + \frac{1}{(3.5 + 16 + 1.2)(10^{-6})},$$

or

$$\text{MTBF}_{sp} \; = \; 909,600 \text{ hr.}$$

This compares with $\text{MTBF}_{ss} = 48,309$ hr for the system with these same three units reliabilitywise in series, or a great increase.

EXAMPLE 2–2

Two unequal units, which have a constant failure rate, are reliability-wise in pure parallel function mode. Do the following:

1. Write down the reliability function.

2. Write down the probability density function.

3. Write down the failure rate function.

4. Write down the mean life equation.

5. Calculate the reliability as a function of mission time if $\lambda_1 = 6.0$ fr/10^6 hr and $\lambda_2 = 10.0$ fr/10^6 hr, and plot the results.

6. Calculate the probability density as a function of age and plot the results on the same figure as that for the reliability.

7. Calculate the failure rate as a function of age and plot the results on the same figure as that for the reliability.

8. Calculate the mean life of this configuration.

9. Discuss in great detail the behavior of the results obtained.

SOLUTIONS TO EXAMPLE 2–2

1. The reliability function is

$$R(T) = e^{-\lambda_1 T} + e^{-\lambda_2 T} - e^{-(\lambda_1 + \lambda_2)T}.$$

2. The probability density function is

$$f(T) = \lambda_1 e^{-\lambda_1 T} + \lambda_2 e^{-\lambda_2 T} - (\lambda_1 + \lambda_2)e^{-(\lambda_1 + \lambda_2)T}.$$

3. The failure rate function is

$$\lambda(T) = \frac{f(T)}{R(T)}.$$

4. The mean life is given by

$$m = \frac{1}{\lambda_1} + \frac{1}{\lambda_2} - \frac{1}{\lambda_1 + \lambda_2}.$$

5. Table 2.2, Column 2, gives the calculated reliability values, and Fig. 2.2 the plot thereof.

6. Table 2.2, Column 3, gives the calculated probability density values, and Fig. 2.2 the plot thereof.

7. Table 2.2, Column 4, gives the calculated failure rate values, and Fig. 2.2 the plot thereof.

8. The mean life of this configuration is given by

$$m = \frac{1}{\lambda_1} + \frac{1}{\lambda_2} - \frac{1}{\lambda_1 + \lambda_2},$$

$$= \frac{1}{(6.0)(10^{-6})} + \frac{1}{(10.0)(10^{-6})} - \frac{1}{(6.0)(10^{-6}) + (10.0)(10^{-6})},$$

$$= (1.667)(10^5) + (1.0)(10^5) - (0.625)(10^5),$$

or

$$m = 204,200 \text{ hr.}$$

9. It may be seen from the reliability values in Table 2.2, Column 2, and their plot in Fig. 2.2 that the reliability does not drop very sharply in the beginning. The plot has a knee in the beginning, indicative of a wear-out life characteristic, even though each unit in parallel has a constant failure rate! After about $T = 10,000$ hr the reliability drops sharply, and it is practically zero by $T = 800,000$ hr.

It may be seen from the probability density values in Table 2.2, Column 3, and their plot in Fig. 2.2 that the probability density starts at zero for $T = 0$, increases up to about $T = 9 \times 10^4$ hr, and then decreases, again indicative of a wear-out life characteristic, even though both units have a constant failure rate!

It may be seen from the failure rate values in Table 2.2, Column 4, and their plot in Fig. 2.2 that the failure rate starts at zero for $T = 0$, increases thereafter at a decreasing rate, reaches a maximum value of about 6.6 fr/10^6 hr at about $T = 300,000$ hr, and decreases thereafter and stabilizes at the value of $\lambda = 6.0$ fr /10^6 hr, or at the failure rate of the surviving unit which has the lower failure rate, with the unit having the higher failure rate having failed already, again indicative of a wear-out life characteristic, even though both units have a constant failure rate!

EXAMPLE 2–3

Figure 2.3 gives the reliability block diagram for two series-parallel systems. Each system has $N \cdot X$ identical units with a unit reliability of 0.75. The first configuration is named "system redundancy"; i.e., there are X groups of units reliabilitywise in parallel, and in each group there are N identical units reliabilitywise in series. The second configuration is named "element redundancy"; i.e., there are N groups

TABLE 2.2 – $R(T)$, $f(T)$ and $\lambda(T)$ for two exponential units reliabilitywise in pure parallel with $\lambda_1 = 6$ fr/10^6 hr and $\lambda_2 = 10$ fr/10^6 hr.

1	2	3	4
$T \times 10^{-3}$	$R(T)$	$f(T) \times 10^6$	$\lambda(T) \times 10^6$
0	1.00000	0.00000	0.00000
1	0.99994	0.11857	0.11858
3	0.99947	0.34728	0.34746
5	0.99856	0.56511	0.56592
7	0.99722	0.77245	0.77460
9	0.99546	0.96970	0.97411
10	0.99446	1.06466	1.07059
30	0.95731	2.51927	2.63163
50	0.89802	3.32095	3.69808
70	0.82735	3.68766	4.45718
80	0.79008	3.75739	4.75574
90	0.75239	3.77134	5.01248
100	0.71479	3.74132	5.02341
110	0.67768	3.67710	5.42603
120	0.64134	3.58674	5.59258
150	0.53898	3.21923	5.97280
170	0.47740	2.93641	6.15079
300	0.20686	1.35799	6.56489
700	0.01589	0.09887	6.22089
1000	0.00252	0.01532	6.07150
1100	0.00138	0.00833	6.04835
1200	0.00075	0.00454	6.03259
3500	0.00000	0.00000	6.00000

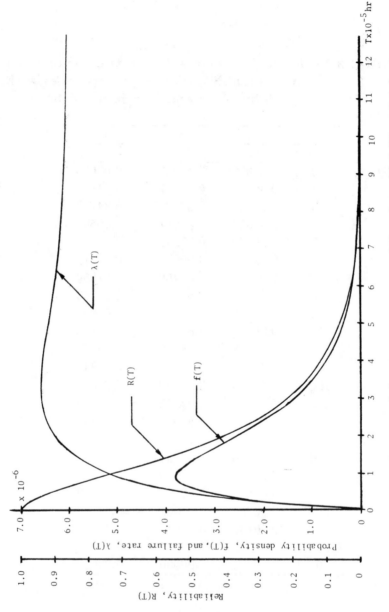

Fig. 2.2 – $R(T)$, $f(T)$, and $\lambda(T)$ for two exponential units reliabilitywise in pure parallel with $\lambda_1 = 6 \text{ fr}/10^6$ hr and $\lambda_2 = 10 \text{ fr}/10^6$ hr.

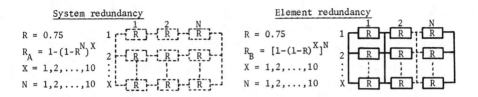

Fig. 2.3 – The configurations of element and system redundancy of Example 2–3.

in series with X units reliabilitywise in parallel in each group. Compare the reliabilities of these two configurations.

SOLUTION TO EXAMPLE 2–3

Let R_A denote the system's reliability with system redundancy; then

$$R_A = 1 - (1 - R^N)^X.$$

And let R_B denote the system's reliability with element redundancy; then

$$R_B = [1 - (1 - R)^X]^N.$$

R is the reliability for each unit. Substituting $R = 0.75$, $N = 1, 3, 5, 7$, and $X = 1, 2, \ldots, 10$ into the equations for R_A and R_B yields the results in Table 2.3, where it is obvious that for the same value of N and X the element redundancy configuration has a higher reliability than does the system redundancy configuration, except when $X = 1$ and $N = 1$.

PROBLEMS

2-1. Two equal units, which have a constant failure rate, are reliabilitywise in pure parallel function mode. Do the following:

    (1) Write down the reliability function.

    (2) Write down the probability density function.

    (3) Write down the failure rate function.

TABLE 2.3 – Comparison of relative effectiveness of element and system redundancy in Example 2–3.

X	N = 1		N = 3		N = 5		N = 7	
	R_A	R_B	R_A	R_B	R_A	R_B	R_A	R_B
1	0.7500	0.7500	0.4219	0.4219	0.2373	0.2373	0.1335	0.1335
2	0.9375	0.9375	0.6658	0.8240	0.4183	0.7242	0.2491	0.6365
3	0.9844	0.9844	0.8068	0.9539	0.5563	0.9243	0.3494	0.8956
4	0.9961	0.9961	0.8883	0.9883	0.6616	0.9806	0.4362	0.9730
5	0.9990	0.9990	0.9354	0.9971	0.7419	0.9951	0.5115	0.9932
6	0.9998	0.9998	0.9627	0.9993	0.8032	0.9988	0.5767	0.9983
7	0.9999	0.9999	0.9784	0.9998	0.8499	0.9997	0.6332	0.9996
8	1.0000	1.0000	0.9875	1.0000	0.8855	0.9999	0.6822	0.9999
9	1.0000	1.0000	0.9928	1.0000	0.9127	1.0000	0.7246	1.0000
10	1.0000	1.0000	0.9958	1.0000	0.9334	1.0000	0.7613	1.0000

(4) Write down the mean life equation.

(5) Calculate the reliability as a function of mission time if each unit's failure rate is $\lambda = 6.0$ fr/10^6 hr, and plot the results.

(6) Calculate the probability density as a function of age and plot the results on the same figure as that for the reliability.

(7) Calculate the failure rate as a function of age and plot the results on the same figure as that for the reliability.

(8) Calculate the mean life of this configuration.

(9) Discuss in great detail the behavior of the results obtained.

2-2. (1) Find the reliability of a system consisting of a generator, a diesel engine driving this generator, and the required controls, if they have the following failure rates:
Generator: 15 fr/10^6 hr.
Diesel engine: 25 fr/10^6 hr.
Controls: 10 fr/10^6 hr.
The mission time is 400 hr. Draw a reliability block diagram.

(2) If two sets of controls are used in parallel and an overall controls reliability of 99% is desired, what should the reliability of each set of controls be? Draw a reliability block diagram for this whole system.

2-3. Given the system in Fig. 2.4, calculate its useful life reliability for a 100-hr mission. All given failure rates are in failures per million hours of operation.

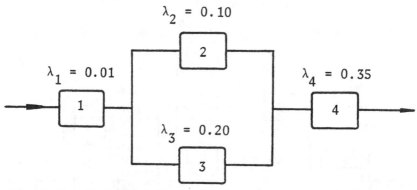

Fig. 2.4 — Reliability block diagram of the system in Problem 2-3.

2-4. The system of Fig. 2.5 is made up of ten components. Components 3, 4, and 5 are unequal and have the reliabilities of R_3, R_4, and R_5, as indicated. Components 8, 9, and 10 are equal and have a reliability of R_8. In component Group III, the components are equal. What is the system's reliability in terms of the respective R's?

2-5. Each system in Fig. 2.6 contains four identical units. Determine the following:

   (1) Which system is the most reliable? Give proof using inequalities, as well as numerical values.

   (2) Which system is the least reliable? Give proof using inequalities, as well as numerical values.

2-6. Draw the reliability block diagram for the complex system of units whose reliability is given by

$$R_S = R_1 \cdot [1 - (1 - R_2)^2] \cdot [1 - (1 - R_3 R_4 R_5)(1 - R_6 R_7 R_8)]$$
$$\cdot R_9 \cdot [1 - (1 - R_{10})^5].$$

2-7. Find (1) the reliability and (2) the MTBF of the system whose reliability block diagram is given in Fig. 2.7.

2-8. Determine the reliability and the MTBF of the system given in Fig. 2.8.

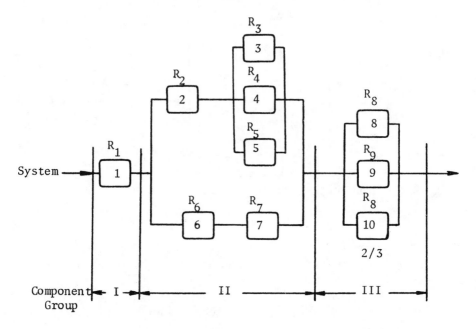

Fig. 2.5 – Reliability block diagram of the complex system
             of Problem 2-4.

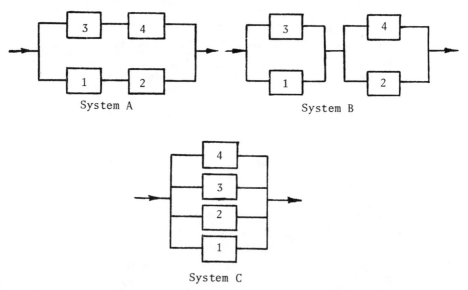

Fig. 2.6 – The three systems of four identical units each of
             Problem 2-5.

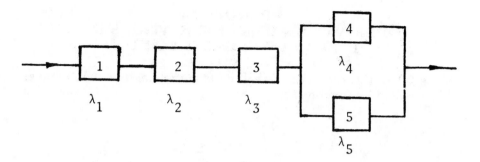

Fig. 2.7 – The reliabilitywise series-parallel configuration of the system of Problem 2-7.

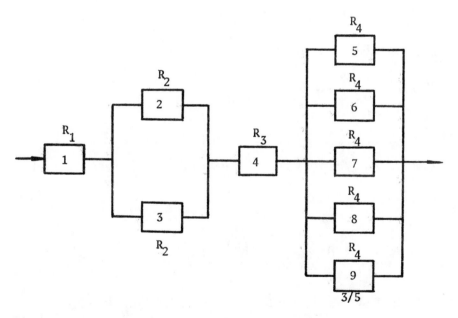

Fig. 2.8 – Reliability block diagram for Problem 2-8.

APPENDIX 2A
DERIVATION OF THE EQUATIONS IN
TABLE 2.1 FOR CASES 2 AND 4

Case 2 – In this case, two Weibull units are reliabilitywise in parallel. Their reliability functions are

$$R_1(T) = \begin{cases} e^{-[(T-\gamma_1)/\eta_1]^{\beta_1}}, & T > \gamma_1, \\ 1, & T \le \gamma_1, \end{cases}$$

and

$$R_2(T) = \begin{cases} e^{-[(T-\gamma_2)/\eta_2]^{\beta_2}}, & T > \gamma_2, \\ 1, & T \le \gamma_2. \end{cases} \tag{A2.1}$$

This system's reliability function is given by

$$R_{sp}(T) = 1 - [1 - R_1(T)][1 - R_2(T)],$$

or

$$R_{sp}(T) = R_1(T) + R_2(T) - R_1(T)R_2(T).$$

Then the MTTF is given by

$$\text{MTTF} = \int_0^\infty R_{sp}(T)\, dT,$$

or

$$\text{MTTF} = \int_0^\infty R_1(T)\, dT + \int_0^\infty R_2(T)\, dT - \int_0^\infty R_1(T)R_2(T)\, dT. \tag{A2.2}$$

Using the results of Cases 3 and 4 in Table 1.1 and assuming $\gamma_2 < \gamma_1$, the MTTF equation can be written as

$$\begin{aligned} \text{MTTF} = \ & \gamma_1 + \eta_1 \Gamma\left(\frac{1}{\beta_1} + 1\right) + \gamma_2 + \eta_2 \Gamma\left(\frac{1}{\beta_2} + 1\right) \\ & - \left[\gamma_2 + \frac{\eta_2}{\beta_2} G\left[\frac{1}{\beta_2}, \left(\frac{\gamma_1 - \gamma_2}{\eta_2}\right)^{\beta_2}\right] \right. \\ & \left. + \int_{\gamma_1}^\infty e^{-\{[(T-\gamma_1)/\eta_1]^{\beta_1} + [(T-\gamma_2)/\eta_2]^{\beta_2}\}}\, dT \right], \end{aligned}$$

or

$$\begin{aligned}
\text{MTTF} \;=\;& \gamma_1 + \eta_1 \Gamma(\frac{1}{\beta_1} + 1) + \eta_2 \Gamma(\frac{1}{\beta_2} + 1) \\
& - \frac{\eta_2}{\beta_2} G\left[\frac{1}{\beta_2}, (\frac{\gamma_1 - \gamma_2}{\eta_2})^{\beta_2}\right] \\
& - \int_{\gamma_1}^{\infty} e^{-\{[(T-\gamma_1)/\eta_1]^{\beta_1} + [(T-\gamma_2)/\eta_2]^{\beta_2}\}} \, dT,
\end{aligned} \qquad (A2.3)$$

where

$\Gamma(n)$ = gamma function,

and

$G(n, z) = \int_0^z e^{-x} x^{n-1} dx$, is the incomplete gamma function.

In the special case where $\gamma_1 = \gamma_2 = \gamma$ and $\beta_1 = \beta_2 = \beta$, Eq. $(A2.3)$ becomes

$$\begin{aligned}
\text{MTTF} \;=\;& \gamma + \eta_1 \Gamma(\frac{1}{\beta} + 1) + \eta_2 \Gamma(\frac{1}{\beta} + 1) - 0 \\
& - \int_{\gamma}^{\infty} e^{-\left[(1/\eta_1)^{\beta} + (1/\eta_2)^{\beta}\right](T-\gamma)^{\beta}} \, dT, \\
=\;& \gamma + [\Gamma(\frac{1}{\beta} + 1)](\eta_1 + \eta_2) - \Gamma(\frac{1}{\beta} + 1) \\
& \cdot \left\{ \frac{1}{\left[(1/\eta_1)^{\beta} + (1/\eta_2)^{\beta}\right]^{1/\beta}} \right\},
\end{aligned}$$

or

$$\text{MTTF} = \gamma + \Gamma(\frac{1}{\beta} + 1)\left\{ \eta_1 + \eta_2 - \frac{1}{\left[(1/\eta_1)^{\beta} + (1/\eta_2)^{\beta}\right]^{1/\beta}} \right\}.$$

$$(A2.4)$$

Case 4 – In this case, n Weibull units are reliabilitywise in parallel, and each unit has the same shape parameter, β, and the same location parameter, γ. This system's reliability function is given by

$$R_{sp}(T) = 1 - [1 - R_1(T)] \cdot [1 - R_2(T)] \ \ldots \ [1 - R_n(T)],$$

or

$$R_{sp}(T) = R_{s1}(T) - R_{s2}(T) + \cdots + (-1)^{n+1} R_{sn}(T),$$

where

$$R_{s1}(T) = R_1(T) + R_2(T) + \cdots + R_n(T),$$
$$R_{s2}(T) = R_1(T)R_2(T) + R_1(T)R_3(T) + \cdots + R_{n-1}(T)R_n(T),$$

and

$$R_{sn}(T) = \prod_{i=1}^{n} R_i(T).$$

Then the MTTF is given by

$$
\begin{aligned}
\text{MTTF} &= \int_0^\infty R_{sp}(T)\, dT, \\
&= \int_0^\infty R_{s1}(T)\, dT - \int_0^\infty R_{s2}(T)\, dT \\
&\quad + (-1)^{n+1} \int_0^\infty R_{sn}(T)\, dT.
\end{aligned}
$$

Using the results of Cases 3, 4 and 6 in Table 1.1,

$$
\begin{aligned}
\int_0^\infty R_{s1}(T)\, dT &= [\gamma + \eta_1 \Gamma(\tfrac{1}{\beta} + 1)] + [\gamma + \eta_2 \Gamma(\tfrac{1}{\beta} + 1)] \\
&\quad + \cdots + [\gamma + \eta_n \Gamma(\tfrac{1}{\beta} + 1)],
\end{aligned}
$$

or

$$
\int_0^\infty R_{s1}(T)\, dT = n\gamma + \Gamma(\tfrac{1}{\beta} + 1) \sum_{i=1}^{n} \eta_i,
$$

$$
\begin{aligned}
\int_0^\infty R_{s2}(T)\, dT &= \gamma + \frac{\Gamma[(1/\beta) + 1]}{[(1/\eta_1)^\beta + (1/\eta_2)^\beta]^{1/\beta}} + \gamma \\
&\quad + \frac{\Gamma[(1/\beta) + 1]}{[(1/\eta_1)^\beta + (1/\eta_3)^\beta]^{1/\beta}} + \cdots + \gamma \\
&\quad + \frac{\Gamma[(1/\beta) + 1]}{\{[1/(\eta_{n-1})]^\beta + (1/\eta_n)^\beta\}^{1/\beta}},
\end{aligned}
$$

or

$$
\begin{aligned}
\int_0^\infty R_{s2}(T)\, dT &= \frac{n!\gamma}{(n-2)!2} + \Gamma(\tfrac{1}{\beta} + 1)\Bigg\{ \frac{1}{[(1/\eta_1)^\beta + (1/\eta_2)^\beta]^{1/\beta}} \\
&\quad + \frac{1}{[(1/\eta_1)^\beta + (1/\eta_3)^\beta]^{1/\beta}} + \cdots
\end{aligned}
$$

$$+ \frac{1}{\{[1/(\eta_{n-1})]^\beta + (1/\eta_n)^\beta\}^{1/\beta}}\Bigg\},$$

and

$$\int_0^\infty R_{sn}(T)\, dT \;=\; \gamma + \Gamma(\frac{1}{\beta}+1)\frac{1}{\Big[\sum\limits_{i=1}^{n}(1/\eta_i)^\beta\Big]^{1/\beta}}.$$

Consequently, the MTTF can be written as

$$
\begin{aligned}
\text{MTTF} \;=\; & \gamma + \Gamma(\frac{1}{\beta}+1)\Bigg(\sum_{i=1}^{n}\eta_i - \Bigg\{\frac{1}{[(1/\eta_1)^\beta + (1/\eta_2)^\beta]^{1/\beta}} \\
& + \frac{1}{[(1/\eta_1)^\beta + (1/\eta_3)^\beta]^{1/\beta}} + \cdots \\
& + \frac{1}{\{[1/(\eta_{n-1})]^\beta + (1/\eta_n)^\beta\}^{1/\beta}}\Bigg\} + \cdots \\
& + (-1)^{n+1}\frac{1}{\Big[\sum\limits_{i=1}^{n}(1/\eta_i)^\beta\Big]^{1/\beta}}\Bigg).
\end{aligned}
\qquad (A2.5)
$$

Chapter 3

RELIABILITY OF STANDBY SYSTEMS

3.1   WHAT IS A STANDBY SYSTEM?

A system is said to have units which are reliabilitywise in standby when there is an active unit or subsystem to which are attached units or subsystems which stand by idly during the mission, either in a quiescent, nonoperating or warm-up mode, until they are called upon to operate at the specified output level by a sensing and switching subsystem when the active unit or subsystem, which operates from the start of the mission, fails before the mission is completed. The function of the sensing subsystem is to detect a failure in the active unit or subsystem and command the switching subsystem to switch in the standby unit or subsystem.

This is unlike the reliabilitywise parallel system in which all units start to operate simultaneously at the beginning of the mission, and the system succeeds when at least one unit is left which completes the mission successfully.

3.2   RELIABILITY OF A TWO-UNIT STANDBY SYSTEM

A two-unit standby system succeeds when (1) the active unit does not fail, the sensing subsystem does not fail, and the switch does not fail open during the mission, or (2) the active unit fails before the end of the mission, the sensing and switching subsystems do not fail, and the standby unit not having already failed succeeds for the remainder of

41

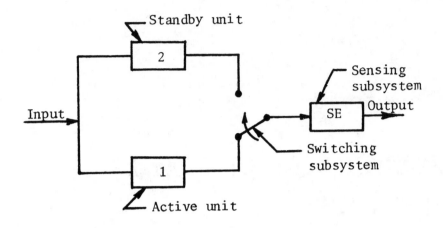

Fig. 3.1– Two-unit standby system with sensing and switching subsystems.

the mission. The reliability block diagram of such a standby system is shown in Fig. 3.1. Probabilistically, the reliability of this system is the probability that

1.　Unit 1 succeeds for the whole mission of T duration, *and* the sensing subsystem does not fail, *and* the switch does not fail open for the whole mission;

or that

2.　Unit 1 fails in time interval $(T_1, T_1 + \Delta T_1)$ prior to T, *and* the sensing subsystem does not fail by T_1, and the switching subsytem does not fail in the quiescent mode by T_1, *and* the switch successfully switches in the standby unit in the energized mode with a one-cycle operation, at which time the sensing subsystem is turned off, *and* the standby unit does not fail by T_1 in the quiescent mode *and* functions successfully for the remainder of the mission in the energized mode, *and* the switch does not fail open until the end of the mission.

Assuming that the failures of different units and the failures due to different failure modes are independent, the probability of the first event is given mathematically by

$$R_{1E}(T) \cdot R_{SE}(T) \cdot R_{SWO}(T).$$

For the second event, let

$A_1 = $ [Unit 1 fails in the time interval $(T_1, T_1 + dT_1)$ prior to T],

$A_2 = $ (sensing subsystem does not fail by T_1),

$A_3 = $ [(switch does not fail in the quiescent mode by T_1) and (switch does not fail open from T_1 to T)],

$A_4 = $ (switch successfully switches in the standby unit in the energized mode with a single-cycle operation),

and

$A_5 = $ [(Unit 2 does not fail by T_1 in the quiescent mode) and (Unit 2 functions successfully from T_1 to T)],

or

$$A_5 = [(A_{51}) \cap (A_{52})],$$

where

$A_{51} = $ (Unit 2 does not fail by T_1 in the quiescent mode),

and

$A_{52} = $ (Unit 2 functions successfully from T_1 to T).

Then

$$
\begin{aligned}
P(A_1) &= f_1(T_1)\, dT_1, \\
P(A_2) &= R_{SE}(T_1), \\
P(A_3) &= P[(\text{switch does not fail open from 0 to } T_1 \text{ and} \\
& \quad \text{switch does not fail closed from 0 to } T_1), \text{ and} \\
& \quad (\text{switch does not fail open from } T_1 \text{ to T})], \\
&= R_{SWO}(T_1) \cdot R_{SWC}(T_1) \cdot R_{SWO}(T_1, T - T_1) \\
&= R_{SWQ}(T_1) \cdot R_{SWO}(T_1, T - T_1),
\end{aligned}
$$

and

$$P(A_4) = R_{SWE}(1 \text{ cycle}).$$

The probability of event A_5, consisting of two events, can be obtained as follows:

$$
\begin{aligned}
P(A_5) &= P(A_{51} \cap A_{52}), \\
&= P(A_{51}) \cdot P(A_{52}|A_{51}),
\end{aligned}
$$

where

$$P(A_{51}) \;=\; R_{2Q}(T_1).$$

It must be pointed out that

$$P(A_{52}|A_{51}) \neq R_{2E}(T_1, T - T_1),$$

because the time-to-failure distributions in the quiescent mode and in the energized mode are different. Let

$$P(A_{52}|A_{51}) = R_{2E}(T_{1e}, T - T_1),$$

where T_{1e} is an equivalent time such that, from the conservation of reliability principle,

$$R_{2Q}(T_1) = R_{2E}(T_{1e}), \tag{3.1}$$

then

$$P(A_5) = R_{2Q}(T_1) \cdot R_{2E}(T_{1e}, T - T_1).$$

Therefore, the probability of the second event is given by

$$\int_{T_1=0}^{T} f_1(T_1) \cdot R_{SE}(T_1) \cdot R_{SWQ}(T_1) \cdot R_{SWO}(T_1, T - T_1)$$
$$\cdot R_{SWE}(1 \text{ cycle}) \cdot R_{2Q}(T_1) \cdot R_{2E}(T_{1e}, T - T_1) \, dT_1,$$

where T_1 is a variable quantity, because we do not know ahead of time when the active unit will fail. The upper limit of this integral is T because the variable time T_1 may vary from a split second after the mission starts to a split second before the misson is completed, thus requiring that the standby unit be brought in to complete the mission successfully. Consequently, T_1 varies conservatively from 0 to T.

Then the reliability of this standby system, starting the mission at age zero, is given by

$$\begin{aligned}
R_{SB}(T) \;=\;& R_{1E}(T) \cdot R_{SE}(T) \cdot R_{SWO}(T) \\
&+ \int_{T_1=0}^{T} f_1(T_1) \cdot R_{SE}(T_1) \cdot R_{SWQ}(T_1) \\
&\quad \cdot R_{SWO}(T_1, T - T_1) \cdot R_{SWE}(1 \text{ cycle}) \\
&\quad \cdot R_{2Q}(T_1) \cdot R_{2E}(T_{1e}, T - T_1) \, dT_1. \tag{3.2}
\end{aligned}$$

If all units have constant failure rates and Unit 1 has an active failure rate of λ_1, the sensing subsystem has a failure rate of λ_{SE}, the switching subsystem has a failure rate of λ_{SWE} and operates for one cycle, the standby Unit 2 has a quiescent (idling) failure rate of λ_{2Q} and an energized, functioning failure rate of λ_{2E}, and the switch has a failing open failure rate of λ_{SWO}, then in Eq. (3.2)

$$R_{SWO}(T_1, T - T_1) = \frac{R_{SWO}(T_1 + T - T_1)}{R_{SWO}(T_1)} = \frac{R_{SWO}(T)}{R_{SWO}(T_1)},$$

which for the exponential case is

$$R_{SWO}(T_1, T - T_1) = \frac{e^{-\lambda_{SWO}T}}{e^{-\lambda_{SWO}T_1}} = e^{-\lambda_{SWO}(T-T_1)}.$$

Therefore, the time domain of function is $(T - T_1)$.

From Eq. (3.1)

$$e^{-\lambda_{2Q}T_1} = e^{-\lambda_{2E}T_{1e}}.$$

Taking the natural logarithm of both sides yields

$$\lambda_{2Q}T_1 = \lambda_{2E}T_{1e},$$

from which

$$T_{1e} = \frac{\lambda_{2Q}T_1}{\lambda_{2E}}.$$

This equation is to be used when T_{1e} needs to be quantified. However, as will be seen next, its quantification is not required to obtain the solution to this case because Unit 2 is exponential.

The term $R_{2E}(T_{1e}, T - T_1)$ is quantified for the exponential case as follows:

$$\begin{aligned} R_{2E}(T_{1e}, T - T_1) &= \frac{R_{2E}(T_{1e} + T - T_1)}{R_{2E}(T_{1e})}, \\ &= \frac{e^{-\lambda_{2E}(T_{1e}+T-T_1)}}{e^{-\lambda_{2E}T_{1e}}}, \end{aligned}$$

or

$$R_{2E}(T_{1e}, T - T_1) = e^{-\lambda_{2E}(T-T_1)}.$$

Now Eq. (3.2) becomes

$$\begin{aligned} R_{SB} &= e^{-\lambda_1 T} \cdot e^{-\lambda_{SE}T} \cdot e^{-\lambda_{SWO}T} \\ &\quad + \int_0^T \lambda_1 e^{-\lambda_1 T_1} \cdot e^{-\lambda_{SE}T_1} \cdot e^{-\lambda_{SWQ}T_1} \cdot e^{-\lambda_{SWO}(T-T_1)} \\ &\quad \cdot e^{-\lambda_{SWE} \cdot 1} \cdot e^{-\lambda_{2Q}T_1} \cdot e^{-\lambda_{2E}(T-T_1)} dT_1, \qquad (3.1') \end{aligned}$$

or

$$\begin{aligned} R_{SB} &= e^{-(\lambda_1+\lambda_{SE}+\lambda_{SWO})T} + \lambda_1 \cdot e^{-\lambda_{SWE}} \cdot e^{-\lambda_{SWO}T} \cdot e^{-\lambda_{2E}T} \\ &\quad \cdot \int_0^T e^{-(\lambda_1+\lambda_{SE}+\lambda_{SWQ}-\lambda_{SWO}+\lambda_{2Q}-\lambda_{2E})T_1} dT_1. \end{aligned}$$

Let

$$\lambda_A = \lambda_1 + \lambda_{SE} + \lambda_{SWO},$$
$$\lambda_B = \lambda_{2E} + \lambda_{SWO},$$

and

$$\lambda_C = \lambda_1 + \lambda_{SE} + \lambda_{SWQ} - \lambda_{SWO} + \lambda_{2Q} - \lambda_{2E};$$

then

$$R_{SB}(T) = e^{-\lambda_A T} + \frac{\lambda_1 e^{-\lambda_{SWE}} e^{-\lambda_B T}}{\lambda_C}(1 - e^{-\lambda_C T}). \qquad (3.3)$$

All of these failure rates except λ_{SWE} are in units of failures per hour, but λ_{SWE} is in failures per cycle of operation.

If

$$\lambda_{SWE} = \lambda_{SWQ} = \lambda_{SWO} = \lambda_{SE} = \lambda_{2Q} = 0, \qquad (3.4)$$

then $\lambda_{2E} = \lambda_2$ and Eq. (3.3) becomes

$$R_{SB}(T) = e^{-\lambda_1 T} + \frac{\lambda_1 e^{-\lambda_2 T}}{\lambda_1 - \lambda_2}[1 - e^{-(\lambda_1 - \lambda_2)T}],$$

or

$$R_{SB}(T) = e^{-\lambda_1 T} + \frac{\lambda_1}{\lambda_1 - \lambda_2}(e^{-\lambda_2 T} - e^{-\lambda_1 T}). \qquad (3.4')$$

If

$$\lambda_1 = \lambda_2 = \lambda,$$

the second term of Eq. (3.4') becomes indeterminate; consequently, substituting in $\lambda_1 = \lambda_2 = \lambda$ and the conditions in Eq. (3.4) in Eq. (3.1') yields

$$R_{SB}(T) = e^{-\lambda T} + \lambda T e^{-\lambda T},$$

or

$$R_{SB}(T) = e^{-\lambda T}(1 + \lambda T). \qquad (3.5)$$

This is the reliability of a system with one functioning unit and one standby unit, both units having the same failure rate, and with sensing, switching and quiescent unit reliability of 100%.

Furthermore, for a standby system with a total of n identical units where one unit is functioning and $(n - 1)$ units are in standby, the reliability is given by

$$R_{SB}(T) = e^{-\lambda T}[1 + \lambda T + \frac{(\lambda T)^2}{2!} + \cdots + \frac{(\lambda T)^{(n-1)}}{(n-1)!}] = P(N_f \leq n - 1), \qquad (3.6)$$

where N_f is the number of units that fail.

Equation (3.6) also gives the reliability of an exponential unit for a mission of T duration with $(n - 1)$ spares, such that if the unit fails it can be replaced by a spare.

The mean time between failures of such a system is given by

$$\text{MTBF}_{SB} = \frac{n}{\lambda} = nm,$$

or n times the MTBF of a single unit, and thereby lies the reason for the high reliability of a standby system when the sensing and switching reliability is 1.

If the units in a standby system are Weibullian, then from Eq. (3.1)

$$e^{-\left(\frac{T_1 - \gamma_{2Q}}{\eta_{2Q}}\right)^{\beta_{2Q}}} = e^{-\left(\frac{T_{1e} - \gamma_{2E}}{\eta_{2E}}\right)^{\beta_{2E}}},$$

or

$$\left(\frac{T_1 - \gamma_{2Q}}{\eta_{2Q}}\right)^{\beta_{2Q}} = \left(\frac{T_{1e} - \gamma_{2E}}{\eta_{2E}}\right)^{\beta_{2E}}. \qquad (3.7)$$

Taking the logarithm to the base e of Eq. (3.7) yields

$$\beta_{2Q} \log_e\left(\frac{T_1 - \gamma_{2Q}}{\eta_{2Q}}\right) = \beta_{2E} \log_e\left(\frac{T_{1e} - \gamma_{2E}}{\eta_{2E}}\right). \qquad (3.8)$$

Solving Eq. (3.8) for T_{1e} yields

$$T_{1e} = \gamma_{2E} + \eta_{2E}\left(\frac{T_1 - \gamma_{2Q}}{\eta_{2Q}}\right)^{\frac{\beta_{2Q}}{\beta_{2E}}}. \qquad (3.9)$$

Then in Eq. (3.2)

$$R_{2E}(T_{1e}, T - T_1) = \frac{R_{2E}(T_{1e} + T - T_1)}{R_{2E}(T_{1e})},$$

and

$$R_i(T) = e^{-\left(\frac{T - \gamma_i}{\eta_i}\right)^{\beta_i}},$$

where the subscript i identifies the corresponding parameters of the Weibull *pdf* for the unit involved. Numerical integration of the second term of Eq. (3.2) will then yield the system's reliability.

EXAMPLE 3–1

A system consists of one active and one standby unit with an energized constant failure rate of 1,053.50 fr/10^6 hr each and a quiescent failure rate of 105.50 fr/10^6 hr; a sensing subsytem with a failure rate of 2.85 fr/10^6 hr; and a switch with a quiescent failure rate of 0.25 fr/10^6 hr, a switching failure rate of 0.55 fr/10^6 cycles, and a failing open failure rate of 0.15 fr/10^6 hr. Determine the reliability of the standby system for a function period of 100 hr.

SOLUTION TO EXAMPLE 3–1

In Eq. (3.3) the following are known:

$$\lambda_1 = 1,053.50 \cdot 10^{-6} \text{ fr/hr}, \lambda_{2Q} = 105.50 \cdot 10^{-6} \text{ fr/hr},$$
$$\lambda_{2E} = 1,053.50 \cdot 10^{-6} \text{ fr/hr}, \lambda_{SE} = 2.85 \cdot 10^{-6} \text{ fr/hr},$$
$$\lambda_{SWQ} = 0.25 \cdot 10^{-6} \text{ fr/hr}, \lambda_{SWE} = 0.55 \cdot 10^{-6} \text{ fr/cycle},$$
$$\lambda_{SWO} = 0.15 \cdot 10^{-6} \text{ fr/hr, and } T = 100 \text{ hr.}$$

Then

$$\lambda_A = \lambda_1 + \lambda_{SE} + \lambda_{SWO} = (1,053.50 + 2.85 + 0.15) \cdot 10^{-6},$$

or

$$\lambda_A = 1,056.50 \cdot 10^{-6} \text{ fr/hr},$$
$$\lambda_B = \lambda_{2E} + \lambda_{SWO} = (1,053.50 + 0.15) \cdot 10^{-6},$$

or

$$\lambda_B = 1,053.65 \cdot 10^{-6} \text{ fr/hr},$$
$$\lambda_C = \lambda_1 + \lambda_{SE} + \lambda_{SWQ} + \lambda_{2Q} - \lambda_{2E} - \lambda_{SWO},$$
$$\lambda_C = (1,053.50 + 2.85 + 0.25 + 105.50 - 1,053.50$$
$$- 0.15) \cdot 10^{-6},$$

or

$$\lambda_C = 108.45 \cdot 10^{-6} \text{ fr/hr.}$$

Substitution of these quantities into Eq. (3.3) yields

$$R_{SB}(T = 100 \text{ hr}) = e^{-1,056.50 \cdot 10^{-6} \cdot 100}$$
$$+ \frac{1,053.50 \cdot 10^{-6} e^{-0.55 \cdot 10^{-6}} e^{-1,053.65 \cdot 10^{-6} \cdot 100}}{108.45 \cdot 10^{-6}}$$
$$\cdot (1 - e^{-108.45 \cdot 10^{-6} \cdot 100}),$$

or

$$R_{SB}(T = 100 \text{ hr}) = 0.994042.$$

EXAMPLE 3–2

Determine the reliability of the three two-unit systems given in Fig. 3.2 and compare them. Each unit has a reliability of 90% for a 100-hr mission.

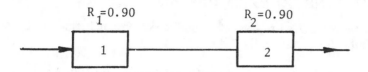

(a) Reliabilitywise in series.

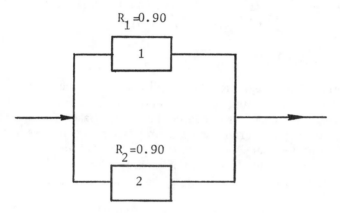

(b) Reliabilitywise in parallel.

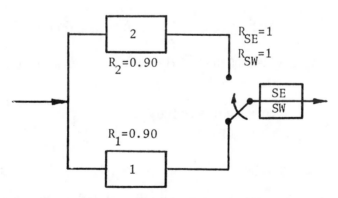

(c) Reliabilitywise in standby.

Fig. 3.2– Two-unit system of Example 3-2.

SOLUTIONS TO EXAMPLE 3–2

1. Figure 3.2(a) represents a series system such that if at least one unit fails the system fails. Then

$$R_{SS} = R_1 R_2 = (0.90)(0.90) = 0.8100.$$

2. Figure 3.2(b) represents a parallel system such that if at least one unit survives the system survives. Then

$$
\begin{aligned}
R_{SP} &= 1 - (Q_1 Q_2) = 1 - (1 - R_1)(1 - R_2), \\
&= 1 - (0.10)(0.10) = 1 - 0.0100,
\end{aligned}
$$

or

$$R_{SP} = 0.9900.$$

3. Figure 3.2(c) represents a standby system such that if at least one unit survives, starting with the active unit, the system survives its mission. Two cases are covered here. In Case 1, $\lambda_{2Q} = 0$ and $R_{SE} = R_{SW} = 1$, or perfect sensing and switching are assumed, and in Case 2 imperfect sensing and switching are assumed with $\lambda_{2Q} = 0$.

Case 1 – First, the failure rate of each unit needs to be determined such that its reliability is 90% for a 100-hr mission; then

$$R = 0.90 = e^{-\lambda T} = e^{-\lambda(100)},$$

$$\log_e 0.90 = -\lambda(100),$$

or

$$\lambda = \frac{\log_e 0.90}{-100} = \frac{-0.10536}{-100} = 0.0010536 \text{ fr/hr.}$$

Therefore, $\lambda_1 = \lambda_2 = 0.0010536$ fr/hr. The system's reliability then is

$$
\begin{aligned}
R_{SB} &= e^{-\lambda T} + \lambda T e^{-\lambda T}, \\
&= 0.90 + 0.0010536(100)(0.90), \\
&= 0.90 + 0.094824,
\end{aligned}
$$

or

$$= 0.994824.$$

Case 2–In this case the following failure rates are chosen:

$$
\begin{aligned}
\lambda_1 &= \lambda_{2E} = 1{,}053.6 \text{ fr}/10^6 \text{ hr}, \ \lambda_{2Q} = 0, \\
\lambda_{SE} &= 2.0 \text{ fr}/10^6 \text{ hr}, \ \lambda_{SWQ} = 1.5 \text{ fr}/10^6 \text{ hr}, \\
\lambda_{SWE} &= 0.5 \text{ fr}/10^6 \text{ cycles, and } \lambda_{SWO} = 1.0 \text{ fr}/10^6 \text{ hr.}
\end{aligned}
$$

Then, from Eq. (3.3),

$$\lambda_A = \lambda_1 + \lambda_{SE} + \lambda_{SWO},$$
$$= 1,053.6 + 2.0 + 1.0 = 1,055.6 \text{ fr}/10^6 \text{ hr},$$
$$\lambda_B = \lambda_{2E} + \lambda_{SWO},$$
$$= 1,053.6 + 1.0 = 1054.6 \text{ fr}/10^6 \text{ hr},$$

and

$$\lambda_C = \lambda_1 + \lambda_{SE} + \lambda_{SWQ} + \lambda_{2Q} - \lambda_{2E} - \lambda_{SWO},$$
$$= 1,053.6 + 2.0 + 1.5 + 0.0 - 1,053.6 - 1.0,$$
$$= 2.5 \text{ fr}/10^6 \text{ hr}.$$

Therefore,

$$\begin{aligned} R_{SB} &= e^{-1,055.6 \cdot 10^6 \cdot 100} \\ &+ \frac{1,055.6 \cdot 10^{-6} \cdot e^{-0.5 \cdot 10^{-6}} \cdot e^{-1,054.6 \cdot 10^{-6} \cdot 100}}{2.5 \cdot 10^{-6}} \\ &\cdot (1 - e^{-2.5 \cdot 10^{-6} \cdot 100}), \\ &= 0.8998204 + (379.98)(0.0002), \\ &= 0.8998204 + 0.075996, \end{aligned}$$

or

$$R_{SB} = 0.9759.$$

It may be seen that the most reliable configuration is the standby Case 1 with $R_{SB} = 0.994824$, the next best is the parallel with $R_{SP} = 0.9900$, the third best is the standby Case 2 with $R_{SB} = 0.9759$, and the least reliable is the series with $R_{SS} = 0.8100$. It must pointed out that a two-unit standby system is not always more reliable than a two-unit parallel system, particularly if the sensing and switching subsystem is not very reliable. Before a decision is made to use the same number of units in standby versus in parallel, the impact of the sensing and switching subsystem's reliability on the overall system's reliability must be evaluated and then the more reliable system configuration is chosen.

3.3   COMPLEX STANDBY SYSTEMS

For complex standby systems, a good approach is to prepare a table of the number of system success combinations, the units involved in the system, their modes of function, the system success function modes, and the time domain of function of each unit. This is best explained by the examples that follow.

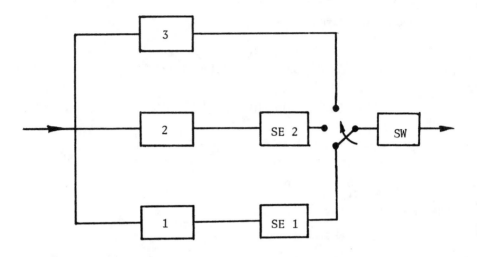

Fig. 3.3– Complex standby system of Example 3–3 with two standby units.

EXAMPLE 3–3

The standby system is shown in Fig. 3.3 and the system success combinations are listed in Table 3.1. Unit 1 is the active unit and Units 2 and 3 are the standby units. SE1 and SE2 are the sensing subsystems and SW is a three-position switch. Unit 1 is up 70% of the operating time with a failure rate of λ_{1E} and down 30% of the time with a failure rate of λ_{1Q}. When Units 2 and 3 are in their quiescent function mode, they have failure rates λ_{2Q} and λ_{3Q}; and when they are energized, they have failure rates λ_{2E} and λ_{3E}. However, during this period they are up 70% of the time and down 30% of the time. The failure-sensing devices have quiescent failure rates of λ_{SEQ1} and λ_{SEQ2}, and energized failure rates of λ_{SEE1} and λ_{SEE2}. The failure-sensing device SE1 is used to detect the failure of Unit 1 and to command the switch to cut off Unit 1 and bring in Unit 2. The failure-sensing device SE2 performs a similar function for Unit 2. The switching device has a switching failure rate of λ_{SWE}, a quiescent failure rate of λ_{SWQ}, and a switch failing-open failure rate of λ_{SWO}. Do the following:

1. Write in probability language the system's reliability.

2. Write the complete reliability equation for the standby system, and its solution.

Table 3.1– Three-unit standby system of Example 3-3.

Number of system success combination	Units, modes of function, system success modes, and time domains of function														
	1	SEE1	SWO	SWQ	SWE	SEQ2	SEE2	2Q	2	SWO	SWQ	SWE	3Q	3	SWO
1	G	G	G												
	T	T	T												
	B	G													
2	T_1	T_1		G	G	G	G	G	G	G					
	B	G		T_1	$1C$	T_1	$T-T_1$	T_1	$T-T_1$	$T-T_1$					
3	T_1	T_1		G	G	G	G	G	B		G	G	G	G	G
	B	G		T_1	$1C$	T_1	T_2-T_1	T_1	T_2-T_1		T_2-T_1	$1C$	T_2	$T-T_2$	$T-T_2$
4	T_1	T_1		T_1	$1C$	T_1		$< T_1$	B			$1C$	T_1	$T-T_1$	$T-T_1$

3. Given the following failure rates, find the system's reliability for a mission of 100 hr:

$$\lambda_{1E} = \lambda_{2E} = \lambda_{3E} = 1000.0 \text{ fr}/10^6 \text{ hr},$$
$$\lambda_{1Q} = \lambda_{2Q} = \lambda_{3Q} = 50.0 \text{ fr}/10^6 \text{ hr},$$
$$\lambda_{SEQ1} = \lambda_{SEQ2} = 0.01 \text{ fr}/10^6 \text{ hr},$$
$$\lambda_{SEE1} = \lambda_{SEE2} = 2.0 \text{ fr}/10^6 \text{ hr},$$
$$\lambda_{SWQ} = 0.5 \text{ fr}/10^6 \text{ hr},$$
$$\lambda_{SWE} = 0.5 \text{ fr}/10^6 \text{ cycles},$$

and

$$\lambda_{SWO} = 0.5 \text{ fr}/10^6 \text{ hr}.$$

SOLUTIONS TO EXAMPLE 3–3

1. The system's reliability is given by the following:

(1) The probability that Unit 1 does not fail during the mission *and* sensing SE1 does not fail up to the end of the mission *and* the switch does not fail open for the whole mission;

or

(2) Unit 1 fails in the time interval $(T_1, T_1 + \Delta T_1)$ prior to the end of the mission, *and* sensing SE1 is good by T_1, *and* the switch does not fail in the quiescent mode by T_1 *and* is good for 1 cycle of switching, *and* sensing SE2 does not fail in the quiescent mode by T_1 *and* functions well thereafter for the rest of the mission, *and* Unit 2 is good in the quiescent mode up to T_1 *and* is good in the energized mode during the rest of the mission or in period $T - T_1$, *and* the switch does not fail open during the rest of the mission or during $T - T_1$;

or

(3) Unit 1 fails in time interval $(T_1, T_1 + \Delta T_1)$ prior to the end of the mission, *and* sensing SE1 is good by T_1, *and* the switch does not fail in the quiescent mode by T_1 *and* is good for 1 cycle of swiching, *and* sensing SE2 does not fail in the quiescent mode by T_1 *and* works well from T_1 to T_2, *and* Unit 2 does not fail in the quiescent mode up to T_1, *and* works well from T_1 to T_2 *and* then fails in time interval $(T_2, T_2 + \Delta T_2)$ prior to T, *and* the switch

does not fail in the quiescent mode from T_1 to T_2 *and* is good when switching for the second time, *and* Unit 3 does not fail in the quiescent mode by T_2 *and* is good energized during period T_2 to T, *and* the switch does not fail open during the rest of the mission or during $T - T_2$;

or

(4) Unit 1 fails in time interval $(T_1, T_1 + \Delta T_1)$ prior to the end of the mission, and sensing SE1 does not fail by T_1, *and* the switch does not fail in the quiescent mode by T_1, *and* is good for one cycle of switching, *and* sensing Unit SE2 does not fail in the quiescent mode by T_1, *and* Unit 2 fails in the quiescent mode by T_1, and the switch is good for the second cycle of switching, *and* Unit 3 does not fail in the quiescent mode until T_1, *and* is good in period $T - T_1$, *and* the switch does not fail open for the rest of the mission or during $T - T_1$.

2. Let

$$\lambda_1 = 0.70\lambda_{1E} + 0.30\lambda_{1Q} = (0.70)(0.001) + (0.30)(0.00005),$$
$$= 0.000715 \text{ fr/hr},$$
$$\lambda_2 = 0.70\lambda_{2E} + 0.30\lambda_{2Q} = (0.70)(0.001) + (0.30)(0.00005),$$
$$= 0.000715 \text{ fr/hr},$$

and

$$\lambda_3 = 0.70\lambda_{3E} + 0.30\lambda_{3Q} = (0.70)(0.001) + (0.30)(0.00005),$$
$$= 0.000715 \text{ fr/hr}.$$

Then, mathematically, the system's reliability can be writen as follows:

$$
\begin{aligned}
R(T) = \ & e^{-\lambda_1 T} e^{-\lambda_{SEE1} T} e^{-\lambda_{SWO} T} \\
& + \int_{T_1=0}^{T} \lambda_1 e^{-\lambda_1 T_1} e^{-\lambda_{SEE1} T_1} e^{-\lambda_{SWQ} T_1} e^{-\lambda_{SWE} \cdot 1} \\
& \cdot e^{-\lambda_{SEQ2} T_1} e^{-\lambda_{SEE2}(T-T_1)} e^{-\lambda_{2Q} T_1} e^{-\lambda_2(T-T_1)} \\
& \cdot e^{-\lambda_{SWO}(T-T_1)} \, dT_1, \\
& + \int_{T_2=0}^{T} \int_{T_1=0}^{T_2} \lambda_1 e^{-\lambda_1 T_1} e^{-\lambda_{SEE1} T_1} e^{-\lambda_{SWQ} T_1} \\
& \cdot e^{-\lambda_{SWE} \cdot 1} e^{-\lambda_{SEQ2} T_1} e^{-\lambda_{SEE2}(T_2-T_1)} e^{-\lambda_{2Q} T_1} \lambda_2 \\
& \cdot e^{-\lambda_2(T_2-T_1)} e^{-\lambda_{SWQ}(T_2-T_1)} \\
& \cdot e^{-\lambda_{SWE} \cdot 1} e^{-\lambda_{3Q} T_2} e^{-\lambda_3(T-T_2)} e^{-\lambda_{SWO}(T-T_2)} \, dT_1 \, dT_2,
\end{aligned}
$$

$$+ \int_{T_1=0}^{T} \lambda_1 e^{-\lambda_1 T_1} e^{-\lambda_{SEE1} T_1} e^{-\lambda_{SWQ} T_1}$$
$$\cdot e^{-\lambda_{SWE} \cdot 1} e^{-\lambda_{SEQ2} T_1} (1 - e^{-\lambda_{2Q} T_1}) e^{-\lambda_{SWE} \cdot 1}$$
$$\cdot e^{-\lambda_{3Q} T_1} e^{-\lambda_3 (T-T_1)} e^{-\lambda_{SWO} (T-T_1)} \, dT_1.$$

Note that in the above expression's fourth term $Q(T_1)$, which is the probability that Unit 2 fails in its quiescent mode before T_1, needs to be quantified. This may be done as follows:

$$Q_{2Q}(T1) = \int_{T=0}^{T_1} \lambda_{2Q} e^{-\lambda_{2Q} T} \, dT,$$

or

$$Q_{2Q}(T_1) = 1 - e^{-\lambda_{2Q} T_1}.$$

This expression has been inserted in this term.

Then

$$R(T) = e^{-(\lambda_1 + \lambda_{SEE1} + \lambda_{SWO})T} + \lambda_1 e^{-[\lambda_{SWE} + (\lambda_{SEE2} + \lambda_2 + \lambda_{SWO})T]}$$
$$\cdot \int_{T_1=0}^{T} e^{-(\lambda_1 + \lambda_{SEE1} + \lambda_{SWQ} + \lambda_{SEQ2} - \lambda_{SEE2} + \lambda_{2Q} - \lambda_2 - \lambda_{SWO})T_1} \, dT_1$$
$$+ \lambda_1 \lambda_2 e^{-[2\lambda_{SWE} + (\lambda_3 + \lambda_{SWO})T]}$$
$$\cdot \int_{T_2=0}^{T} \int_{T_1=0}^{T_2} e^{-(\lambda_{SEE2} + \lambda_2 + \lambda_{SWQ} + \lambda_{3Q} - \lambda_3 - \lambda_{SWO})T_2}$$
$$\cdot e^{-(\lambda_1 + \lambda_{SEE1} + \lambda_{SWQ} + \lambda_{SEQ2} - \lambda_{SEE2} + \lambda_{2Q} - \lambda_2 - \lambda_{SWQ})T_1} \, dT_1 \, dT_2,$$
$$+ \lambda_1 e^{-[2 \cdot \lambda_{SWE} + (\lambda_3 + \lambda_{SWO})T]}$$
$$\cdot \int_{T_1=0}^{T} (1 - e^{-\lambda_{2Q} T_1})$$
$$\cdot e^{-(\lambda_1 + \lambda_{SEE1} + \lambda_{SWQ} + \lambda_{SEQ2} + \lambda_{3Q} - \lambda_3 - \lambda_{SWO})T_1} \, dT_1.$$

Denote

$$
\begin{aligned}
A &= \lambda_1 + \lambda_{SEE1} + \lambda_{SWO}, \\
B &= \lambda_{SEE2} + \lambda_2 + \lambda_{SWO}, \\
C &= \lambda_1 + \lambda_{SEE1} + \lambda_{SWQ} + \lambda_{SEQ2} - \lambda_{SEE2} + \lambda_{2Q} - \lambda_2 \\
  &\quad - \lambda_{SWO}, \\
D &= \lambda_{SEE2} + \lambda_2 + \lambda_{SWQ} + \lambda_{3Q} - \lambda_3 - \lambda_{SWO}, \\
E &= \lambda_1 + \lambda_{SEE1} + \lambda_{SWQ} + \lambda_{SEQ2} - \lambda_{SEE2} + \lambda_{2Q} \\
  &\quad - \lambda_2 - \lambda_{SWQ}, \\
F &= \lambda_1 + \lambda_{SEE1} + \lambda_{SWQ} + \lambda_{SEQ2} + \lambda_{3Q} - \lambda_3 - \lambda_{SWO},
\end{aligned}
$$

and

$$G = \lambda_3 + \lambda_{SWO}.$$

Then

$$R(T) = e^{-AT} + \lambda_1 e^{-(\lambda_{SWE}+BT)} \int_{T_1=0}^{T} e^{-CT_1} \, dT_1$$

$$+ \lambda_1 \lambda_2 e^{-[2\lambda_{SWE}+GT]} \int_{T_2=0}^{T} \int_{T_1=0}^{T_2} e^{-DT_2} e^{-ET_1} \, dT_1 \, dT_2$$

$$+ \lambda_1 e^{-[2\lambda_{SWE}+GT]} \int_{T_1=0}^{T} e^{-FT_1} (1 - e^{-\lambda_{2Q}T_1}) \, dT_1,$$

or

$$R(T) = e^{-AT} + \frac{\lambda_1 e^{-(\lambda_{SWE}+BT)}}{C}(1 - e^{-CT})$$

$$+ \lambda_1 \lambda_2 e^{-[2\lambda_{SWE}+GT]} \{ \frac{1}{DE}(1 - e^{-DT})$$

$$- \frac{1}{E(D+E)}[1 - e^{-(D+E)T}] \}$$

$$+ \lambda_1 e^{-[2\lambda_{SWE}+GT]} \{ \frac{1}{F}(1 - e^{-FT})$$

$$- \frac{1}{F+\lambda_{2Q}}[1 - e^{-(F+\lambda_{2Q})T}] \}.$$

3. Substitution of the values of the failure rates into the expressions of A, B, C, D, E, F and G yields

$$A = (715.0 + 2.0 + 0.5)(10^{-6}) = 0.7175 \cdot 10^{-3},$$

$$B = (2.0 + 715.0 + 0.5)(10^{-6}) = 0.7175 \cdot 10^{-3},$$

$$C = (715.0 + 2.0 + 0.5 + 0.01 - 2.0 + 50.0 - 715.0$$
$$- 0.5)(10^{-6}),$$
$$= 0.05 \cdot 10^{-3},$$

$$D = (2.0 + 715.0 + 0.5 + 50.0 - 715.0$$
$$- 0.5)(10^{-6}),$$
$$= 0.052 \cdot 10^{-3},$$

$$E = (715.0 + 2.0 + 0.5 + 0.01 - 2.0 + 50.0 - 715.0$$
$$- 0.5)(10^{-6}),$$
$$= 0.05 \cdot 10^{-3},$$

$$F = (715.0 + 2.0 + 0.5 + 0.01 + 50.0 - 715.0 - 0.5)(10^{-6}),$$
$$= 0.052001 \cdot 10^{-3},$$

and

$$G = (715.0 + 0.5) = 0.7155 \cdot 10^{-3}.$$

Then

$$
\begin{aligned}
R(T) = \ & e^{-(0.0007175)T} \\
& + \frac{(715.0 \cdot 10^{-6})e^{-((0.5 \cdot 10^{-6}) + (0.0007175)T)}}{0.00005} \\
& \cdot (1 - e^{-(-0.00005T)}) \\
& + (715.0 \cdot 10^{-6})(715.0 \cdot 10^{-6}) \\
& \cdot e^{-[2(0.5 \cdot 10^{-6}) + (0.0007175)T]} \\
& \cdot \{ \frac{1 - e^{-(0.000052)T}}{(0.000052)(0.00005)} \\
& - \frac{1 - e^{-(0.000052 + 0.00005)T}}{(0.00005)(0.000052 + 0.00005)} \} \\
& + (715.0 \cdot 10^{-6})e^{-[2(0.5 \cdot 10^{-6}) + (0.0007155)T]} \\
& \cdot \{ \frac{1}{0.000052001}(1 - e^{-0.000052001T}) \\
& - \frac{1}{(0.000052001) + (50.0 \cdot 10^{-6})} \\
& \cdot [1 - e^{-((0.000052001) + (50.0 \cdot 10^{-6}))T}] \} \},
\end{aligned}
$$

or

$$R(T = 100 \text{ hr}) = 0.99968.$$

EXAMPLE 3–4

In the standby system shown in Fig. 3.4, Units 1 and 2 are functioning in parallel. The failure-sensing device has a failure rate of λ_{SE}. Assume that as soon as the standby Unit 4 is brought into function the sensing subsystem is no longer needed. The switching device contains three switches with energized failure rates of λ_{SWE1}, λ_{SWE2}, and λ_{SWE3}, quiescent failure rates of λ_{SWQ1}, λ_{SWQ2}, λ_{SWQ3}, and failing open failure rates of λ_{SWO1}, λ_{SWO2}, and λ_{SWO3}. The standby units have quiescent failure of λ_{3Q} and λ_{4Q} and energized failure rates of λ_{3E} and λ_{4E}. If both Units 1 and 2 fail, Unit 3 is switched in, and if Unit 3 fails, Unit 4 is switched in. Do the following:

1. Prepare a table of system success function modes.

2. Write the complete reliability equation in the form of integrals.

3. Integrate out completely the first single integral in the solution.

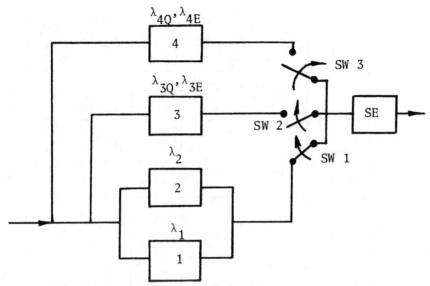

Fig. 3.4– Reliability block diagram for two Units in parallel and two Units in standby with sensing and switching Units of Example 3–4.

4. Integrate out completely the first double integral in the solution.

5. Write out the complete solution.

SOLUTIONS TO EXAMPLE 3–4

1. Table 3.2 gives the system success function modes.
2. The system's reliability, in integrals form, is given by

$$R(T) = [e^{-\lambda_1 T} + e^{-\lambda_2 T} - e^{-(\lambda_1+\lambda_2)T}]e^{-\lambda_{SE}T}e^{-\lambda_{SWO1}T}$$

$$+ \int_{T_1=0}^{T} \{[\lambda_1 e^{-\lambda_1 T_1} + \lambda_2 e^{-\lambda_2 T_1} - (\lambda_1 + \lambda_2)e^{-(\lambda_1+\lambda_2)T_1}]$$

$$\cdot e^{-\lambda_{SE}T}e^{-\lambda_{SWO1}T_1}e^{-\lambda_{SWQ2}T_1}e^{-\lambda_{SWE2}\cdot 1}e^{-\lambda_{3Q}T_1}$$

$$\cdot e^{-\lambda_{3E}(T-T_1)}e^{-\lambda_{SWO2}(T-T_1)}\} dT_1$$

$$+ \int_{T_2=0}^{T}\int_{T_1=0}^{T_2} \{[\lambda_1 e^{-\lambda_1 T_1} + \lambda_2 e^{-\lambda_2 T_1} - (\lambda_1 + \lambda_2)e^{-(\lambda_1+\lambda_2)T_1}]$$

$$\cdot e^{-\lambda_{SE}T_2}e^{-\lambda_{SWO1}T_1}e^{-\lambda_{SWQ2}T_1}e^{-\lambda_{SWE2}\cdot 1}e^{-\lambda_{3Q}T_1}e^{-\lambda_{3E}(T_2-T_1)}$$

$$\cdot \lambda_{SWO2}e^{-\lambda_{SWO2}(T_2-T_1)}e^{-\lambda_{SWQ3}T_2}e^{-\lambda_{SWE3}\cdot 1}e^{-\lambda_{4Q}T_2}$$

$$\cdot e^{-\lambda_{4E}(T-T_2)}e^{-\lambda_{SWO3}(T-T_2)}\} dT_1\, dT_2$$

Table 3.2– Tabular solution to Example 3-4.

Number of system success combination	Units, modes of function, system success modes, and time domains of function												
	1&2	SE	SWO1	SWQ2	SWE2	3Q	3E	SWO2	SWQ3	SWE3	4Q	4E	SWO3
1	G / T	G / T	G / T										
2	B / T_1	G / T	G / T_1	G / T_1	G / $1C$	G / T_1	G / $T-T_1$	G / $T-T_1$	G / T_2	G / $1C$	G / T_2	G / $T-T_2$	G / $T-T_2$
3	B / T_1	G / T_2	G / T_1	G / T_1	G / $1C$	G / T_1	G / T_2-T_1	B / T_2-T_1	G / T_2	G / $1C$	G / T_2	G / $T-T_2$	G / $T-T_2$
4	B / T_1	G / T_1	G / T_1	G / T_1	B / $1C$				G / T_1	G / $1C$	G / T_1	G / $T-T_1$	G / $T-T_1$
5	B / T_1	G / T_2	G / T_1	G / T_1	G / $1C$	G / T_1	B / T_2-T_1	G / T_2-T_1	G / T_2	G / $1C$	G / T_2	G / $T-T_2$	G / $T-T_2$
6	B / T_1	G / T_1	G / T_1	G / T_1	G / $1C$	B / $<T_1$			G / T_1	G / $1C$	G / T_1	G / $T-T_1$	G / $T-T_1$
7	B / T_1	G / T_1	G / T_1	B / $<T_1$					G / T_1	G / $1C$	G / T_1	G / $T-T_1$	G / $T-T_1$
8	G / T_1	G / T	B / T_1	G / T_1	G / $1C$	G / T_1	G / $T-T_1$	G / $T-T_1$	G / T_2	G / $1C$	G / T_2	G / $T-T_2$	G / $T-T_2$
9	G / T_1	G / T_2	B / T_1	G / T_1	G / $1C$	G / T_1	G / T_2-T_1	B / T_2-T_1	G / T_2	G / $1C$	G / T_2	G / $T-T_2$	G / $T-T_2$
10	G / T_1	G / T_1	B / T_1	G / T_1	B / $1C$				G / T_1	G / $1C$	G / T_1	G / $T-T_1$	G / $T-T_1$
11	G / T_1	G / T_2	B / T_1	G / T_1	G / $1C$	G / T_1	B / T_2-T_1	G / T_2-T_1	G / T_2	G / $1C$	G / T_2	G / $T-T_2$	G / $T-T_2$
12	G / T_1	G / T_1	B / T_1	G / T_1	G / $1C$	B / $<T_1$			G / T_1	G / $1C$	G / T_1	G / $T-T_1$	G / $T-T_1$
13	G / T_1	G / T_1	B / T_1	B / $<T_1$					G / T_1	G / $1C$	G / T_1	G / $T-T_1$	G / $T-T_1$

$$+ \int_{T_1=0}^{T} \{[\lambda_1 e^{-\lambda_1 T_1} + \lambda_2 e^{-\lambda_2 T_1} - (\lambda_1 + \lambda_2)e^{-(\lambda_1+\lambda_2)T_1}]$$

$$\cdot e^{-\lambda_{SE}T_1} e^{-\lambda_{SWO1}T_1} e^{-\lambda_{SWQ2}T_1}(1 - e^{-\lambda_{SWE2}\cdot 1})e^{-\lambda_{SWQ3}T_1}$$

$$\cdot e^{-\lambda_{SWE3}\cdot 1} e^{-\lambda_4 Q T_1} e^{-\lambda_4 E(T-T_1)} e^{-\lambda_{SWO3}(T-T_1)} \} \, dT_1$$

$$+ \int_{T_2=0}^{T} \int_{T_1=0}^{T_2} \{[\lambda_1 e^{-\lambda_1 T_1} + \lambda_2 e^{-\lambda_2 T_1} - (\lambda_1 + \lambda_2)e^{-(\lambda_1+\lambda_2)T_1}]$$

$$\cdot e^{-\lambda_{SE}T_2} e^{-\lambda_{SWO1}T_1} e^{-\lambda_{SWQ2}T_1} e^{-\lambda_{SWE2}\cdot 1} e^{-\lambda_{3Q}T_1} \lambda_{3E}$$

$$\cdot e^{-\lambda_{3E}(T_2-T_1)} e^{-\lambda_{SWO2}(T_2-T_1)} e^{\lambda_{SWQ3}T_2} e^{-\lambda_{SWE3}\cdot 1} e^{-\lambda_4 Q T_2}$$

$$\cdot e^{-\lambda_4 E(T-T_2)} e^{-\lambda_{SWO3}(T-T_2)} \} \, dT_1 \, dT_2$$

$$+ \int_{T_1=0}^{T} \{[\lambda_1 e^{-\lambda_1 T_1} + \lambda_2 e^{-\lambda_2 T_1} - (\lambda_1 + \lambda_2)e^{-(\lambda_1+\lambda_2)T_1}]$$

$$\cdot e^{-\lambda_{SE}T_1} e^{-\lambda_{SWO1}T_1} e^{-\lambda_{SWQ2}T_1} e^{-\lambda_{SWE2}\cdot 1}(1 - e^{-\lambda_{3Q}T_1})$$

$$\cdot e^{-\lambda_{SWQ3}T_1} e^{-\lambda_{SWE3}\cdot 1} e^{-\lambda_4 Q T_1} e^{-\lambda_4 E(T-T_1)}$$

$$\cdot e^{-\lambda_{SWO3}(T-T_1)} \} \, dT_1$$

$$+ \int_{T_1=0}^{T} \{[\lambda_1 e^{-\lambda_1 T_1} + \lambda_2 e^{-\lambda_2 T_1} - (\lambda_1 + \lambda_2)e^{-(\lambda_1+\lambda_2)T_1}]$$

$$\cdot e^{-\lambda_{SE}T_1} e^{-\lambda_{SWO1}T_1}(1 - e^{-\lambda_{SWQ2}T_1})e^{-\lambda_{SWQ3}T_1} e^{-\lambda_{SWE3}\cdot 1}$$

$$\cdot e^{-\lambda_4 Q T_1} e^{-\lambda_4 E(T-T_1)} e^{-\lambda_{SWO3}(T-T_1)} \} \, dT_1$$

$$+ \int_{T_1=0}^{T} \{[e^{-\lambda_1 T_1} + e^{-\lambda_2 T_1} - e^{-(\lambda_1+\lambda_2)T_1}]e^{-\lambda_{SE}T_1} \lambda_{SWO1}$$

$$\cdot e^{-\lambda_{SWO1}T_1} e^{-\lambda_{SWQ2}T_1} e^{-\lambda_{SWE2}\cdot 1} e^{-\lambda_{3Q}T_1}$$

$$\cdot e^{-\lambda_{3E}(T-T_1)} e^{-\lambda_{SWO2}(T-T_1)} \} \, dT_1$$

$$+ \int_{T_2=0}^{T} \int_{T_1=0}^{T_2} \{[e^{-\lambda_1 T_1} + e^{-\lambda_2 T_1} - e^{-(\lambda_1+\lambda_2)T_1}]e^{-\lambda_{SE}T_2}$$

$$\cdot \lambda_{SWO1} e^{-\lambda_{SWO1}T_1} e^{-\lambda_{SWQ2}T_1} e^{-\lambda_{SWE2}\cdot 1} e^{-\lambda_{3Q}T_1} e^{-\lambda_{3E}(T_2-T_1)}$$

$$\cdot \lambda_{SWO2} e^{-\lambda_{SWO2}(T_2-T_1)} e^{-\lambda_{SWQ3}T_2} e^{-\lambda_{SWE3}\cdot 1} e^{-\lambda_4 Q T_2}$$

$$\cdot e^{-\lambda_4 E(T-T_2)} e^{-\lambda_{SWO3}(T-T_2)} \} \, dT_1 \, dT_2$$

$$+ \int_{T_1=0}^{T} \{[e^{-\lambda_1 T_1} + e^{-\lambda_2 T_1} e^{-(\lambda_1+\lambda_2)T_1}]e^{-\lambda_{SE}T_1} \lambda_{SWO1}$$

$$\cdot e^{-\lambda_{SWO1}T_1} e^{-\lambda_{SWQ2}T_1}(1 - e^{-\lambda_{SWE2}})e^{-\lambda_{SWQ3}T_1} e^{-\lambda_{SWE3}\cdot 1}$$

$$\cdot e^{-\lambda_4 Q T_1} e^{-\lambda_4 E(T-T_1)} e^{-\lambda_{SWO3}(T-T_1)} \} \, dT_1$$

$$+ \int_{T_2=0}^{T} \int_{T_1=0}^{T_2} \{[e^{-\lambda_1 T_1} + e^{-\lambda_2 T_1} - e^{-(\lambda_1+\lambda_2)T_1}]e^{-\lambda_{SE}T_2}$$

$$\cdot \lambda_{SWO1} e^{-\lambda_{SWO1}T_1} e^{-\lambda_{SWQ2}T_1} e^{-\lambda_{SWE2}\cdot 1} e^{-\lambda_{3Q}T_1} \lambda_{3E}$$

$$\cdot\, e^{-\lambda_{3E}(T_2-T_1)}e^{-\lambda_{SWO2}(T_2-T_1)}e^{-\lambda_{SWQ3}T_2}e^{-\lambda_{SWE3}\cdot1}e^{-\lambda_{4Q}T_2}$$

$$\cdot\, e^{-\lambda_{4E}(T-T_2)}e^{-\lambda_{SWO3}(T-T_2)}\}\,dT_1\,dT_2$$

$$+\int_{T_1=0}^{T}\{[e^{-\lambda_1 T_1}+e^{-\lambda_2 T_1}e^{-(\lambda_1+\lambda_2)T_1}]e^{-\lambda_{SE}T_1}\lambda_{SWO1}$$

$$\cdot\, e^{-\lambda_{SWO1}T_1}e^{-\lambda_{SWQ2}T_1}e^{-\lambda_{SWE2}\cdot1}(1-e^{-\lambda_{3Q}T_1})e^{-\lambda_{SWQ3}T_1}$$

$$\cdot\, e^{-\lambda_{SWE3}\cdot1}e^{-\lambda_{4Q}T_1}e^{-\lambda_{4E}(T-T_1)}e^{-\lambda_{SWO3}(T-T_1)}\}\,dT_1$$

$$+\int_{T_1=0}^{T}\{[e^{-\lambda_1 T_1}+e^{-\lambda_2 T_1}-e^{-(\lambda_1+\lambda_2)T_1}]e^{-\lambda_{SE}T_1}$$

$$\cdot\, \lambda_{SWO1}e^{-\lambda_{SWO1}T_1}(1-e^{-\lambda_{SWQ2}T_1})e^{-\lambda_{SWQ3}T_1}e^{-\lambda_{SWE3}\cdot1}$$

$$\cdot\, e^{-\lambda_{4Q}T_1}e^{-\lambda_{4E}(T-T_1)}e^{-\lambda_{SWO3}(T-T_1)}\}\,dT_1.$$

3. The first single integral in the solution is

$$\int_{T_1=0}^{T}\{[\lambda_1 e^{-\lambda_1 T_1}+\lambda_2 e^{-\lambda_2 T_1}-(\lambda_1+\lambda_2)e^{-(\lambda_1+\lambda_2)T_1}]e^{-\lambda_{SE}T}$$

$$\cdot\, e^{-\lambda_{SWO1}T_1}e^{-\lambda_{SWQ2}T_1}e^{-\lambda_{SWE2}\cdot1}e^{-\lambda_{3Q}T_1}$$

$$\cdot\, e^{-\lambda_{3E}(T-T_1)}e^{-\lambda_{SWO2}(T-T_1)}\}\,dT_1$$

$$=e^{-\lambda_{SWE2}+(\lambda_{SE}+\lambda_{3E}+\lambda_{SWO2})T}$$

$$\{\int_{T_1=0}^{T}\lambda_1 e^{-(\lambda_1+\lambda_{SWO1}+\lambda_{SWQ2}+\lambda_{3Q}-\lambda_{3E}-\lambda_{SWO2})T_1}\,dT_1$$

$$+\int_{T_1=0}^{T}\lambda_2 e^{-(\lambda_2+\lambda_{SWO1}+\lambda_{SWQ2}+\lambda_{3Q}-\lambda_{3E}-\lambda_{SWO2})T_1}\,dT_1$$

$$-\int_{T_1=0}^{T}(\lambda_1+\lambda_2)e^{-(\lambda_1+\lambda_2+\lambda_{SWO1}+\lambda_{SWQ2}+\lambda_{3Q}-\lambda_{3E}-\lambda_{SWO2})T_1}\,dT_1\},$$

$$=S_1(T)\{\frac{\lambda_1}{A_1}(1-e^{-A_1 T})+\frac{\lambda_2}{B_1}(1-e^{-B_1 T})-\frac{\lambda_1+\lambda_2}{C_1}(1-e^{-C_1 T})\},$$

where

$$
\begin{aligned}
S_1(T) &= e^{-\lambda_{SWE2}+(\lambda_{SE}+\lambda_{3E}+\lambda_{SWO2})T},\\
A_1 &= \lambda_1+\lambda_{SWO1}+\lambda_{SWQ2}+\lambda_{3Q}-\lambda_{3E}-\lambda_{SWO2},\\
B_1 &= \lambda_2+\lambda_{SWO1}+\lambda_{SWQ2}+\lambda_{3Q}-\lambda_{3E}-\lambda_{SWO2},\\
C_1 &= \lambda_1+\lambda_2+\lambda_{SWO1}+\lambda_{SWQ2}+\lambda_{3Q}-\lambda_{3E}-\lambda_{SWO2}.
\end{aligned}
$$

4. The first double integral in the solution is

$$\int_{T_2=0}^{T} \int_{T_1=0}^{T_2} \{[\lambda_1 e^{-\lambda_1 T_1} + \lambda_2 e^{-\lambda_2 T_1} - (\lambda_1 + \lambda_2)e^{-(\lambda_1+\lambda_2)T_1}]$$

$$\cdot e^{-\lambda_{SE}T_2} e^{-\lambda_{SWO1}T_1} e^{-\lambda_{SWQ2}T_1} e^{-\lambda_{SWE2}\cdot 1} e^{-\lambda_{3Q}T_1}$$

$$\cdot e^{-\lambda_{3E}(T_2-T_1)} \lambda_{SWO2} e^{-\lambda_{SWO2}(T_2-T_1)} e^{-\lambda_{SWQ3}T_2} e^{-\lambda_{SWE3}\cdot 1}$$

$$\cdot e^{-\lambda_{4Q}T_2} e^{-\lambda_{4E}(T-T_2)} e^{-\lambda_{SWO3}(T-T_2)}\} \, dT_1 \, dT_2$$

$$= \lambda_{SWO2} e^{-[\lambda_{SWE2}+\lambda_{SWE3}+(\lambda_{4E}+\lambda_{SWO3})T]}$$

$$\int_{T_2=0}^{T} e^{-(\lambda_{SE}+\lambda_{3E}+\lambda_{SWO2}+\lambda_{SWQ3}+\lambda_{4Q}-\lambda_{4E}-\lambda_{SWO3})T_2}$$

$$[\int_{T_1=0}^{T_2} \lambda_1 e^{-(\lambda_1+\lambda_{SWO1}+\lambda_{SWQ2}+\lambda_{3Q}-\lambda_{3E}-\lambda_{SWO2})T_1} \, dT_1$$

$$+ \int_{T_1=0}^{T_2} \lambda_2 e^{-(\lambda_2+\lambda_{SWO1}+\lambda_{SWQ2}+\lambda_{3Q}-\lambda_{3E}-\lambda_{SWO2})T_1} \, dT_1$$

$$- \int_{T_1=0}^{T_2} (\lambda_1 + \lambda_2)e^{-(\lambda_1+\lambda_2+\lambda_{SWO1}+\lambda_{SWQ2}+\lambda_{3Q}-\lambda_{3E}-\lambda_{SWO2})T_1} \, dT_1] dT_2,$$

$$= \lambda_{SWO2} e^{-[\lambda_{SWE2}+\lambda_{SWE3}+(\lambda_{4E}+\lambda_{SWO3})T]}$$

$$\int_{T_2=0}^{T} e^{-D_2 T_2} [\int_{T_1=0}^{T_2} \lambda_1 e^{-A_2 T_1} \, dT_1$$

$$+ \int_{T_1=0}^{T_2} \lambda_2 e^{-B_2 T_1} dT_1 - \int_{T_1=0}^{T_2} (\lambda_1 + \lambda_2)e^{-C_2 T_1} \, dT_1] dT_2,$$

$$= \lambda_{SWO2} e^{-[\lambda_{SWE2}+\lambda_{SWE3}+(\lambda_{4E}+\lambda_{SWO3})T]}$$

$$\int_{T_2=0}^{T} e^{-D_2 T_2} [\frac{\lambda_1}{A_2}(1 - e^{-A_2 T_2}) + \frac{\lambda_2}{B_2}(1 - e^{-B_2 T_2})$$

$$- \frac{\lambda_1 + \lambda_2}{C_2}(1 - e^{-C_2 T_2})] dT_2,$$

$$= S_2(T)\{(\frac{\lambda_1}{A_2 D_2} + \frac{\lambda_2}{B_2 D_2} - \frac{\lambda_1 + \lambda_2}{C_2 D_2})(1 - e^{-D_2 T})$$

$$- \frac{\lambda_1}{(A_2 + D_2)A_2}(1 - e^{-(A_2+D_2)T})$$

$$- \frac{\lambda_2}{(B_2 + D_2)B_2}(1 - e^{-(B_2+D_2)T})$$

$$+ \frac{\lambda_1 + \lambda_2}{(C_2 + D_2)C_2}(1 - e^{-(C_2+D_2)T)})\},$$

where

$$S_2(T) = \lambda_{SWO2} e^{-\lambda_{SWE2}+\lambda_{SWE3}+(\lambda_{4E}+\lambda_{SWO3})T},$$

$$D_2 = \lambda_{SE} + \lambda_{3E} + \lambda_{SWO2} + \lambda_{SWQ3} + \lambda_{4Q} - \lambda_{4E}$$
$$- \lambda_{SWO3},$$
$$A_2 = \lambda_1 + \lambda_{SWO1} + \lambda_{SWQ2} + \lambda_{3Q} - \lambda_{3E} - \lambda_{SWO2},$$
$$B_2 = \lambda_2 + \lambda_{SWO1} + \lambda_{SWQ2} + \lambda_{3Q} - \lambda_{3E} - \lambda_{SWO2},$$

and

$$C_2 = \lambda_1 + \lambda_2 + \lambda_{SWO1} + \lambda_{SWQ2} + \lambda_{3Q} - \lambda_{3E} - \lambda_{SWO2}.$$

5. Using the procedures given in Cases 3 and 4 and integrating the solution given in Case 1 term by term yields the following complete solution:

$$R(T) = e^{-(\lambda_{SE}+\lambda_{SWO1})T}[e^{-\lambda_1 T} + e^{-\lambda_2 T} - e^{-(\lambda_1+\lambda_2)T}]$$

$$+ S_1(T)\{\frac{\lambda_1}{A_1}(1 - e^{-A_1 T}) + \frac{\lambda_2}{B_1}(1 - e^{-B_1 T})$$

$$- \frac{\lambda_1 + \lambda_2}{C_1}(1 - e^{-C_1 T})\}$$

$$+ S_2(T)\{(\frac{\lambda_1}{A_2 D_2} + \frac{\lambda_2}{B_2 D_2} - \frac{\lambda_1 + \lambda_2}{C_2 D_2})(1 - e^{-D_2 T})$$

$$- \frac{\lambda_1}{(A_2 + D_2)A_2}[1 - e^{-(A_2+D_2)T}]$$

$$- \frac{\lambda_2}{(B_2 + D_2)B_2}[1 - e^{-(B_2+D_2)T}]$$

$$+ \frac{\lambda_1 + \lambda_2}{(C_2 + D_2)C_2}[1 - e^{-(C_2+D_2)T}]\}$$

$$+ S_3(T)\{\frac{\lambda_1}{A_3}(1 - e^{-A_3 T}) + \frac{\lambda_2}{B_3}(1 - e^{-B_3 T})$$

$$- \frac{\lambda_1 + \lambda_2}{C_3}(1 - e^{-C_3 T})\}$$

$$+ S_4(T)\{(\frac{\lambda_1}{B_4 A_4} + \frac{\lambda_2}{A_4 C_4} - \frac{\lambda_1 + \lambda_2}{A_4 D_4})(1 - e^{-A_4 T})$$

$$- \frac{\lambda_1}{(A_4 + B_4)B_4}[1 - e^{-(B_4+A_4)T}]$$

$$- \frac{\lambda_2}{(A_4 + C_4)C_4}[1 - e^{-(C_4+A_4)T}]$$

$$+ \frac{\lambda_1 + \lambda_2}{(A_4 + D_4)D_4}[1 - e^{-(D_4+A_4)T}]\}$$

$$+ S_5(T)\{[\frac{\lambda_1}{A_5}(1 - e^{-A_5 T}) + \frac{\lambda_2}{B_5}(1 - e^{-B_5 T})$$

$$- \frac{\lambda_1 + \lambda_2}{C_5}(1 - e^{-C_5 T})]$$

$$- \frac{\lambda_1}{A_5 + \lambda_{3Q}}[1 - e^{-(A_5 + \lambda_{3Q})T}]$$

$$- \frac{\lambda_2}{B_5 + \lambda_{3Q}}[1 - e^{-(B_5 + \lambda_{3Q})T}]$$

$$+ \frac{\lambda_1 + \lambda_2}{C_5 + \lambda_{3Q}}[1 - e^{-(C_5 + \lambda_{3Q})T}]\}$$

$$+ S_6(T)\{[\frac{\lambda_1}{A_6}(1 - e^{-A_6 T}) + \frac{\lambda_2}{B_6}(1 - e^{-B_6 T})$$

$$- \frac{\lambda_1 + \lambda_2}{C_6}(1 - e^{-C_6 T})]$$

$$- \frac{\lambda_1}{A_6 + \lambda_{SWQ2}}[1 - e^{-(A_6 + \lambda_{SWQ2})T}]$$

$$- \frac{\lambda_2}{B_6 + \lambda_{SWQ2}}[1 - e^{-(B_6 + \lambda_{SWQ2})T}]$$

$$+ \frac{\lambda_1 + \lambda_2}{C_6 + \lambda_{SWQ2}}[1 - e^{-(C_6 + \lambda_{SWQ2})T}]\}$$

$$+ S_7(T)\{\frac{\lambda_1}{A_7}(1 - e^{-A_7 T}) + \frac{\lambda_2}{B_7}(1 - e^{-B_7 T})$$

$$- \frac{\lambda_1 + \lambda_2}{C_7}(1 - e^{-C_7 T})$$

$$+ S_8(T)\{(\frac{1}{A_8 B_8} + \frac{1}{A_8 C_8} - \frac{1}{A_8 D_8})(1 - e^{-A_8 T})$$

$$- \frac{1}{B_8(A_8 + B_8)}[1 - e^{-(A_8 + B_8)T}]$$

$$- \frac{1}{C_8(A_8 + C_8)}[1 - e^{-(A_8 + C_8)T}]$$

$$+ \frac{1}{D_8(A_8 + D_8)}[1 - e^{-(A_8 + D_8)T}]\}$$

$$+ S_9(T)\{(\frac{1}{A_9}(1 - e^{-A_9 T}) + \frac{1}{B_9}(1 - e^{-B_9 T})$$

$$- \frac{1}{C_9}(1 - e^{-C_9 T})\}$$

$$+ S_{10}(T)\{(\frac{1}{A_{10} B_{10}} + \frac{1}{A_{10} C_{10}} - \frac{1}{A_{10} D_{10}})(1 - e^{-A_{10} T})$$

$$- \frac{1}{B_{10}(A_{10} + B_{10})}[1 - e^{-(A_{10}+B_{10})T}]$$

$$- \frac{1}{C_{10}(A_{10} + C_{10})}[1 - e^{-(A_{10}+C_{10})T}]$$

$$+ \frac{1}{D_{10}(A_{10} + D_{10})}[1 - e^{-(A_{10}+D_{10})T}]\}$$

$$+ S_{11}(T)\{[\frac{1}{A_{11}}(1 - e^{-A_{11}T}) + \frac{1}{B_{11}}(1 - e^{-B_{11}T})$$

$$- \frac{1}{C_{11}}(1 - e^{-C_{11}T})]$$

$$- \frac{1}{A_{11} + \lambda_{3Q}}[1 - e^{-(A_{11}+\lambda_{3Q})T}]$$

$$- \frac{1}{B_{11} + \lambda_{3Q}}[1 - e^{-(B_{11}+\lambda_{3Q})T}]$$

$$+ \frac{1}{C_{11} + \lambda_{3Q}}[1 - e^{-(C_{11}+\lambda_{3Q})T}]\},$$

$$+ S_{12}(T)\{[\frac{1}{A_{12}}(1 - e^{-A_{12}T}) + \frac{1}{B_{12}}(1 - e^{-B_{12}T})$$

$$- \frac{1}{C_{12}}(1 - e^{-C_{12}T})]$$

$$- \frac{1}{A_{12} + \lambda_{SWQ2}}[1 - e^{-(A_{12}+\lambda_{SWQ2})T}]$$

$$- \frac{1}{B_{12} + \lambda_{SWQ2}}[1 - e^{-(B_{12}+\lambda_{SWQ2})T}]$$

$$+ \frac{1}{C_{12} + \lambda_{SWQ2}}[1 - e^{-(C_{12}+\lambda_{SWQ2})T}]\},$$

where

$$S_3(T) = (1 - e^{-\lambda_{SWE2}})e^{-[\lambda_{SWE3}+(\lambda_{4E}+\lambda_{SWO})]T},$$

$$A_3 = \lambda_1 + \lambda_{SE} + \lambda_{SWO1} + \lambda_{SWQ2} + \lambda_{SWQ3} + \lambda_{4Q}$$
$$- \lambda_{4E} - \lambda_{SWO3},$$

$$B_3 = \lambda_2 + \lambda_{SE} + \lambda_{SWO1} + \lambda_{SWQ2} + \lambda_{SWQ3} + \lambda_{4Q}$$
$$- \lambda_{4E} - \lambda_{SWO3},$$

$$C_3 = \lambda_1 + \lambda_2 + \lambda_{SE} + \lambda_{SWO1} + \lambda_{SWQ2} + \lambda_{SWQ3} + \lambda_{4Q}$$
$$- \lambda_{4E} - \lambda_{SWO3},$$

$$S_4(T) = \lambda_{3E}e^{-[\lambda_{SWE2}+\lambda_{SWE3}+(\lambda_{4E}+\lambda_{SWO3})T]},$$

$$A_4 = \lambda_{SE} + \lambda_{3E} + \lambda_{SWO2} + \lambda_{SWQ3} + \lambda_{4Q}$$

$$-\lambda_{4E} - \lambda_{SWO2},$$

$$B_4 = \lambda_1 + \lambda_{SE} + \lambda_{SWO1} + \lambda_{SWQ2} + \lambda_{3Q} - \lambda_{3E}$$
$$- \lambda_{SWO3},$$

$$C_4 = \lambda_2 + \lambda_{SE} + \lambda_{SWO1} + \lambda_{SWQ2} + \lambda_{3Q} - \lambda_{3E}$$
$$- \lambda_{SWO3},$$

$$D_4 = \lambda_1 + \lambda_2 + \lambda_{SE} + \lambda_{SWO1} + \lambda_{SWQ2} + \lambda_{3Q}$$
$$- \lambda_{3E} - \lambda_{SWO3},$$

$$S_5(T) = e^{-[\lambda_{SWE2}+\lambda_{SWE3}+(\lambda_{4E}+\lambda_{SWO3})T]},$$

$$A_5 = \lambda_1 + \lambda_{SE} + \lambda_{SWO1} + \lambda_{SWQ2} + \lambda_{SWQ3} + \lambda_{4Q} - \lambda_{4E}$$
$$- \lambda_{SWO3},$$

$$B_5 = \lambda_2 + \lambda_{SE} + \lambda_{SWO1} + \lambda_{SWQ2} + \lambda_{SWQ3} + \lambda_{4Q} - \lambda_{4E}$$
$$- \lambda_{SWO3},$$

$$C_5 = \lambda_1 + \lambda_2 + \lambda_{SE} + \lambda_{SWO1} + \lambda_{SWQ2} + \lambda_{SWQ3} + \lambda_{4Q}$$
$$- \lambda_{4E} - \lambda_{SWO3},$$

$$S_6(T) = e^{-[\lambda_{SWE3}+(\lambda_{4E}+\lambda_{SWO3})T]},$$

$$A_6 = \lambda_1 + \lambda_{SE} + \lambda_{SWO1} + \lambda_{SWQ3} + \lambda_{4Q} - \lambda_{4E} - \lambda_{SWO3},$$
$$B_6 = \lambda_2 + \lambda_{SE} + \lambda_{SWO1} + \lambda_{SWQ3} + \lambda_{4Q} - \lambda_{4E} - \lambda_{SWO3},$$
$$C_6 = \lambda_1 + \lambda_2 + \lambda_{SE} + \lambda_{SWO1} + \lambda_{SWQ3} + \lambda_{4Q} - \lambda_{4E}$$
$$- \lambda_{SWO3},$$

$$S_7(T) = \lambda_{SWO1}e^{-[\lambda_{SWE2}+(\lambda_{3E}+\lambda_{SWO2})T]},$$

$$A_7 = \lambda_1 + \lambda_{SE} + \lambda_{SWO1} + \lambda_{SWQ2} + \lambda_{3Q} - \lambda_{3E} - \lambda_{SWO2},$$
$$B_7 = \lambda_2 + \lambda_{SE} + \lambda_{SWO1} + \lambda_{SWQ2} + \lambda_{3Q} - \lambda_{3E} - \lambda_{SWO2},$$
$$C_7 = \lambda_1 + \lambda_2 + \lambda_{SE} + \lambda_{SWO1} + \lambda_{SWQ2} + \lambda_{3Q}$$
$$- \lambda_{3E} - \lambda_{SWO2},$$

$$S_8(T) = \lambda_{SWO1}\lambda_{SWO2}e^{-[(\lambda_{SWE2}+\lambda_{SWE3}+(\lambda_{4E}+\lambda_{SWO3})T]},$$

$$A_8 = \lambda_{SE} + \lambda_{3E} + \lambda_{SWO2} + \lambda_{SWQ3} + \lambda_{4Q} - \lambda_{4E}$$
$$- \lambda_{SWO3},$$

$$B_8 = \lambda_1 + \lambda_{SWO1} + \lambda_{SWO2} + \lambda_{SWQ2} + \lambda_{3Q} - \lambda_{3E},$$
$$C_8 = \lambda_2 + \lambda_{SWO1} + \lambda_{SWO2} + \lambda_{SWQ2} + \lambda_{3Q} - \lambda_{3E},$$
$$D_8 = \lambda_1 + \lambda_2 + \lambda_{SWO1} + \lambda_{SWO2} + \lambda_{SWQ2} + \lambda_{3Q} - \lambda_{3E},$$

$$S_9(T) = \lambda_{SWO1}(1 - e^{-\lambda_{SWE2}})e^{-[\lambda_{SWE3}+(\lambda_{4E}+\lambda_{SWO3})T]},$$

$$A_9 = \lambda_1 + \lambda_{SE} + \lambda_{SWO1} + \lambda_{SWQ2} + \lambda_{SWQ3} + \lambda_{4Q}$$
$$- \lambda_{4E} - \lambda_{SWO3},$$

$$B_9 = \lambda_2 + \lambda_{SE} + \lambda_{SWO1} + \lambda_{SWQ2} + \lambda_{SWQ3} + \lambda_{4Q}$$
$$- \lambda_{4E} - \lambda_{SWO3},$$

$$C_9 = \lambda_1 + \lambda_2 + \lambda_{SE} + \lambda_{SWO1} + \lambda_{SWQ2} + \lambda_{SWQ3} + \lambda_{4Q} - \lambda_{4E} - \lambda_{SWO3},$$

$$S_{10}(T) = \lambda_{SWO1}\lambda_{3E}e^{-[\lambda_{SWE2}+\lambda_{SWE3}+(\lambda_{4E}+\lambda_{SWO3})T]},$$

$$A_{10} = \lambda_{SE} + \lambda_{3E} + \lambda_{SWO2} + \lambda_{SWQ3} + \lambda_{4Q} - \lambda_{4E} - \lambda_{SWO3},$$

$$B_{10} = \lambda_1 + \lambda_{SWO1} + \lambda_{SWO2} + \lambda_{SWQ2} + \lambda_{3Q} - \lambda_{3E},$$

$$C_{10} = \lambda_2 + \lambda_{SWO1} + \lambda_{SWO2} + \lambda_{SWQ2} + \lambda_{3Q} - \lambda_{3E},$$

$$D_{10} = \lambda_1 + \lambda_2 + \lambda_{SWO1} + \lambda_{SWO2} + \lambda_{SWQ2} + \lambda_{3Q} - \lambda_{3E},$$

$$S_{11}(T) = \lambda_{SWO1}e^{-[\lambda_{SWE2}+\lambda_{SWE3}+(\lambda_{4E}+\lambda_{SWO3})T]},$$

$$A_{11} = \lambda_1 + \lambda_{SE} + \lambda_{SWO1} + \lambda_{SWQ2} + \lambda_{SWQ3} + \lambda_{4Q} - \lambda_{4E} - \lambda_{SWO3},$$

$$B_{11} = \lambda_2 + \lambda_{SE} + \lambda_{SWO1} + \lambda_{SWQ2} + \lambda_{SWQ3} + \lambda_{4Q} - \lambda_{4E} - \lambda_{SWO3},$$

$$C_{11} = \lambda_1 + \lambda_2 + \lambda_{SE} + \lambda_{SWO1} + \lambda_{SWQ2} + \lambda_{SWQ3} + \lambda_{4Q} - \lambda_{4E} - \lambda_{SWO3},$$

$$S_{12}(T) = \lambda_{SWO}e^{-[\lambda_{SWE3}+(\lambda_{4E}+\lambda_{SWO3})T]},$$

$$A_{12} = \lambda_1 + \lambda_{SE} + \lambda_{SWO1} + \lambda_{SWQ3} + \lambda_{4Q} - \lambda_{4E} - \lambda_{SWO3},$$

$$B_{12} = \lambda_2 + \lambda_{SE} + \lambda_{SWO1} + \lambda_{SWQ3} + \lambda_{4Q} - \lambda_{4E} - \lambda_{SWO3},$$

and

$$C_{12} = \lambda_1 + \lambda_2 + \lambda_{SE} + \lambda_{SWO1} + \lambda_{SWQ3} + \lambda_{4Q} - \lambda_{4E} - \lambda_{SWO3}.$$

EXAMPLE 3-5

Rework Example 3-3 assuming that there is only one sensing unit in the system and its function stops as soon as the standby Unit 3 is brought into service. The system's reliability block diagram is given in Fig. 3.5. Find the system's reliability and compare the result with that obtained in Example 3-3.

SOLUTION TO EXAMPLE 3-5

Table 3.3 gives the system success combinations, the units involved in the system, their modes of function, the system success function modes, and the time domain of function of each unit. The function times given under SE in Column 3 are based on the assumption that as soon as the standby Unit 3 is brought into function the sensing subsystem is no longer needed. The system's reliability is then given

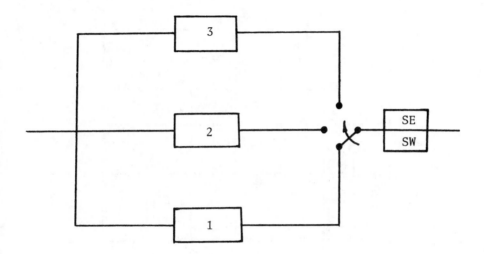

Fig. 3.5– Complex standby system of Example 3–5 with
only one sensing unit.

by

$$R(T) = e^{-(\lambda_1 + \lambda_{SE} + \lambda_{SWO})T} + \int_{T_1=0}^{T} \lambda_1 e^{-\lambda_1 T_1} e^{-\lambda_{SE} T} e^{-\lambda_{SWO}(T-T_1)}$$
$$\cdot\, e^{-\lambda_{SWQ} T_1} e^{-\lambda_{SWE}\cdot 1} e^{-\lambda_{2Q} T_1} e^{-\lambda_2(T-T_1)} \, dT_1$$
$$+ \int_{T_2=0}^{T} \int_{T_1=0}^{T_2} \lambda_1 e^{-\lambda_1 T_1} e^{-\lambda_{SE} T_2} e^{-\lambda_{SWO}(T-T_2)} e^{-\lambda_{SWQ} T_1}$$
$$\cdot\, e^{-\lambda_{SWE}\cdot 1} e^{-\lambda_{2Q} T_1} \lambda_2 e^{-\lambda_2(T_2-T_1)} e^{-\lambda_{SWQ}(T_2-T_1)}$$
$$\cdot\, e^{-\lambda_{SWE}\cdot 1} e^{-\lambda_{3Q} T_2} e^{-\lambda_3(T-T_2)} \, dT_1 \, dT_2$$
$$+ \int_{T_1=0}^{T_2} \lambda_1 e^{-\lambda_1 T_1} e^{-\lambda_{SE} T_1} e^{-\lambda_{SWO}(T-T_1)} e^{-\lambda_{SWQ} T_1} e^{-\lambda_{SWE}\cdot 1}$$
$$\cdot\, (1 - e^{-\lambda_{2Q} T_1}) e^{-\lambda_{SWE}\cdot 1} e^{-\lambda_{3Q} T_1} e^{-\lambda_3(T-T_1)} \, dT_1,$$
$$= e^{-(\lambda_1 + \lambda_{SE} + \lambda_{SWO})T}$$
$$+ \lambda_1 e^{-[\lambda_{SWE} + (\lambda_{SE} + \lambda_{SWO} + \lambda_2)T]}$$
$$\cdot \int_{T_1=0}^{T} e^{-(\lambda_1 - \lambda_{SWO} + \lambda_{SWQ} + \lambda_{2Q} - \lambda_2)T_1} \, dT_1$$
$$+ \lambda_1 \lambda_2 e^{-[2\lambda_{SWE} + (\lambda_{SWO} + \lambda_3)T]}$$

Table 3.3– Three-unit standby system of Example 3-5.

Number of system success combination	Units, modes of function, system success modes, and time domains of function										
	1	SE	SWO	SWQ	SWE	2Q	2	SWQ	SWE	3Q	3
1	G	G	G								
	T	T	T								
2	B	G	G	G	G	G	G				
	T_1	T	$T - T_1$	T_1	$1C$	T_1	$T - T_1$				
3	B	G	G	G	G	G	B	G	G	G	G
	T_1	T_2	$T - T_2$	T_1	$1C$	T_1	$T_2 - T_1$	$T_2 - T_1$	$1C$	T_2	$T - T_2$
4	B	G	G	G	G	B			G	G	G
	T_1	T_1	$T - T_1$	T_1	$1C$	$< T_1$			$1C$	T_1	$T - T_1$

70

$$\cdot \int_{T_2=0}^{T} e^{-(\lambda_{SE}-\lambda_{SWO}+\lambda_2+\lambda_{SWQ}+\lambda_{3Q}-\lambda_3)T_2}$$

$$\cdot \{ \int_{T_1=0}^{T_2} e^{-(\lambda_1+\lambda_{2Q}-\lambda_2)T_1} \, dT_1 \} \, dT_2$$

$$+ \lambda_1 e^{-[2\lambda_{SWE}+(\lambda_{SWO}+\lambda_3)T]}$$

$$\cdot \int_{T_1=0}^{T} e^{-(\lambda_1+\lambda_{SE}-\lambda_{SWO}+\lambda_{SWQ}+\lambda_{3Q}-\lambda_3)T_1}$$

$$\cdot (1 - e^{-\lambda_{2Q}T_1}) \, dT_1,$$

$$= \quad e^{-AT} + \frac{\lambda_1 e^{-(\lambda_{SWE}+BT)}}{C}(1 - e^{-CT})$$

$$+ \lambda_1 \lambda_2 e^{-(2\lambda_{SWE}+GT)}$$

$$\cdot \{ \frac{1}{DE}(1 - e^{-DT}) - \frac{1}{E(D+E)}[1 - e^{-(D+E)T}] \}$$

$$+ \lambda_1 e^{-(2\lambda_{SWE}+GT)}$$

$$\cdot \{ \frac{1}{F}(1 - e^{-FT}) - \frac{1}{F+\lambda_{2Q}}[1 - e^{-(F+\lambda_{2Q})T}] \},$$

where

$$
\begin{aligned}
A &= \lambda_1 + \lambda_{SE} + \lambda_{SWO}, \\
B &= \lambda_{SE} + \lambda_{SWO} + \lambda_2, \\
C &= \lambda_1 - \lambda_{SWO} + \lambda_{SWQ} + \lambda_{2Q} - \lambda_2, \\
D &= \lambda_{SE} + \lambda_2 + \lambda_{SWQ} - \lambda_3 - \lambda_{SWO}, \\
E &= \lambda_1 + \lambda_{2Q} - \lambda_2, \\
F &= \lambda_1 + \lambda_{SE} - \lambda_{SWO} + \lambda_{SWQ} + \lambda_{3Q} - \lambda_3,
\end{aligned}
$$

and

$$G = \lambda_{SWO} + \lambda_3.$$

Substituting the values of the failure rates into the expressions of A, B, C, D, E, and F yields

$$
\begin{aligned}
A &= (715.0 + 2.0 + 0.5)(10^{-6}) = 0.7175 \cdot 10^{-3}, \\
B &= (2.0 + 0.5 + 715.0)(10^{-6}) = 0.7175 \cdot 10^{-3}, \\
C &= (715.0 + 0.5 - 0.5 + 50.0 - 715.0)(10^{-6}) = 0.05 \cdot 10^{-3}, \\
D &= (2.0 + 715.0 + 0.5 - 715.0 - 0.5)(10^{-6}) = 0.002 \cdot 10^{-3}, \\
E &= (715.0 + 50.0 - 715.0)(10^{-6}) = 0.05 \cdot 10^{-3}, \\
F &= (715.0 + 2.0 + 0.5 - 0.5 + 50.0 - 715.0)(10^{-6}), \\
&= 0.052 \cdot 10^{-3},
\end{aligned}
$$

and

$$G = (0.5 + 715.0)(10^{-6}) = 0.7155 \cdot 10^{-3}.$$

Then

$$
\begin{aligned}
R(T) = \ & e^{-(0.0007175)T} \\
& + \frac{(715.0 \cdot 10^{-6})e^{-[(0.5 \cdot 10^{-6})+(0.0007175)T]}}{0.00005} \\
& \cdot (1 - e^{-0.00005T}) \\
& + (715.0 \cdot 10^{-6})(715.0 \cdot 10^{-6})e^{-[2(0.5 \cdot 10^{-6})+(0.0007155)T]} \\
& \cdot \{\frac{1}{(0.000002)(0.00005)}(1 - e^{-(0.000002)T}) \\
& - \frac{1}{(0.00005)(0.000002 + 0.00005)} \\
& \cdot [1 - e^{-(0.000002+0.00005)T}]\} \\
& + (715.0 \cdot 10^{-6})e^{-[2(0.5 \cdot 10^{-6})+(0.0007155)T]} \\
& \cdot \{\frac{1}{0.000052}(1 - e^{-0.000052T}) \\
& - \frac{1 - e^{(0.000052+50.0 \cdot 10^{-6})T}}{0.000052 + 50.0 \cdot 10^{-6}}\},
\end{aligned}
$$

or

$$R(T = 100 \text{ hr}) = 0.99969.$$

It may be seen that the system reliability of this example is almost the same as the result obtained in Example 3–3. This is due to the fact that these two systems have very close configurations. Further investigations on these two systems were conducted by increasing the failure rates of the sensing and switching units in both systems. The results, given in Table 3.4, show that they still have relatively close reliabilities.

PROBLEMS

3-1. What is the reliability of the system with one unit in standby to two units functioning reliabilitywise in parallel as shown in Fig. 3.6. The nomenclature in Fig. 3.6 corresponds to the following:

λ_1 = failure rate of Unit 1,

λ_2 = failure rate of Unit 2,

TABLE 3.4– Comparison of system reliabilities for Examples 3–3 and 3–5 with different sensing and switching failure rates.

Case	Failure rates, $fr/10^6$ hr	System reliability	
		Example 3–3	Example 3–5
1	$\lambda_{SWE} = 1,000$ $\lambda_{SWQ} = 50.0$ $\lambda_{SWO} = 50.0$ $\lambda_{SE} = 2.0$ $\lambda_{SEQ} = 0.01$	0.99467	0.99468
2	$\lambda_{SWE} = 0.5$ $\lambda_{SWQ} = 0.5$ $\lambda_{SWO} = 0.5$ $\lambda_{SE} = 1,000$ $\lambda_{SEQ} = 50.0$	0.90465	0.90481
3	$\lambda_{SWE} = 1,000$ $\lambda_{SWQ} = 50.0$ $\lambda_{SWO} = 50.0$ $\lambda_{SE} = 1,000$ $\lambda_{SEQ} = 50.0$	0.90012	0.90028

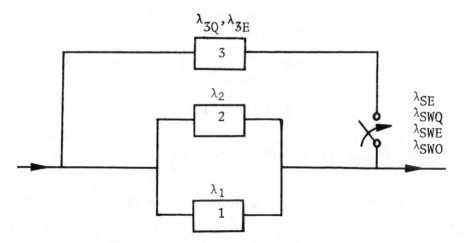

Fig. 3.6– The complex system of Problem 3-1.

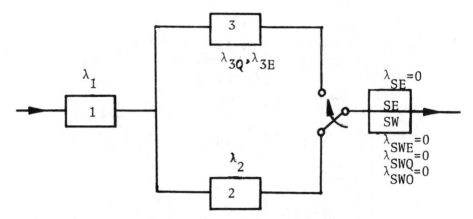

Fig. 3.7– The reliability block diagram of the system in Problem 3–2.

λ_{3Q} = quiescent failure rate of Unit 3,

λ_{3E} = energized failure rate of Unit 3,

λ_{SE} = sensing failure rate of sensing subsystem,

λ_{SWQ} = quiescent switching subsystem failure rate,

λ_{SWE} = energized switching subsystem failure rate, considering only a *one-cycle* operation,

and

λ_{SWO} = failing open failure rate of the switching subsystem, in fr/hr.

All failure rates are in fr/hr, except that of λ_{SWE}, which is in fr/cycle. Prepare a table of all units with all modes of function, system success modes, and time domains of function of each unit. Then proceed with the mathematical solution.

3-2. Find (1) the reliability and (2) the MTBF of the system whose reliability block diagram is given in Fig. 3.7. All *pdf*'s involved are exponential.

3-3. The reliability block diagram of a standby system is shown in Fig. 3.8.

(1) Prepare a table of system success combinations.

(2) Write out the reliability math model with the respective failure rates substituted.

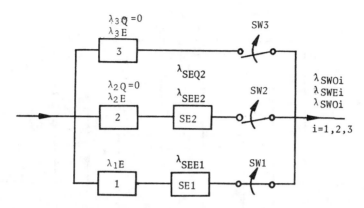

Fig. 3.8 – The complex standby system of Problem 3-3 with two separate sensing and three separate switching subsystems.

(3) Integrate out completely and give the worked out answer in terms of all of the λ's.

3-4. In the standby system of Fig. 3.9, Units 1 and 2 are the active units, and Units 3 and 4 are the standby units. The failure-sensing devices have Weibull energized *pdf*'s with β_{sei}, η_{sei}, and γ_{sei}, and the switching devices have Weibull quiescent *pdf*'s with β_{swqi}, η_{swqi}, and γ_{swqi} and Weibull energized *pdf*'s with β_{swei}, η_{swei}, and γ_{swei}. The switch failing open *pdf*'s Weibull parameters are β_{swoi}, η_{swoi}, and γ_{swoi} after the standby units are switched into operation. Each standby unit has a Weibull quiescent *pdf* with β_{qi}, η_{qi}, and γ_{qi} and an energized Weibull *pdf* with β_{ei}, η_{ei}, and γ_{ei}. If *at least one unit is required for system success,* what is the system's reliability when the active units have Weibull *pdfs*? Give all of the system success function combinations and the complete reliability equation in terms of all of the integrals and the probability density functions involved.

3-5. Figure 3.10 shows three equal functioning units with a failure rate of λ_1 and a standby unit with a de-energized failure rate of λ_{4D} and an energized failure rate of λ_{4E}. Consider a sensing failure rate of λ_{SE}, a de-energized switching failure rate of λ_{SWD}, an energized switching failure rate of λ_{SWE}, and a switch failing open (after closing) failure rate of λ_{SWD}. *At least one unit is required at all times for system success. What is the reliability of the system?* Prepare a complete table of successful function modes of the system. Write out the system's reliability in terms

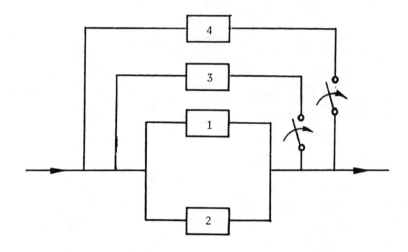

Fig. 3.9– The standby system of Problem 3–4.

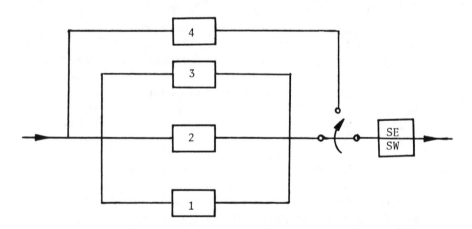

Fig. 3.10 – The system of Problem 3–5 with three equal units reliabilitywise in parallel and one unit in standby to them.

of the R's, then in terms of the λ's, and finally in terms of the integrals. Do not integrate out!

3-6. What is the reliability of the system whose reliability block diagram is given in Fig. 3.11? Consider energized active units, a quiescent and energized standby unit, sensing energized, switch quiescent and energized for one cycle, and switch failing open. The switch can fail open anytime during the mission. Solve the following four cases:

    Case 1: Take Units 1 and 2 together as one system. See Fig. 3.11.

    Case 2: Either Unit 1 or 2 is replaced by the standby unit whenever any one of them fails. See Fig. 3.12.

    Case 3: The standby unit replaces both Units 1 and 2 whenever any one of them fails. See Fig. 3.13.

    Case 4: Compare and discuss the results of these three cases for the following two situations:

    (1) When $\lambda_{SE_1} = \lambda_{SE_2} = \lambda_{SE_3} = \lambda_{SE}$.

    (2) When $\lambda_{SE_1} \neq \lambda_{SE_2} \neq \lambda_{SE_3}$.

Work out all answers completely.

3-7. Draw a reliability block diagram of the system whose reliability is given by

$$R_S = R_1 R_2 R_3 [1 - (1 - R_A)^2]^2 (R_6^4 + 4R_6^3 Q_6 + 6R_6^2 Q_6^2)$$
$$\cdot [e^{-\lambda_7 t}(1 + \lambda_7 t + \frac{1}{2}\lambda_7^2 t^2)],$$

where

$$R_A = R_4[1 - (1 - R_5)^2].$$

3-8. Prove that the reliability of a standby system of n units, one active and $n - 1$ in standby, with perfect sensing and switching and a quiescent reliability of 1, is given by

$$R(T) = \frac{\lambda_2 \lambda_3 \lambda_4 \ \ldots \ \lambda_n e^{-\lambda_1 T}}{(\lambda_2 - \lambda_1)(\lambda_3 - \lambda_1) \ \ldots \ (\lambda_n - \lambda_1)}$$

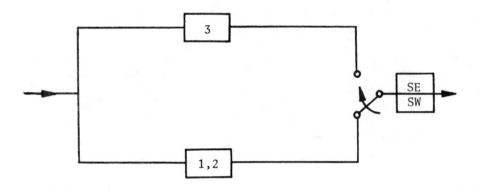

Fig. 3.11– Schematic of the system of Problem 3-6, Case 1.

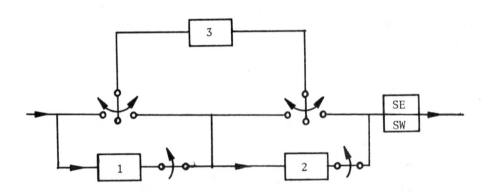

Fig. 3.12 – Schematic of the system of Problem 3-6 when either Unit 1 or 2 is replaced by the standby unit whenever any one of them fails (Case 2).

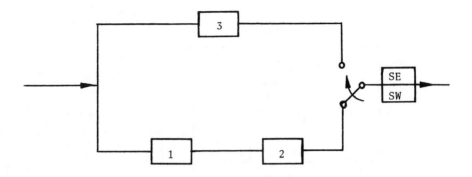

Fig. 3.13 – Schematic of the system of Problem 3-6 when the standby unit replaces both Units 1 and 2 whenever any one of them fails (Case 3).

$$+ \frac{\lambda_1 \lambda_3 \lambda_4 \ \cdots \ \lambda_n e^{-\lambda_2 T}}{(\lambda_1 - \lambda_2)(\lambda_3 - \lambda_2) \ \cdots \ (\lambda_n - \lambda_2)}$$

$$+ \cdots + \frac{\lambda_1 \lambda_2 \ \cdots \ \lambda_{i-1} \lambda_{i+1} \ \cdots \ \lambda_n e^{-\lambda_i T}}{(\lambda_1 - \lambda_i) \ \cdots \ (\lambda_{i-1} - \lambda_i)(\lambda_{i+1} - \lambda_i) \ \cdots \ (\lambda_n - \lambda_i)}$$

$$+ \cdots + \frac{\lambda_1 \lambda_2 \lambda_3 \ \cdots \ \lambda_{n-1} e^{-\lambda_n T}}{(\lambda_1 - \lambda_n)(\lambda_2 - \lambda_n) \ \cdots \ (\lambda_{n-1} - \lambda_n)}.$$

3-9. Figure 3.14 shows three functioning units with a failiure rate of λ_1 and a standby unit with a de-energized failure rate of λ_{4Q} and an energized failure rate of λ_{4E}. Consider a sensing failure rate of λ_{SE}, a de-energized switching failure rate of λ_{SWQ}, an energized switching failure rate of λ_{SWE}, and a switch failing open failure rate of λ_{SWO} after closing. If *two units are required at all times for system success, what is the reliability of the system?* Prepare a complete table of successful function modes of the system. Write out the system's reliability in terms of the integrals and then integrate out!

3-10. What is the reliability of the system with one active unit and two standby units shown in Fig. 3.15? The nomenclature in Fig. 3.15 corresponds to the following:

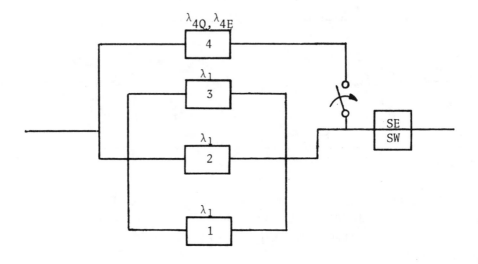

Fig. 3.14-- The system of Problem 3–9 with three equal units
in parallel and one unit in standby to them.

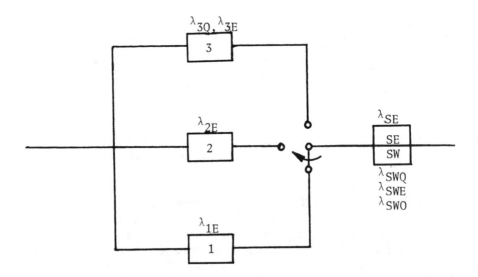

Fig. 3.15– The complex standby system of Problem 3–10.

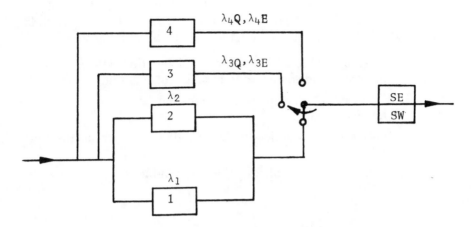

Fig. 3.16– Reliability block diagram for two units in parallel and two units in standby, with sensing and switching units, for Problem 3–11.

λ_{1E} = energized failure rate of Unit 1,

λ_{2E} = energized failure rate of Unit 2,

λ_{3Q} = quiescent failure rate of Unit 3,

λ_{3E} = energized failure rate of Unit 3,

λ_{SE} = failure rate of sensing subsystem,

λ_{SWQ} = quiescent failure rate of switching subsystem,

λ_{SWE} = energized failure rate of switching subsystem, considering only a one-cycle operation,

and

λ_{SWO} = failing open failure rate of the switching subsystem.

All failure rates are in fr/hr, except for λ_{SWE}, which is in fr/cycle.

3-11. In the standby system shown in Fig. 3.16, Units 1 and 2 are functioning reliabilitywise in parallel. The failure-sensing device has an energized failure rate of λ_{SE}, and the switching device has a switching failure rate of λ_{SWE}. The switch failing open failure rate is λ_{SWO} after Units 3 and 4 are switched into operation, respectively. The standby units have quiescent failure rates of λ_{3Q} and λ_{4Q} and energized failure rates of λ_{3E} and λ_{4E}. If both

Units 1 and 2 fail, Unit 3 is switched in, and if Unit 3 also fails, Unit 4 is switched in. What is this system's reliability?

(1) Prepare a table of units, modes of function, system success function modes, and time domains of function.

(2) Write out the complete reliability equation in the form of integrals.

(3) Integrate out completely the first single integral in the solution.

(4) Integrate out completely the first double integral in the solution.

Chapter 4

APPLICATIONS OF THE BINOMIAL AND POISSON DISTRIBUTIONS TO SYSTEM RELIABILITY PREDICTION

4.1 THE BINOMIAL DISTRIBUTION

4.1.1 IDENTICAL UNITS

The binomial expansion is given by

$$
(R + Q)^n = R^n + \frac{n}{1!}R^{n-1}Q + \frac{n(n-1)}{(1!)(2!)}R^{n-2}Q^2
$$

$$
+ \ ... \ + \frac{n(n-1)}{(1!)(2!)}R^2Q^{n-2} + \frac{n}{1!}RQ^{n-1} + Q^n,
$$

or

$$
(R + Q)^n = \sum_{k=0}^{n} \frac{n!}{k!(n-k)!}R^{n-k}Q^k,
$$

where n is the total number of units in a system. As we know that $R + Q = 1$, then also $(R + Q)^n = 1$. Each term in the expanded form of $(R+Q)^n$ may also be obtained from the following pdf of the discrete binomial distribution for k failures:

$$
f(k) = \frac{n!}{k!(n-k)!}R^{n-k}Q^k, \ k = 1, 2, ..., n,
$$

which gives the probability of exactly k failures. The cumulative binomial function is given by

$$F(k) = \sum_{i=0}^{k} f(i) = P(k \text{ or less failures}).$$

Let's take a system with three units in it; then $n = 3$ and

$$(R + Q)^3 = R^3 + 3R^2Q + 3RQ^2 + Q^3 = 1.$$

This says that the sum of the probabilities of all outcomes in a mission of length t, for which the R's and Q's are calculated, is equal to 1, as it should be.

What are these outcomes and their probabilities? The binomial distribution gives these to us very conveniently. The first term, R^3, gives the probability of all three units succeeding, or surviving, the mission. The second term without the coefficient, R^2Q, gives the probability of two out of the three units surviving and the third failing, or RRQ. With the coefficient, it is $3R^2Q$ because all probabilities of two units surviving and one unit failing need to be determined; i.e., $RRQ = R^2Q$, $RQR = R^2Q$, and $QRR = R^2Q$, for a total probability of $3R^2Q$. The third term, $3RQ^2$, similarly, is all probabilities of only one unit surviving and two units failing, and the last term, Q^3, is the probability of all three units failing.

The probability of no unit or of only one unit failing is then given by

$$P[\text{one or fewer (or none) units fail}] = R^3 + 3R^2Q.$$

This then would be the reliability of a system with three equal units, where all units surviving or any one unit failing and the remaining two units surviving give system success.

The probability of no unit, or of only one unit, or of only two units failing is given by

$$P[\text{two or fewer units fail}] = R^3 + 3R^2Q + 3RQ^2.$$

This is also the reliability of a three-unit, reliabilitywise parallel system, because then

$$R^3 + 3R^2Q + 3RQ^2 = 1 - (1 - R)^3,$$

where the right side is the expression for the reliability of a three-unit, reliabilitywise parallel system.

4.1.2 DIFFERENT UNITS

If the three units are different then

$$(R_1 + Q_1)(R_2 + Q_2)(R_3 + Q_3) = 1,$$

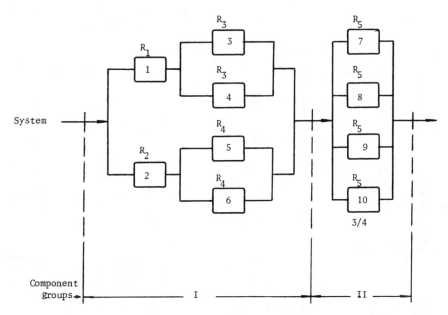

Fig. 4.1 – A system with ten components in series-parallel and conditional function configuration.

and expanding it yields

$$R_1 R_2 R_3 \quad + \quad (R_1 R_2 Q_3 + R_1 Q_2 R_3 + Q_1 R_2 R_3)$$
$$+ \quad (R_1 Q_2 Q_3 + Q_1 Q_2 R_3 + Q_1 R_2 Q_3) + Q_1 Q_2 Q_3 = 1.$$

Here the first term gives the probability of all three units surviving a mission, the terms in the first set of parentheses give the probability of two units surviving and one unit failing, the terms in the second set of parentheses give the probability of one unit surviving and two units failing, and the last term gives the probability of all three units failing.

A similar analysis may be used for any number of units, n, identical or different, by using the proper number for n in the binomial expansion.

EXAMPLE 4–1

What is the reliability of the system given in Fig. 4.1 when three out of the four of Units 7 through 10 are needed for system success?

SOLUTION TO EXAMPLE 4–1

The system's reliability block diagram, given in Fig. 4.1, can be split into two groups, Groups I and II; then the system's reliability is

$$R_S = R_I R_{II},$$

where

$$R_I = 1 - Q_{top}Q_{bottom} = 1 - (1 - R_{top})(1 - R_{bottom}),$$
$$R_{top} = R_1[1 - (1 - R_3)^2],$$

and

$$R_{bottom} = R_2[1 - (1 - R_4)^2].$$

Also,

$$R_{II} = R_5^3(4 - 3R_5).$$

Therefore, the system's reliability is

$$R_S = \{1 - [1 - R_1(1 - Q_3^2)][1 - R_2(1 - Q_4^2)]\}[R_5^3(4 - 3R_5)].$$

EXAMPLE 4-2

The system of Fig. 4.2 is given.   It is made up of 13 components. Components 3, 4, and 5 are unequal and have the reliabilities of R_3, R_4, and R_5, as indicated. Components 6 and 8 have a reliability of R_6, and Components 7 and 9 have a reliability of R_7. Components 10 through 13 have a reliability of R_8.

In Component Group III, it is required that at least two out of the four components function satisfactorily for Group III success. What is the system's reliability in terms of the individual R's?

SOLUTION TO EXAMPLE 4-2

This system's reliability is given by

$$R_S = R_I R_{II} R_{III},$$

where

$$R_I = R_1,$$
$$R_{II} = 1 - [Q_{II-top}Q_{II-bottom}],$$
$$Q_{II-top} = 1 - R_2[1 - (1 - R_3)(1 - R_4)(1 - R_5)],$$
$$Q_{II-bottom} = (1 - R_6R_7)^2,$$

and by substitution

$$R_{II} = 1 - \{1 - R_2[1 - (1 - R_3)(1 - R_4)(1 - R_5)]\}(1 - R_6R_7)^2,$$

$$R_{III} = \sum_{x=2}^{4} \binom{4}{x} R_8^x (1 - R_8)^{4-x},$$

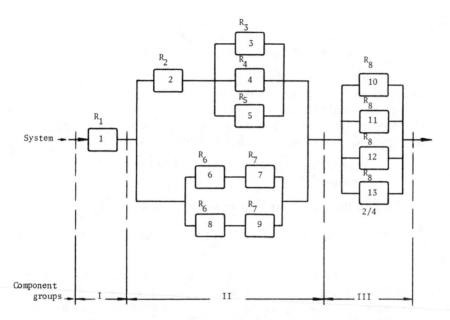

Fig. 4.2 – The system with 13 components in series-parallel and conditional function configuration for Example 4–4.

$$= \binom{4}{2} R_8^2 (1 - R_8)^2 + \binom{4}{3} R_8^3 (1 - R_8)^1 + \binom{4}{4} R_8^4,$$

$$= 6R_8^2 (1 - R_8)^2 + 4R_8^3 (1 - R_8) + R_8^4,$$

or

$$R_{III} = 6R_8^2 - 8R_8^3 + 3R_8^4.$$

Consequently, substitution of R_I, R_{II}, and R_{III} into R_S yields

$$R_S = R_1 \Big[1 - \{1 - R_2[1 - (1 - R_3)(1 - R_4)(1 - R_5)]\}[(1 - R_6 R_7)^2] \Big]$$
$$\cdot (6R_8^2 - 8R_8^3 + 3R_8^4).$$

4.2 THE POISSON DISTRIBUTION

The Poisson distribution may be obtained from the identity

$$e^{-x} e^{+x} = 1.$$

Also,

$$e^{-x} \left(1 + \frac{x}{1!} + \frac{x^2}{2!} + \cdots + \frac{x^n}{n!} + \cdots \right) = 1.$$

Then

$$e^{-x} + \frac{x}{1}e^{-x} + \frac{x^2}{2}e^{-x} + \cdots + \frac{x^n}{n!}e^{-x} + \cdots = 1.$$

Each term represents a probability, and the sum of all these probabilities is equal to 1.   Hence, each term is a term of a *pdf*, and in this case of the Poisson *pdf*.

The interpretation of this distribution is as follows: If x is taken to be the expected, or average, number of occurrences of an event, then

e^{-x} = probability that that event will not occur if x remains constant,

xe^{-x} = probability that that event will occur exactly once,

$\frac{x^2}{2!}e^{-x}$ = probability that that event will occur exactly twice,

and so on.

In reliability, the event of concern is failure, and the average number of failures in time t is given by $x = \lambda t$ when the λ is constant. Consequently, $e^{-x} = e^{-\lambda t}$, which is $R(t)$ for a single system having a constant failure rate, λ, gives the probability that no failure will occur in time t. $xe^{-x} = \lambda t e^{-\lambda t}$ is the probability of exactly one failure occurring in time t, $[x^2/2!]e^{-x} = [(\lambda t)^2/2!]e^{-\lambda t}$ is the probability of exactly two failures occurring in time t, and so on.

Therefore, the probability of exactly k failures occurring in t is given by

$$f(k) = e^{-\lambda t}\frac{(\lambda t)^k}{k!}, \ k = 0, 1, 2, \ldots.$$

This is the discrete Poisson distribution.

The cumulative Poisson distribution is given by

$$F(k) = P(k \text{ or less failures}),$$

or

$$F(k) = \sum_{j=0}^{k} e^{-\lambda t}\frac{(\lambda t)^j}{j!}.$$

These results may be used to determine the probability of occurrence of a specific number of failures during a mission, to determine standby reliability, or to calculate the number of spares required when units have constant and identical failure rates. These applications are given in the examples that follow.

EXAMPLE 4-3

Given is a system exhibiting a *constant* failure rate of 150 fr/10^6 hr and operating for a mission of 100 hr.   Find the following probabilities:

1. No failures occur during this mission.

2. One failure occurs during this mission.

3. Two failures occur during this mission.

4. Two or fewer failures occur during this mission.

SOLUTIONS TO EXAMPLE 4-3
1. The probability that no failures occur during this mission, or $f(0)$, is the system's reliability for this mission, or

$$R_S = f(0) = e^{-\lambda t},$$
$$R_S = e^{-(150)(10^{-6})(100)} = e^{-0.0150},$$
$$R_S = 0.98511.$$

2. The probability that one failure occurs during this mission is given by

$$f(1) = e^{-\lambda t}\frac{(\lambda t)^1}{1!},$$
$$f(1) = \lambda t e^{-\lambda t},$$
$$f(1) = (150)(10^{-6})(100)e^{-(150)(10^{-6})(100)},$$
$$f(1) = (0.015)(0.9851),$$

or

$$f(1) = 0.01477,$$

a relatively low probability, fortunately!
3. The probability that two failures occur during this mission is given by

$$f(2) = e^{-\lambda t}\frac{(\lambda t)^2}{2!},$$
$$f(2) = e^{-0.015}\frac{(0.015)^2}{2},$$
$$f(2) = (0.9851)\frac{0.000225}{2},$$

or

$$f(2) = 0.00011,$$

a much lower probability than for one failure.

4. The probability that two or fewer failures occur during this mission is given by

$$P(k \le 2) \; = \; F(2) = \sum_{k=0}^{2} e^{-\lambda t}\frac{(\lambda t)^k}{k!} = f(0) + f(1) + f(2),$$

$$F(2) \; = \; 0.98511 + 0.01477 + 0.00011,$$

or

$$F(2) \; = \; 0.99999.$$

EXAMPLE 4–4

In a system there exists a very critical unit which requires spares to attain a specified unit reliability of 99%, for a period of 250 hr. The unit has an MTBF of 1,250 hr and exhibits a constant failure rate characteristic.

How many spares would be required if the unit is easily accessible and can be replaced almost immediately, by successfully plugging in an identical spare when the functioning unit fails, to increase its reliability from 81.87% to 99%?

SOLUTION TO EXAMPLE 4–4

The solution may be found by using the Poisson distribution and answering the question, "How many failures, equal to the number of spares, can be tolerated to attain the reliability of 99%?" or the question, "How many standby (spare) units are required to attain the reliability of 99%?" Therefore,

$$F(k) \; = \; \sum_{j=0}^{k} e^{-\lambda t}\frac{(\lambda t)^j}{j!},$$

$$F(k) \; = \; e^{-\lambda t}[1 + \frac{\lambda t}{1!} + \frac{(\lambda t)^2}{2!} + \cdots + \frac{(\lambda t)^k}{k!}],$$

$$\lambda t \; = \; (\frac{1}{MTBF})(t) = (\frac{1}{1,250})(250) = 0.2.$$

Consequently,

$$0.99 = e^{-0.2}[1 + 0.2 + \frac{(0.2)^2}{2} + \cdots + \frac{(0.2)^k}{k!}].$$

This equation should be solved for the nearest integer, k, which satisfies the equality; then the required number of spares is

$$k = 2.$$

With two spare units, the actual reliability is

$$R_S = F(2) = 0.99885,$$

whereas with one spare it would be

$$R_S = F(1) = 0.98248.$$

Consequently, $k = 2$ is the right answer.

EXAMPLE 4-5

A battery has an expected failure rate of 0.01 fr/hr and is used 24 hr per day.

1. How many spares will be required for a three-calendar-month period (assume 30 days per month) for a 95% probability (adequacy, assurance, or confidence) that there will be a sufficient number of spares?

2. What would the battery reliability be for a 24-hr period?

3. If a battery adequacy probability of 95% is required for a 24-hr period, how many spares would be required for a three-calendar-month period, assuming the replacement of the failed batteries is immediate?

4. Compare and discuss the results obtained in Cases 1 and 3.

SOLUTIONS TO EXAMPLE 4-5

1. The average number of spares required for a three-calendar-month period, assuming operation during useful life and no preventive maintenance other than immediate replacement of a battery whenever it fails, is

$$\bar{N}_F = \lambda t,$$

where

$$t = \text{total operating time,}$$
$$t = (\frac{24 \text{ hr}}{\text{day}})(\frac{30 \text{ days}}{\text{month}})(3 \text{ months}),$$
$$t = 2,160 \text{ hr,}$$

and

$$\lambda = 0.01 \text{ fr/hr.}$$

Then

$$\bar{N}_F = (0.01)(2,160),$$

or
$$\bar{N}_F = 21.6 \quad \text{failures.}$$

For a 95% assurance of having sufficient spares

$$0.95 \leq \sum_{j=0}^{N_S} e^{-\bar{N}_F} \frac{(\bar{N}_F)^j}{j!},$$

and with $\bar{N}_F = 21.6$, $N_S = 30$ spares.

2. The battery reliability for a 24-hr period is

$$R(t) \quad = \quad e^{-\lambda t} = e^{(-0.01)(24)},$$

or

$$R(t) \quad = \quad 0.7866.$$

3. For a 24-hr period and 95% assurance,

$$\bar{N}_F = (0.01)(24) = 0.24 \quad \text{failures.}$$

Then

$$0.95 \quad \leq \quad \sum_{j=0}^{N_S} e^{-\bar{N}_F} \frac{(\bar{N}_F)^j}{j!},$$

$$\leq \quad e^{-\bar{N}_F} [1 + \frac{\bar{N}_F}{1!} + \frac{(\bar{N}_F)^2}{2!} + \cdots],$$

or, for $\bar{N}_F = 1$,

$$0.95 \leq e^{-0.24}(1 + 0.24),$$
$$0.95 < 0.9754.$$

Therefore, one spare will be required for a 24-hr operating period. Extending this to a three-calendar-month period, or to 90 days, yields

$$N_S = (1)(90) = 90 \quad \text{spares.}$$

4. In Case 1 the 95% assurance requirement is for a period of 3 months, whereas in Case 3 it is for a 24-hr period; consequently, Case 1 requires only 30 spares, whereas Case 3 requires 90 spares, i.e., substantially more!

With 90 spares a daily battery adequacy probability of 97.54% is achieved, whereas only 95% is required. This means that money can be saved by having fewer spares that give an adequacy probability closer to 95%. To find the number of spares x that are necessary, we can interpolate as follows:

$$x(0.9754) + (90 - x)(0.7866) = (90)(0.95),$$

where

$(90)(0.95)$ = average number of days the battery completes a 24-hr mission for a total of 90 days,

$x(0.9754)$ = average number of days the battery completes a 24-hr mission for a total of x days,

$(90 - x)(0.7866)$ = average number of days the battery completes a 24-hr mission for a total of $(90 - x)$ days,

0.9754 = probability that there is one or fewer battery failures during a 24-hr period,

and

0.7866 = probability that there is no failure during a 24-hr period.

Therefore, the left side of the equation is the average number of days the battery completes a 24-hr mission, for a total of 90 days, and so is the right side. Now solving for x yields

$$x(0.9754 - 0.7866) = (90)(0.95 - 0.7866),$$

$$x = \frac{(90)(0.95 - 0.7866)}{(0.9754 - 0.7866)},$$

or

$$x = 77.88, \text{ or 78 spares.}$$

Therefore, 78 spares would be sufficient to achieve a daily battery adequacy probability of 95% for a 3-month period.

PROBLEMS

4-1. Find the probability of hitting a target with at least one missile when three missiles are fired, each with a probability of 0.833 of scoring a hit.

4-2. A 100-hr MTBF unit needs to be replaced with spares during a mission of 270 hr to attain a mission reliability of 0.9995. How many such spare units would be required?

4-3. Spares need to be provided for a critical unit in a system for 500 cumulative hours of operation with an assurance of 95%. If the unit's failure rate is 0.010 failures per hour, how many spares should be provided for this period of operation?

4-4. A 500-hr MTBF unit needs to be replaced with spares during a mission of 1,100 hr to attain a mission reliability of 0.995. How many such spare units would be required?

4-5. Draw the reliability block diagram of the complex system of components whose reliability is given by

$$R_S = R_1 \cdot [1 - (1 - R_2)^3(1 - R_3)^3] \cdot (4R_4^3 - 3R_4^4)$$
$$\cdot \{1 - [(1 - R_5^2)(1 - R_6)]\}.$$

4-6. Draw a reliability block diagram of the system whose reliability is given by

$$R_S = R_1 R_2 R_3 [1 - (1 - R_A)^2]^2 (R_6^4 + 4R_6^3 Q_6 + 6R_6^2 Q_6^2$$
$$+ 4R_6 Q_6^3) \cdot [e^{-\lambda_7 t}(1 + \lambda_7 t + \frac{1}{2}\lambda_7^2 t^2 + \frac{1}{6}\lambda_7^3 t^3)],$$

where $R_A = R_4[1 - (1 - R_5)^4]$.

Chapter 5

METHODS OF RELIABILITY PREDICTION FOR COMPLEX SYSTEMS

5.1   BAYES' THEOREM METHOD

Bayes' theorem may be stated as follows: The reliability of a system is equal to the reliability of the system, given that a chosen unit, say Unit A, is good, $R_S|A_G$, times the reliability of Unit A, plus the reliability of the system, given that Unit A is bad, $R_S|A_B$, times the unreliability of Unit A. Mathematically,

$$R_S = (R_S|A_G)R_A + (R_S|A_B)Q_A.$$

Bayes' theorem can also be used to determine a system's unreliability, as follows:

$$Q_S = (Q_S|A_G)R_A + (Q_S|A_B)Q_A,$$

where

$$A_G = \text{ event that Unit } A \text{ is good,}$$

and

$$A_B = \text{ event that Unit } A \text{ is bad.}$$

Bayes' theorem can be applied to all combinations of units in a system; however, if the combination, as given by the reliability block diagram of the system, is that of series, parallel, and standby units, then the procedures discussed in previous chapters may be used more

95

expediently. Nevertheless, if there is any doubt about the reliability-wise combination of the units, then Bayes' theorem may be used to determine the system's reliability or unreliability very conveniently.

If the terms $R_S|A_G$, $R_S|A_B$, $Q_S|A_G$, or $Q_S|A_B$ cannot be written out directly, then another unit, say Unit B, may be chosen, and these terms may be expanded as follows:

$$R_S|A_G = (R_S|B_G|A_G)R_B + (R_S|B_B|A_G)Q_B.$$

This procedure can be continued until each term can be written out directly in terms of the reliabilities or unreliabilities of all of the system's units involved.

Generally,

$$R_S = \sum_{i=1}^{n}(R_S|\text{Unit } A \text{ is in State } S_i) \cdot P(\text{Unit } A \text{ is in State } S_i),$$

where

n =all possible modes of function of Unit A

and

$$\sum_{i=1}^{n} P(S_i) = 1.$$

Also,

$$Q_S = \sum_{i=1}^{n}(Q_S|\text{Unit } A \text{ is in State } S_i) \cdot P(\text{Unit } A \text{ is in State } S_i).$$

EXAMPLE 5–1

A system's schematic is given in Fig. 5.1(a). Find the reliability of this system using Bayes' theorem when system success requires that *at least one of the following paths is good*: AD, CE or BE.

SOLUTIONS TO EXAMPLE 5–1
Solution 1:

Choosing Unit C,

$$R_S = (R_S|C_G)R_C + (R_S|C_B)(1 - R_C).$$

With Unit C good, the system whose reliability is sought is given in Fig. 5.1(b). Hence the expression for $R_S|C_G$ is

$$R_S|C_G = 1 - (1 - R_A R_D)(1 - R_E).$$

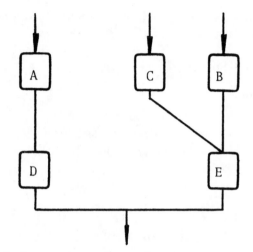

(a) System's schematic for Example 5-1.

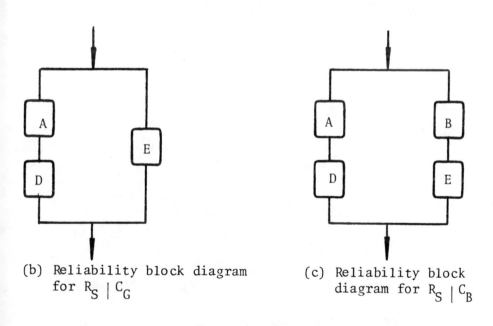

(b) Reliability block diagram
    for $R_S \mid C_G$

(c) Reliability block
    diagram for $R_S \mid C_B$

Fig. 5.1 – Schematic and reliability block diagrams for
        Example 5-1.

With Unit C bad, the system whose reliability is sought is given in Fig. 5.1(c). Hence the expression for $R_S|C_B$ is

$$R_S|C_B = 1 - (1 - R_A R_D)(1 - R_B R_E).$$

Substitution yields

$$
\begin{aligned}
R_S &= [1 - (1 - R_A R_D)(1 - R_E)]R_C \\
&\quad + [1 - (1 - R_A R_D)(1 - R_B R_E)](1 - R_C),
\end{aligned}
$$

or

$$
\begin{aligned}
R_S &= R_A R_D + R_B R_E + R_C R_E - R_B R_C R_E \\
&\quad - R_A R_C R_D R_E - R_A R_B R_D R_E + R_A R_B R_C R_D R_E.
\end{aligned}
$$

Solution 2:
Unit E may be chosen; then

$$R_S = (R_S|E_G)R_E + (R_S|E_B)(1 - R_E),$$

or

$$R_S = [1 - Q_B Q_C(1 - R_A R_D)]R_E + R_A R_D(1 - R_E).$$

Expanded this gives the same result as in Solution 1.

Solution 3:
For this system, Fig. 5.1(a) is the reliability block diagram. Hence it is a series-parallel configuration, and the system's reliability can be written directly, or

$$
\begin{aligned}
R_S &= 1 - (1 - R_A R_D)\{1 - \{R_E[1 - (1 - R_C)(1 - R_B)]\}\}, \\
&= 1 - (1 - R_A R_D)(1 - R_B R_E - R_C R_E + R_B R_C R_E),
\end{aligned}
$$

or

$$
\begin{aligned}
R_S &= R_A R_D + R_B R_E + R_C R_E - R_B R_C R_E \\
&\quad - R_A R_C R_D R_E - R_A R_B R_D R_E + R_A R_B R_C R_D R_E,
\end{aligned}
$$

the same result as in Solution 1.

EXAMPLE 5–2
Using Bayes' theorem, determine the reliability of the system given in Fig. 5.2 when *at least two of the three functioning units are required* for system success.

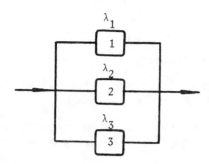

Fig. 5.2 – Schematic for Example 5–2.

SOLUTIONS TO EXAMPLE 5–2
Solution 1
The system's reliability, using Bayes' theorem, is

$$
\begin{aligned}
R_S &= (R_S|\text{No. 1 good})R_1 + (R_S|\text{No. 1 bad})Q_1, \\
&= [1 - (1 - R_2)(1 - R_3)]R_1 + R_2 R_3(1 - R_1), \\
&= (R_2 + R_3 - R_2 R_3)R_1 + R_2 R_3 - R_1 R_2 R_3, \\
&= R_1 R_2 + R_1 R_3 - R_1 R_2 R_3 + R_2 R_3 - R_1 R_2 R_3,
\end{aligned}
$$

or

$$
R_S = R_1 R_2 + R_2 R_3 + R_1 R_3 - 2R_1 R_2 R_3.
$$

Solution 2
Using the binomial method

$$
(R_1 + Q_1)(R_2 + Q_2)(R_3 + Q_3) = 1,
$$

or

$$
\begin{aligned}
R_1 R_2 R_3 + R_1 R_2 Q_3 + R_1 Q_2 R_3 + Q_1 R_2 R_3 \\
+ R_1 Q_2 Q_3 + Q_1 R_2 Q_3 + Q_1 Q_2 R_3 + Q_1 Q_2 Q_3 = 1.
\end{aligned}
$$

The system's reliability is given by those terms which contain two or more R's, or

$$
\begin{aligned}
R_S &= R_1 R_2 R_3 + R_1 R_2 Q_3 + R_1 Q_2 R_3 + Q_1 R_2 R_3, \\
&= R_1 R_2 R_3 + R_1 R_2(1 - R_3) + R_1(1 - R_2)R_3 + (1 - R_1)R_2 R_3, \\
&= R_1 R_2 R_3 + R_1 R_2 - R_1 R_2 R_3 + R_1 R_3 - R_1 R_2 R_3 + R_2 R_3 \\
&\quad - R_1 R_2 R_3, \\
&= R_1 R_2 + R_2 R_3 + R_3 R_1 - 2R_1 R_2 R_3,
\end{aligned}
$$

or the same as in Solution 1.

EXAMPLE 5–3
Figure 5.3(a) is the schematic of a system.

1. What is this system's reliability if system success requires that *at least one of paths ACF, AEF*, or *DEF* is good?

2. What is this system's reliability if system success requires that *at least two of the paths ACF, AEF*, or *DEF* are good?

SOLUTIONS TO EXAMPLE 5–3
1. Solution 1
Choosing Unit A,

$$R_S = (R_S|A_G)R_A + (R_S|A_B)(1 - R_A).$$

With Unit A good, the system's reliability block diagram is as in Fig. 5.3(b). Since Unit D does not affect the reliability of the subsystem in Fig. 5.3(b), the conditional system reliability, given Unit A is good, is

$$R_S|A_G = [1 - (1 - R_C)(1 - R_E)]R_F.$$

With Unit A bad, the system's reliability block diagram is as in Fig. 5.3(c). It is a reliabilitywise series system. The conditional reliability of the system, given Unit A is bad, is

$$R_S|A_B = R_D R_E R_F.$$

This term may also be found by expanding the term $R_S|A_B$ as follows:

$$R_S|A_B = (R_S|D_G|A_B)R_D + (R_S|D_B|A_B)(1 - R_D).$$

It may be seen that the value of $R_S|D_G|A_B$ is $R_E R_F$, because from Fig. 5.3(c) for system success with A_B and D_G we need E and F to be good to have a successful path. Also, $R_S|D_B|A_B$ is zero, because from Fig. 5.3(a), if both A and D are bad, there is no input and the system fails. Consequently, we may now write

$$R_S|A_B = (R_E R_F)R_D + (0)(1 - R_D),$$

which yields the same result as before, or

$$R_S|A_B = R_E R_F R_D.$$

A conclusion that may be drawn is that, if one is not sure of the quantification of any term, it may be expanded continuously until each term has a value of 1 or 0.

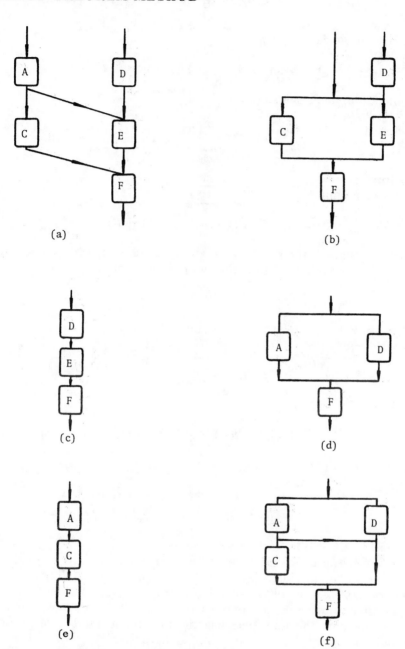

Fig. 5.3 – Schematic and reliability block diagrams of the system and subsystems for Example 5–3.

Substitution yields

$$R_S = \{[1 - (1 - R_C)(1 - R_E)]R_F\}R_A + R_D R_E R_F(1 - R_A).$$

Simplification yields the solution

$$\begin{aligned} R_S \;=\;\; & R_A R_C R_F + R_A R_E R_F + R_D R_E R_F \\ & - R_A R_C R_E R_F - R_A R_D R_E R_F. \end{aligned}$$

Solution 2

Choosing Unit E,

$$R_S = (R_S|E_G)R_E + (R_S|E_B)(1 - R_E).$$

With Unit E good, the system's reliability block diagram is as in Fig. 5.3(d), and the conditional reliability of the system is

$$R_S|E_G = [1 - (1 - R_A)(1 - R_D)]R_F.$$

With Unit E bad, the system's reliability block diagram is as in Fig. 5.3(e); therefore,

$$R_S|E_B = R_A R_C R_F.$$

Substitution yields

$$R_S = \{[1 - (1 - R_A)(1 - R_D)]R_F\}R_E + R_A R_C R_F(1 - R_E).$$

Simplification yields the solution

$$\begin{aligned} R_S \;=\;\; & R_A R_C R_F + R_A R_E R_F + R_D R_E R_F \\ & - R_A R_C R_E R_F - R_A R_D R_E R_F, \end{aligned}$$

the same result as obtained in Solution 1.

It may be noticed that, starting with a different unit, the calculations are different, but they lead to the same result. Therefore, a faster solution may be obtained when a proper unit is chosen to start the solution, especially for complex systems. In general, choose as the starting unit the unit that has the maximum number of inputs plus outputs, or the unit that reduces the system to reliabilitywise series-parallel subsystems, because their reliability can be written out directly.

2. Choosing Unit E, the system's reliability is

$$R_S = (R_S|E_G)R_E + (R_S|E_B)(1 - R_E).$$

With Unit E good, the system's reliability block diagram is as in Fig. 5.3(f), since at least two paths are required for system success. The conditional reliability, given that Unit E is good, is

$$R_S|E_G = (R_S|C_G|E_G)R_C + (R_S|C_B|E_G)(1 - R_C),$$

where Unit C is chosen in the expansion of terms. With Units C and E good, the conditional reliability of the system is

$$R_S|C_G|E_G = R_A R_F,$$

and with Units C bad and E good,

$$R_S|C_B|E_G = R_A R_D R_F.$$

Then

$$R_S|E_G = (R_A R_F)R_C + R_A R_D R_F(1 - R_C).$$

If Unit E is bad, only one path is left and the system fails, or

$$R_S|E_B = 0.$$

Consequently, substitution into the first equation yields

$$R_S = [R_A R_F R_C + R_A R_D R_F(1 - R_C)]R_E + (0)(1 - R_E),$$

or

$$R_S = R_A R_C R_E R_F + R_A R_D R_E R_F - R_A R_C R_D R_E R_F.$$

5.2 BOOLEAN TRUTH TABLE METHOD

For a complex system which is composed of n independently functioning units, the system's reliability can be calculated using the Boolean truth table method. It is known that a unit in a system has only two states: functioning or failed. In this method, all units are listed in a truth table, as shown later in Table 5.2, with 2^n possible combinations, or 2^n entries, where n is the number of units in the system. The table has a 1 or 0 entry in each column under Units indicating the success or failure, respectively, of each unit. Therefore, all possible combinations of all units succeeding or failing are thus listed. The next step is to examine each row of the truth table and judge whether the combination of units succeeding or failing yields system success or failure.

Let an S or F, respectively, denote system success or failure, and insert S or F in the next column in the table. For each S entry, multiply the respective probabilities for the indicated state of each unit to yield a contribution, R_S, to the probability of system success for that entry, and these R_S are listed in the last column of the truth table. Finally, all probabilities in the R_S column of the truth table are summed to obtain the system's reliability. To illustrate the methodology, a numerical example is given in Example 5-4. From the Boolean truth table, the system's reliability equation can be written, as given by Eq. (5.1).

A similar method is presented in Chapter 6 and applied to cases of more than two function modes of the system's units.

5.3 PROBABILITY MAPS METHOD

The success or failure of a system which is composed of n independently functioning units can be analyzed by probability or truth maps [1]. A probability map is a block diagram, as shown in Figs. 5.4 and 5.7. For an n unit system, there are 2^n cells in its probability map. Each cell indicates one combination of a possible event of the system's success or failure. For example, if a system consists of two units, A and B, then the probability map has $2^2 = 4$ cells, because there are four combinations of the possible events of the system's success or failure. These four cells form a probability map, as may be seen in Fig. 5.4(a). For a three or four-unit system, the probability map has $2^3 = 8$ cells or $2^4 = 16$ cells, and as may be seen in Fig. 5.4(f) and (g), respectively. Figure 5.7 gives the probability map for a five-unit system, and it is composed of two 16-cell maps. In a probability map the cells spanned by a letter, say A, indicate the events which are the intersections of Unit A functioning [for example, the Cells d and c in Fig. 5.4(a)], and the cells not spanned by A indicate the events which are the intersections of Unit A failing [for example, the Cells a and c in Fig. 5.4(a). Therefore, in a probability map, every letter which denotes a unit should span half of the total cells, and every cell indicates an event different than the other, as shown in Figs. 5.4 and 5.7. This is the requirement to construct a probability map. It may be noted that the structure of the probability map for a given system is not unique. For a two-unit system, it may be either in the form of Figs. 5.4(a) through (d) or in the form of Fig. 5.4(e), and all of these maps satisfy the requirements mentioned above.

In Fig. 5.4(a), Cell a represents the event of both Units A and B failing; Cell d, Unit A functioning and Unit B failing; Cell b, Unit A failing and Unit B functioning; and Cell c, both Units A and B functioning.

Let the number 1 indicate the event of the system's success, then for different system configurations the corresponding probability map

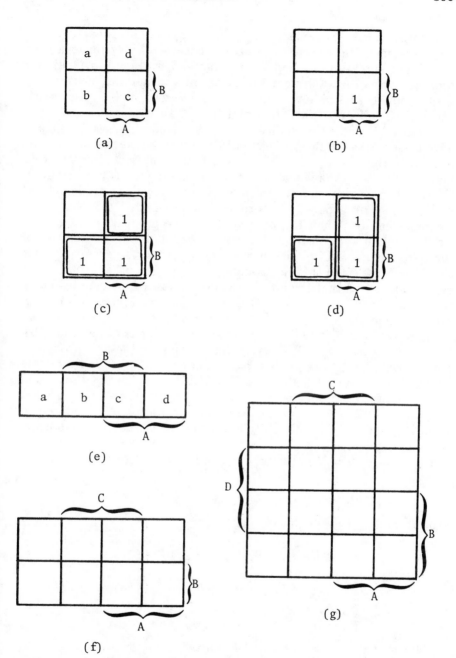

Fig. 5.4 – Probability maps.

can be obtained. For example, Fig. 5.4(b) is the probability map for a system with two units in series, because the system succeeds only when both Units A and B are functioning. Figure 5.4(c) is the probability map for a system with two units in parallel, because the system succeeds when both Units A and B succeed, or Unit A succeeds and Unit B fails, or Unit A fails and Unit B succeeds.

After all the cells that indicate the events leading to system success have been marked with 1s, the system's reliability is equal to the sum of all probabilities of these events, which are represented by the cells marked with 1s. It may seen that there are many terms of probabilities that need to be written out. To simplify these terms, the cells marked with 1s may be grouped. A group can be selected that has some common features; then the probability of the grouped event can be written out in the form which is the product of the reliabilities or the unreliabilities of individual units only. Also, the groups of cells should be so selected that no grouping overlaps any other grouping. It needs to be noted that the group selection is not unique. For example, two groups may be selected, either in the form of Fig. 5.4(c) or Fig. 5.4(d). Then the system's reliability equation can be obtained from the groups of 1s which indicate system success. For example, from Fig. 5.4(d), the two groups of 1s are exclusive, and for the first group the probability of system success is $R_B Q_A$, because that cell indicates the intersection of Unit A failing and Unit B succeeding. For the second group the probability of system success is R_A, because no matter what the state of Unit B is, the system succeeds when Unit A is functioning. Therefore, the system's reliability is the sum of these two probabilities, or

$$R_S = R_B Q_A + R_A = R_B (1 - R_A) + R_A,$$

or

$$R_S = R_A + R_B - R_A R_B.$$

Also, from parallel reliability analysis,

$$R_S = R_B Q_A + R_A R_B + R_A Q_B,$$

$$R_S = R_B Q_A + R_A (R_B + Q_B),$$

or

$$R_S = R_A + R_B - R_A R_B.$$

The probability map method is universal in the sense that it is a uniform technique which can be applied equally well to all system configurations. The general procedure of this method for a system, especially for a complex system, is as follows:

1. Identify all continuous paths which allow system success.

2. Construct a probability map for each successful path, and step by step fill in the cells in these probability maps with 1s if the event denoted by this cell leads to system success.

3. Pile up all maps of each successful path; if a cell already has a 1, do not place another 1 in that cell.

4. Select groups of 1s by rectangles, as shown in Fig. 5.7(e), such that the probability of the grouped event can be written as the product of the reliabilities or the unreliabilities of the individual units, and no one grouping overlaps any other grouping.

5. Write down the probabilities of system success events denoted by each group of 1s. Then the system's reliability is the sum of all of these probabilities.

To illustrate the methodology, a numerical example is given in Example 5-4.

5.4   LOGIC DIAGRAMS METHOD

In the logic diagrams method the reliability block diagram is transformed into a switching network [2], as shown in Table 5.1. A closed contact represents unit success; an open contact, unit failure. Each complete path of contacts represents an alternate mode of system success. Each unit that is required for each alternative mode of system success is identified by a contact along that path. All paths terminate at a common point which represents system success. It has to be noted that the logic diagram is so constructed that all paths are mutually exclusive.

Examples of the logic diagram for simple systems are given in Table 5.1. For complex configurations the procedure is to reduce the reliability block diagram to a simple series or parallel configuration by successively splitting the diagram into subdiagrams by removing one unit and replacing it with a short circuit and an open circuit, as shown in Fig. 5.8.

After the logic diagram is drawn, two approaches can be used to obtain a numerical answer. The first involves writing an equation for the system's reliability by writing down every path's probability with an addition sign joining all paths. For example, for the first logic diagram given in the second row of Table 5.1, the system's reliability equation is

$$R_S = R_A Q_B + Q_A R_B + R_A R_B,$$

TABLE 5.1 – Logic diagrams and related reliability block diagrams.

Reliability block diagram	Logic diagram

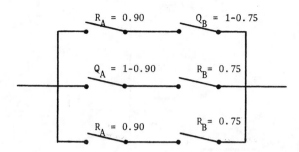

Fig. 5.5 – A logic diagram illustrating the numerical determination of its reliability.

$$= R_A(1 - R_B) + (1 - R_A)R_B + R_AR_B,$$

or

$$R_S = R_A + R_B - R_AR_B.$$

The second method is to insert the values for the various probabilities directly into the logic diagram, multiply the series terms, and add the parallel terms until just one series term remains. For example, if $R_A = 0.90$ and $R_B = 0.75$, then the first logic diagram given in the second row of Table 5.1 can be redrawn as shown in Fig. 5.5. Then the system's reliability is

$$R_S = (0.90)(1 - 0.75) + (1 - 0.90)(0.75) + (0.90)(0.75),$$

or

$$R_S = 0.975.$$

EXAMPLE 5–4

In the system of Fig. 5.6, system success requires that *at least one path be good.* Find the reliability of this system using the following methods:

1. The Bayes' theorem approach.

2. The Boolean truth table method.

3. The probability maps method.

4. The logic diagrams method.

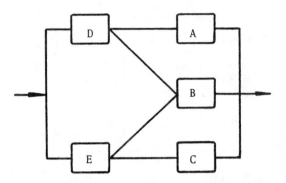

Fig. 5.6 – Schematic of a complex system for Example 5–4.

Assume that $R_A = 0.70$, $R_B = 0.90$, $R_C = 0.85$, $R_D = 0.80$, and $R_E = 0.80$.

SOLUTIONS TO EXAMPLE 5–4

1. *The Bayes' theorem approach:*

$$R_S = (R_S|B_G)R_B + (R_S|B_B)(1 - R_B),$$

where

$$R_S|B_G = 1 - (1 - R_D)(1 - R_E) = R_D + R_E - R_D R_E,$$

and

$$R_S|B_B = 1 - (1 - R_A R_D)(1 - R_C R_E),$$
$$= R_A R_D + R_C R_E - R_A R_C R_D R_E.$$

Substitution yields

$$R_S = (R_D + R_E - R_D R_E)R_B$$
$$+ (R_A R_D + R_C R_E - R_A R_C R_D R_E)(1 - R_B),$$

or

$$R_S = R_B R_D + R_B R_E + R_A R_D + R_C R_E - R_A R_B R_D$$
$$- R_B R_C R_E - R_B R_D R_E - R_A R_C R_D R_E$$
$$+ R_A R_B R_C R_D R_E.$$

Substituting the values of the reliabilities into the system's reliability function yields

$$R_S = (0.90)(0.80) + (0.90)(0.80) + (0.70)(0.80) + (0.85)(0.80)$$

$$- (0.70)(0.90)(0.80) - (0.90)(0.85)(0.80)$$
$$- (0.90)(0.80)(0.80) - (0.70)(0.85)(0.80)(0.80)$$
$$+ ((0.70)(0.90)(0.85)(0.80)(0.80),$$

or

$$R_S = 0.94992.$$

2. *The Boolean truth table method:* The Boolean truth table is given in Table 5.2. The table has $2^5 = 32$ entries and has a 1 or 0 entered in each column indicating the success or failure, respectively, of each unit. The contribution to the system's reliability of each entry is obtained by multiplying the probability of occurrence of each event in that entry. For example, the reliability of Entry 4 is

$$R_{S4} = (1 - R_A)(1 - R_C)(1 - R_D)R_E R_B,$$
$$= (1 - 0.70)(1 - 0.85)(1 - 0.80)(0.80)(0.90),$$

or

$$R_{S4} = 0.00648.$$

Then the system's reliability is equal to the sum of the reliabilities of all entries, or

$$\sum_{i=1}^{32} R_{Si} = 0.94992.$$

The system's reliability equation can be written from the Boolean truth table as

$$
\begin{aligned}
R_S = \ & (1 - R_A)(1 - R_C)(1 - R_D)R_E R_B \\
& + (1 - R_A)(1 - R_C)(1 - R_E)R_D R_B \\
& + (1 - R_A)(1 - R_C)R_D R_E R_B \\
& + (1 - R_A)(1 - R_D)(1 - R_B)R_C R_E \\
& + (1 - R_A)(1 - R_D)R_C R_E R_B + (1 - R_A)(1 - R_E)R_C R_D R_B \\
& + (1 - R_A)(1 - R_B)R_C R_D R_E + (1 - R_A)R_C R_D R_E R_B \\
& + (1 - R_C)(1 - R_D)R_A R_E R_B \\
& + (1 - R_C)(1 - R_E)(1 - R_B)R_A R_D \\
& + (1 - R_C)(1 - R_E)R_A R_D R_B + (1 - R_C)(1 - R_B)R_A R_D R_E \\
& + (1 - R_C)R_A R_C R_E R_B + (1 - R_D)(1 - R_B)R_A R_C R_E \\
& + (1 - R_D)R_A R_C R_D R_B + (1 - R_E)(1 - R_B)R_A R_C R_D \\
& + (1 - R_E)R_A R_C R_D R_B + (1 - R_B)R_A R_C R_D R_E \\
& + R_A R_B R_C R_D R_E.
\end{aligned}
$$

TABLE 5.2 – The Boolean truth table for Example 5-4.

Entry No.	A	C	D	E	B	Success (S) or failure (F)	R_{Si}
1	0	0	0	0	0	F	–
2	0	0	0	0	1	F	–
3	0	0	0	1	0	F	–
4	0	0	0	1	1	S	0.00648
5	0	0	1	0	0	F	–
6	0	0	1	0	1	S	0.00648
7	0	0	1	1	0	F	–
8	0	0	1	1	1	S	0.02592
9	0	1	0	0	0	F	–
10	0	1	0	0	1	F	–
11	0	1	0	1	0	S	0.00408
12	0	1	0	1	1	S	0.03672
13	0	1	1	0	0	F	–
14	0	1	1	0	1	S	0.03672
15	0	1	1	1	0	S	0.01632
16	0	1	1	1	1	S	0.14688
17	1	0	0	0	0	F	–
18	1	0	0	0	1	F	–
19	1	0	0	1	0	F	–
20	1	0	0	1	1	S	0.01512
21	1	0	1	0	0	S	0.00168
22	1	0	1	0	1	S	0.01512
23	1	0	1	1	0	S	0.00672
24	1	0	1	1	1	S	0.06048
25	1	1	0	0	0	F	–
26	1	1	0	0	1	F	–
27	1	1	0	1	0	S	0.00952
28	1	1	0	1	1	S	0.08568
29	1	1	1	0	0	S	0.00952
30	1	1	1	0	1	S	0.08568
31	1	1	1	1	0	S	0.03808
32	1	1	1	1	1	S	0.34272

$$\sum R_{Si} = 0.94992$$

A reduction table, Table 5.3, is constructed which allows us to reduce the 19 Boolean success terms to a simplified expression for the given reliability model [3], because some of the 19 terms represent events that complement each other. These terms can be combined and the final system's reliability expression can be simplified. For example, the first and the third terms of Column 1 in Table 5.3 represent the complementary events of Unit D failing, given that Units A and C fail and Units E and B function, and Unit D functioning, given that Units A and C fail and Units E and B function, respectively. Therefore, the summation of these two terms is equal to the probability of Units A and C failing and Units E and B functioning, which is the first term of Column 2 in Table 5.3. Checking out and combining all of the possible terms in the first column of Table 5.3, listing the combined terms in the second column, checking out and combining all of the possible terms in the second column, listing the combined terms in the third column, and repeating the procedure until no more terms can be combined completes Table 5.3. In Table 5.3 the terms which cannot be combined are underlined. Eventually, there are six underlined terms left; therefore, the system's reliability can be expressed by these six terms, or

$$
\begin{aligned}
R_S \; = \; & (1 - R_A)(1 - R_C)R_E R_B + (1 - R_A)R_D(1 - R_E)R_B \\
& + (1 - R_A)R_C R_E + R_A(1 - R_C)(1 - R_D)R_E R_B \\
& + R_A R_C(1 - R_D)R_E + R_A R_D,
\end{aligned}
$$

or

$$
\begin{aligned}
R_S \; = \; & R_B R_D + R_B R_E + R_A R_D + R_C R_E - R_A R_B R_D - R_B R_C R_E \\
& - R_B R_D R_E - R_A R_C R_D R_E + R_A R_B R_C R_D R_E, \qquad (5.1)
\end{aligned}
$$

the same result as that obtained using the Bayes' theorem approach.

3. *The probability maps method:* Let A, B, C, D, and E denote success, and $\bar{A}$, $\bar{B}$, $\bar{C}$, $\bar{D}$, and $\bar{E}$ denote failure. Then the probability maps method would proceed as follows: First, identify all continuous paths which allow mission success. These would be AD, BD, BE, and CE. As shown in Fig. 5.7, step by step, fill in the cells in the probability maps for five variables with 1s; see Figs. 5.7(a) through (d). Then pile up steps (a) through (d) in step (e). If a cell already has a 1, do not place another 1 in that cell. A total of six groups is selected such that no grouping overlaps any other grouping. Actually, the groups of 1s are chosen arbitrary, but the fewer the number of groups, the fewer the terms in the reliability expression. In Fig. 5.7(e), the eight-cell group gives AD, the four-cell group gives $\bar{A}CE$, because

TABLE 5.3 – The reduction tabulation for the Boolean truth table method solution of Example 5–4.

$(1-R_A)(1-R_C)(1-R_D)R_E R_B$ $(1-R_A)(1-R_C)R_E R_B$

$(1-R_A)(1-R_C)R_D(1-R_E)R_B$ $\underline{(1-R_A)R_D(1-R_E)R_B}$

$(1-R_A)(1-R_C)R_D R_E R_B$ $(1-R_A)R_C(1-R_D)R_E$

$(1-R_A)R_C(1-R_D)R_E(1-R_B)$

$(1-R_A)R_C(1-R_D)R_E R_B$ $\underline{(1-R_A)R_C R_E}$

$(1-R_A)R_C R_D(1-R_E)R_B$

$(1-R_A)R_C R_D R_E(1-R_B)$ $(1-R_A)R_C R_D R_E$

$(1-R_A)R_C R_D R_E R_B$

$R_A(1-R_C)(1-R_D)R_E R_B$

$R_A(1-R_C)R_D(1-R_E)(1-R_B)$

$R_A(1-R_C)R_D(1-R_E)R_B$ $R_A(1-R_C)R_D(1-R_B)$

$R_A(1-R_C)R_D R_E(1-R_B)$ $R_A(1-R_C)R_D R_B$

$R_A(1-R_C)R_D R_E R_B$ $R_A(1-R_C)R_D$

$R_A R_C(1-R_D)R_E(1-R_B)$

$R_A R_C(1-R_D)R_E R_B$ $\underline{R_A R_C(1-R_D)R_E}$

$R_A R_C R_D(1-R_E)(1-R_B)$

$R_A R_C R_D(1-R_E)R_B$ $R_A R_C R_D(1-R_E)$

$R_A R_C R_D R_E(1-R_B)$ $R_A R_C R_D R_E$

$R_A R_C R_D R_E R_B$ $R_A R_C R_D$ $\underline{R_A R_D}$

Units C and E completely span this grouping, the three two-cell groups give $\bar{A}\bar{C}BE$, $\bar{A}BD\bar{E}$, and $AC\bar{D}E$, and the single cell gives $AB\bar{C}\bar{D}E$. Therefore, the system's reliability is

$$
\begin{aligned}
R_S = \ & R_A R_D + (1 - R_A) R_C R_E + (1 - R_A)(1 - R_C) R_B R_E \\
& + (1 - R_A) R_B R_D (1 - R_E) + R_A R_C (1 - R_D) R_E \\
& + R_A R_B (1 - R_C)(1 - R_D) R_E,
\end{aligned}
$$

or

$$
\begin{aligned}
R_S = \ & R_B R_D + R_B R_E + R_A R_D + R_C R_E - R_A R_B R_D \\
& - R_B R_C R_E - R_B R_D R_E - R_A R_C R_D R_E \\
& + R_A R_B R_E R_D R_E,
\end{aligned}
$$

the same expression as that obtained in Methods 1 and 2.

4. *The logic diagrams method:* In this problem the complex configuration can be reduced to a series-parallel configuration by splitting the diagram into subdiagrams by removing Unit B and replacing it with a short circuit and an open circuit. The procedure is given in Fig. 5.8.

Following the logic diagram given in Fig. 5.8, the system's reliability is

$$
\begin{aligned}
R_S = \ & R_B[R_D + (1 - R_D) R_E] \\
& + (1 - R_B)[R_A R_D + R_A (1 - R_D) R_C R_E \\
& + (1 - R_A) R_D R_C R_E + (1 - R_A)(1 - R_D) R_C R_E],
\end{aligned}
$$

or

$$
\begin{aligned}
R_S = \ & R_B R_D + R_B R_E + R_A R_D + R_C R_E \\
& - R_B R_D R_E - R_A R_B R_D - R_B R_C R_E \\
& - R_A R_C R_D R_E + R_A R_B R_C R_D R_E,
\end{aligned}
$$

which is the same as that obtained by previous methods.

The numerical answer is

$$
\begin{aligned}
R_S = \ & (0.90)(0.80) + (0.90)(1 - 0.80)(0.80) \\
& + (1 - 0.90)(0.70)(0.80) \\
& + (1 - 0.90)(1 - 0.70)(0.80)(0.85)(0.80) \\
& + (1 - 0.90)(1 - 0.70)(1 - 0.80)(0.85)(0.80) \\
& + (1 - 0.90)(0.70)(1 - 0.80)(0.85)(0.80),
\end{aligned}
$$

or

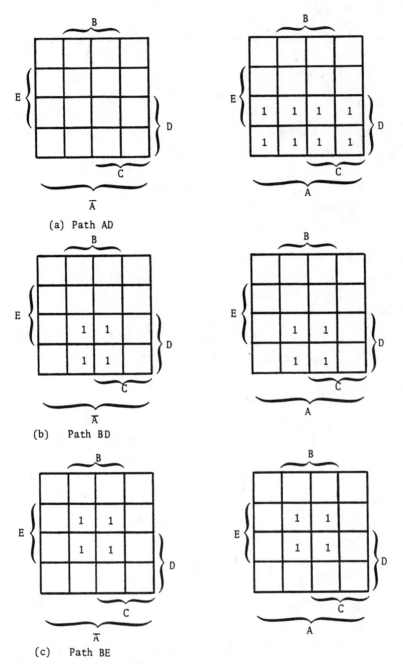

(a) Path AD

(b)   Path BD

(c)   Path BE

Fig. 5.7 – Step by step construction of the probability maps
for Example 5–4.

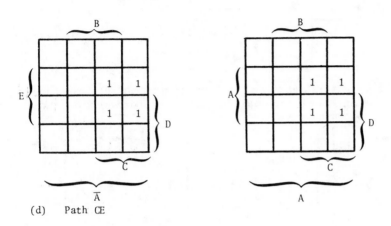

(d)   Path CE

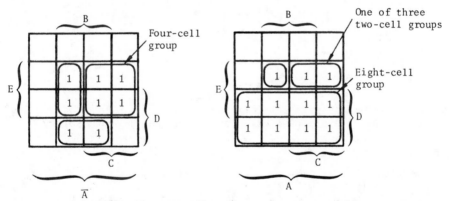

(e)   The sum of paths AD, BD, BE and CE with an arbitrary selection of grouping.

Fig. 5.7 (continued)

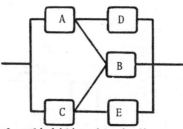

Splitting the original reliability block diagram above into two sub-diagrams by removing Unit B and replacing it by a short circuit and an open circuit, as shown next:

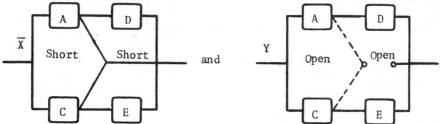

Now the logic diagram of the system can be expressed as shown below:

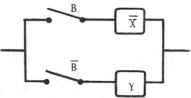

$\bar{X}$ and Y are in series-parallel form and their logic diagram can be drawn directly. Then the system's logic diagram becomes as shown below:

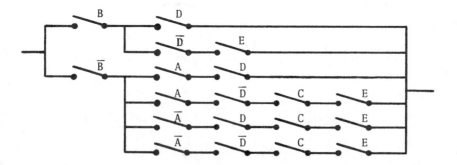

Fig. 5.8 – The procedure of developing the logic diagram of the system in Example 5–4.

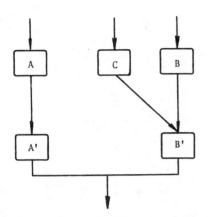

Fig. 5.9 – The system of Problem 5-1.

$$R_S = 0.94992.$$

PROBLEMS

5-1. In the system of Fig. 5.9, *system success requires that at least two of the following paths be good: AA', CB' or BB'.* Find the reliability of this system using Bayes' theorem.

5-2. Given the system in Fig. 5.10, using Bayes' theorem, find its reliability if *system success requires that at least one of the following paths be good: ABG, ACF, DCF, DEF.* Preferably, use Bayes' theorem at least twice to obtain the solution. In the final solution, expand and multiply out all terms, and write it out in terms of the respective R_i's of all units involved.

5-3. Given the system in Fig. 5.10, using Bayes' theorem, find its reliability if *system success requires that at least two of the following paths be good: ABG, ACF, DCF, DEF.* Preferably, use Bayes' theorem at least twice to obtain the solution. In the final solution, expand and multiply out all terms, and write it out in terms of the respective R_i's of all units involved.

5-4. (1) What is the reliability of the system in Fig. 5.11 whose success requires *that at least one of the following paths be good: ABG, ACF, AEF, or DEF.*

    (2) What is the reliability if *at least two good paths are required for success?*

Use Bayes' theorem and give the solutions in terms of the reliabilities of all units involved.

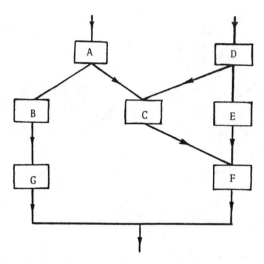

Fig. 5.10 – Schematic of a network of units of Problems 5-2 and 5-3.

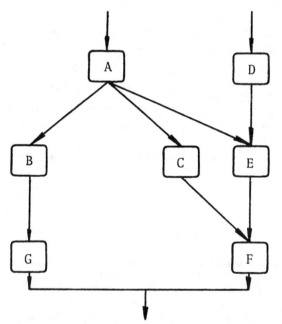

Fig. 5.11 – Schematic of the complex system of Problem 5-4.

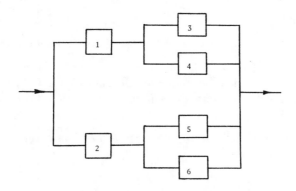

Fig. 5.12 – The phased-array radar configuration for Problem 5-5.

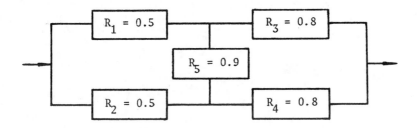

Fig. 5.13 – The schematic diagram of the five units in the system of Problem 5-6.

5-5. The block diagram in Fig. 5.12 is that of a phased-array radar with dual Transmitters 1 and 2, with dual Antennas 3 and 4 and 5 and 6 for each transmitter, respectively. Using Bayes' theorem, determine the system's reliability if only one out of four antenna elements is needed for full system capability. Write out the final answer in terms of R's only.

5-6. Find the system reliability of the arrangement given in Fig. 5.13. At least one of paths $R_1 R_3$, $R_1 R_5 R_4$, $R_2 R_4$, and $R_2 R_5 R_3$ is required for system success.

REFERENCES

1. Hurley, R. B., "Probability Maps," *IEEE Transactions on Reliability*, Vol. 12, No. 3, pp. 39-44, 1963.

2. Case, T., "A Reduction Technique for Obtaining a Simplified Reliability," *IEEE Transactions on Reliability*, Vol. 26, No. 4, pp. 248-249, 1977.

3. MIL–HDBK–217C, *Reliability Prediction of Electronic Equipment*, Appendix A, 9 Apr. 1979.

Chapter 6

RELIABILITY OF SYSTEMS WITH MULTIMODE FUNCTION AND LOGIC

6.1 RELIABILITY PREDICTION METHODOLOGY

When there are units in a system with multiple modes of function such as (1) normal, (2) failed open, and (3) failed short in the case of a resistor, or (1) normal, (2) failed open energized, (3) failed open de-energized, (4) failed shut energized, and (5) failed shut de-energized in the case of a valve, the following procedure is recommended to determine the system's reliability:

1. Define system success.

2. Determine the total number of permutations (P), with repetitions permitted, of all modes of function of all units in the system. This may be calculated from

$$P = M_1^{N_1} \times M_2^{N_2} \times \cdots \times M_n^{N_n},$$

where

M_n = number of modes of function of a group of units in the system,

and

N_n = number of units in the system having M_n modes of function.

For example, if in a system of three units, two units (N_1) have three modes (M_1) of function, and one unit (N_2) has two modes (M_2) of function, then

$$P = M_1^{N_1} \times M_2^{N_2} = 3^2 \times 2^1,$$

or

$$P = 18.$$

3. Prepare a table having the following headings:

     (1) Permutation number.

     (2) Units in system identified by unit number and the mode of function of each unit.

     (3) System output.

     (4) Is system successful?

     (5) Probability of occurrence of the permutation resulting in system success.

4. Fill in the first column in the previous table with the numbers 1 through P.

5. Under "Units and modes of function," enter all possible permutations of modes of function of all system units. Preferably, use vertical rotation for all modes of function of the last unit first, then of the next to the last unit, and then finally of the first unit.

6. Enter the system output in the next column.

7. Under column "Is system successful?" enter whether of not a particular permutation has an output which leads to system success by writing Y (yes) for system success and N (no) for system failure.

8. In the next column, write the probability of success of each successful permutation, or for each permutation with a Y in the previous column. It must be noted that in each success permutation, the probability terms are independent of each other. If not, the correct probability term for the intersection of the probabilities involved must be used.

9. The system's reliability is the sum of all successful permutations. This is so if all permutations are mutually exclusive of each other; consequently, either one successful permutation leads to system success, or the next, or the next, etc., until all the probabilities of all successful permutations are added, as we are dealing with the union of these mutually exclusive success probabilities. It must be noted that the sum of probabilities given by all permutations is equal to 1, because all permutations give the complete sample space of all possible probabilities: successful and unsuccessful permutations of system function.

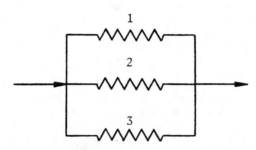

Fig. 6.1 – The schematic of a circuit with three resistors.
Each resistor's resistance is 3 ohms.

For a similar approach, the Boolean truth table method, see Chapter 5.

EXAMPLE 6–1

Figure 6.1 is the schematic of three resistors.

Resistor 1 has three modes of function: N (normal), O (failed open), and S (failed short).

Resistor 2 has three modes of function: N (normal), O (failed open), and S (failed short).

Resistor 3 has two modes of function: N (normal) and S (failed short). Its probability of failing open is practically zero, or $Q_{3o} = 0$.

Circuit success is defined as that combination of all units in the circuit that gives an output of 1 ohm, 1.5 ohms, or 3 ohms. Take the shorted resistance as 0 and the failed open resistance as ∞. Find the reliability of the circuit of these resistors.

SOLUTION TO EXAMPLE 6–1

The number of permutations is

$$P = 3^2 \times 2^1 = 18$$

Table 6.1 gives the solution. For Permutation 3 the circuit's resistance is given by

$$\frac{1}{r_s} = \frac{1}{r_1} + \frac{1}{r_2} + \frac{1}{r_3},$$

or

$$\frac{1}{r_s} = \frac{1}{3} + \frac{1}{\infty} + \frac{1}{3} = \frac{2}{3}.$$

Therefore, r_s, the circuit's resistance, is 1.5 ohms. The remainder of the outputs are found similarly. The next column is filled in with

TABLE 6.1 – Solution to Example 6–1.

Permu-tation number	Units and modes of function 1	2	3	Circuit output, ohms	Is circuit successful? Y - Yes N - No	Probability of circuit success
1	N	N	N	1.0	Y	$R_1 R_2 R_3$
2	N	N	S	0.0	N	0
3	N	O	N	1.5	Y	$R_1 Q_{2o} R_3$
4	N	O	S	0.0	N	0
5	N	S	N	0.0	N	0
6	N	S	S	0.0	N	0
7	O	N	N	1.5	Y	$Q_{1o} R_2 R_3$
8	O	N	S	0.0	N	0
9	O	O	N	3.0	Y	$Q_{1o} Q_{2o} R_3$
10	O	O	S	0.0	N	0
11	O	S	N	0.0	N	0
12	O	S	S	0.0	N	0
13	S	N	N	0.0	N	0
14	S	N	S	0.0	N	0
15	S	O	N	0.0	N	0
16	S	O	S	0.0	N	0
17	S	S	N	0.0	N	0
18	S	S	S	0.0	N	0

$$R_S = \sum_{i=1}^{4} P_i \text{ (of all circuit success permutations)},$$

or

$$R_S = R_1 R_2 R_3 + R_1 Q_{2o} R_3 + Q_{1o} R_2 R_3 + Q_{1o} Q_{2o} R_3.$$

Y or N by comparing the circuit's output with the circuit's success definition. The last column is filled in as follows: For Permutation 3, under "Units and modes of function," we have N, O, N; hence, we want the probability that Unit 1 is functioning normally, *and* the probability that Unit 2 has failed open, *and* the probability that Unit 3 is functioning normally. In probability mathematics, this is written as $R_1 Q_{2o} R_3$. The remaining successful permutation probabilities are written similarly. This circuit's reliability is then given by

$$R_S = R_1 R_2 R_3 + R_1 Q_{2o} R_3 + Q_{1o} R_2 R_3 + Q_{1o} Q_{2o} R_3.$$

If the resistors have constant failure rates λ_1, λ_2, and λ_3, respectively, then

$$R_1 = e^{-\lambda_1 t}, \quad R_2 = e^{-\lambda_2 t}, \quad R_3 = e^{-\lambda_3 t},$$

$$Q_{1o} = 1 - e^{\lambda_{1o} t}, \quad \text{and} \quad Q_{2o} = 1 - e^{-\lambda_{2o} t},$$

where

λ_{1o} = failing open failure rate of Resistor 1,

λ_{2o} = failing open failure rate of Resistor 2,

$$Q_{io} = \int_0^t f_{io}(T)\, dT = \int_0^t \lambda_{io} e^{-\lambda_{io} T}\, dT,$$

or

$$Q_{io}(t) = 1 - e^{-\lambda_{io} t},$$

and

$$\lambda_{io} = \frac{N_{ifo}(\Delta T)}{N_T \times \Delta T}.$$

It must be noted that this procedure can be used for any combination of units in a system. Just substitute for the word *circuit* the word *system*.

EXAMPLE 6–2

In Fig. 6.2, steam comes from a boiler and passes through the stop valves, then through the inlet valves, and then enters a steam turbine. What is the reliability of the subsystem consisting of the four stop valves only? Subsystem success is defined as no steam flowing past the four stop valves when the valves are called upon to shut, so that no steam goes through the inlet valves when the inlet valves happen to stay open, and thus the turbine does not overspeed. The stop valves are assumed to have two modes of function: successful or failed. Only

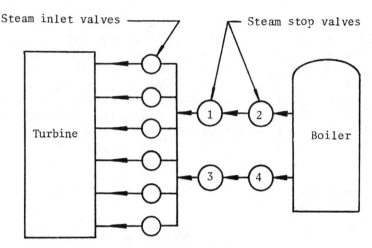

Fig. 6.2 – The schematic of a turbogenerator system including steam-turbine stop valves of Example 6–2.

one mode of success and one mode of failure are assumed to exist in this case.

SOLUTION TO EXAMPLE 6–2

$M = 2$, $N = 4$, and $P = M^N = 2^4 = 16$ permutations. Table 6.2 gives the solution. The sum of the nine favorable permutations yields:

$$R_S = \sum_{i=1}^{9} P_i(\text{of all system success permutations}),$$

or

$$
\begin{aligned}
R_S =\ & R_1 R_2 R_3 R_4 + R_1 R_2 R_3 (1 - R_4) + R_1 R_2 (1 - R_3) R_4 \\
& + R_1 (1 - R_2) R_3 R_4 + R_1 (1 - R_2) R_3 (1 - R_4) \\
& + R_1 (1 - R_2)(1 - R_3) R_4 + (1 - R_1) R_2 R_3 R_4 \\
& + (1 - R_1) R_2 R_3 (1 - R_4) + (1 - R_1) R_2 (1 - R_3) R_4.
\end{aligned}
$$

If $R_1 = R_2 = R_3 = R_4 = R$, then

$$R_S = R^4 + 4R^3(1 - R) + 4R^2(1 - R)^2,$$

or

$$R_S = (2R - R^2)^2 = [1 - (1 - R)^2]^2.$$

TABLE 6.2 – Solution of Example 6–2.

Permu-tation number	Units and modes of function				Is system successful?	Probability of system success*
	1	2	3	4		
1	S	S	S	S	Y	$R_1 R_2 R_3 R_4$
2	S	S	S	F	Y	$R_1 R_2 R_3 (1 - R_4)$
3	S	S	F	S	Y	$R_1 R_2 (1 - R_3) R_4$
4	S	S	F	F	N	0
5	S	F	S	S	Y	$R_1 (1 - R_2) R_3 R_4$
6	S	F	S	F	Y	$R_1 (1 - R_2) R_3 (1 - R_4)$
7	S	F	F	S	Y	$R_1 (1 - R_2)(1 - R_3) R_4$
8	S	F	F	F	N	0
9	F	S	S	S	Y	$(1 - R_1) R_2 R_3 R_4$
10	F	S	S	F	Y	$(1 - R_1) R_2 R_3 (1 - R_4)$
11	F	S	F	S	Y	$(1 - R_1) R_2 (1 - R_3) R_4$
12	F	S	F	F	N	0
13	F	F	S	S	N	0
14	F	F	S	F	N	0
15	F	F	F	S	N	0
16	F	F	F	F	N	0

* It may be seen that as long as there is at least one good stop valve in each leg of steam flow the system is successful.

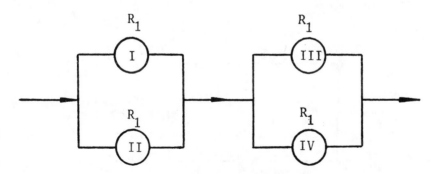

Fig. 6.3 – The reliability block diagram of a parallel-series, four-stop-valve system of the turbogenerator of Example 6–2.

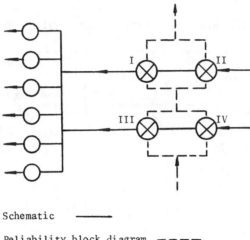

Schematic   ────────

Reliability block diagram ─ ─ ─ ─

Fig. 6.4 – The schematic and the reliability block diagram
of the system of four steam turbine stop valves
of Example 6–2.

But this is the reliability of the system given in Fig. 6.3; therefore, the
valve arrangement's reliability block diagram is the reliability block
diagram given by the dashed lines in Fig. 6.4.

EXAMPLE 6–3
What is the reliability of the system of two steam turbine stop
valves whose physical arrangement is given in Fig. 6.5? Each valve
has the following five modes of function:

1. G = valve good.

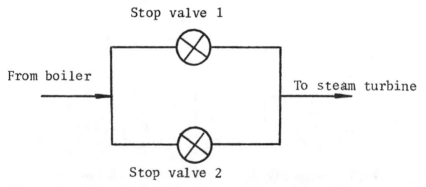

Fig. 6.5 – Schematic of the two stop-valve system of Example 6–3.

2. B_{od} = valve fails open when not called upon to shut (bad open de-energized).

3. B_{oe} = valve fails open when called upon to shut (bad open energized).

4. B_{sd} = valve fails shut when not called upon to shut (bad shut de-energized).

5. B_{se} = valve fails shut when called upon to shut (bad shut energized).

System success is defined as follows:

1. No steam flows to the turbine when the valves are called upon to shut, or when they are energized.

2. Steam flow is not cut off when the valves are not called upon to shut, or when they are de-energized.

SOLUTION TO EXAMPLE 6–3

Here $M^N = 5^2 = 25$, and from Table 6.3

$$R_S = \sum_{i=1}^{8} P_i (\text{of all system success permutations}).$$

Therefore,

$$
\begin{aligned}
R_S = \ & R_1 R_2 + R_1 Q_{2sd} + R_1 Q_{2se} + Q_{1sd} R_2 \\
& + Q_{1se} R_2 + Q_{1sd} Q_{2se} + Q_{1se} Q_{2sd} + Q_{1se} Q_{2se}.
\end{aligned}
$$

The terms $R_1 Q_{2sd}$, $Q_{1sd} R_2$, $Q_{1sd} Q_{2se}$, and $Q_{1se} Q_{2sd}$ will be success terms only if the size of the normally functioning (good) valve is sufficient to allow the necessary quantity of steam to reach the turbine under normal turbogenerator operating conditions.

It may seen that for the success of this system, arranging the valves physically in parallel is a relatively poor arrangement, because of the 25 possible permutations, in this case, only 8 permutations yield success terms. The valves should be arranged physically in series; then 12 permutations yield success terms, or 50% more! Also, the answer could have been written directly by inspection and reliability logic without writing out the full Table 6.3. However, only the accumulation of experience in solving such problems enables this to be done right away!

TABLE 6.3 – Solution to Example 6–3.

Permutation number	Units and modes of function		Is system successful	Probability of system success
	1	2		
1	G	G	Y	$R_1 R_2$
2	G	B_{od}	N	0
3	G	B_{oe}	N	0
4	G	B_{sd}	Y	$R_1 Q_{2sd}$
5	G	B_{se}	Y	$R_1 Q_{2se}$
6	B_{od}	G	N	0
7	B_{oe}	G	N	0
8	B_{sd}	G	Y	$Q_{1sd} R_2$
9	B_{se}	G	Y	$Q_{1se} R_2$
10	B_{od}	B_{od}	N	0
11	B_{od}	B_{oe}	N	0
12	B_{od}	B_{sd}	N	0
13	B_{od}	B_{se}	N	0
14	B_{oe}	B_{od}	N	0
15	B_{oe}	B_{oe}	N	0
16	B_{oe}	B_{sd}	N	0
17	B_{oe}	B_{se}	N	0
18	B_{sd}	B_{sd}	N	0
19	B_{sd}	B_{se}	Y	$Q_{1sd} Q_{2se}$
20	B_{sd}	B_{od}	N	0
21	B_{sd}	B_{oe}	N	0
22	B_{se}	B_{od}	N	0
23	B_{se}	B_{oe}	N	0
24	B_{se}	B_{sd}	Y	$Q_{1se} Q_{2sd}$
25	B_{se}	B_{se}	Y	$Q_{1se} Q_{2se}$

PROBLEMS

6-1. Given is a subsystem of *three unequal valves* connected *physically in series*. Consider three modes of function for these valves: normal, failed open, and failed closed.

    (1) What is this system's reliability if these were stop valves which are normally open, and their function is to stop flow on command?

    (2) What is this system's reliability if these were inlet valves whose function is to govern and control the quantity of flow continuously?

    (3) How would you increase the reliability of the subsystem for these two cases?

Write out the answers in terms of the respective r's, q's and λ's.

6-2. (1) Find the reliability of a subsystem of *two stop valves* arranged *physically in series*. The valves are *normally open*. Their function is to stop the flow of a fluid on command and not to stop the flow of the fluid when not commanded. Consider the following five function modes for each valve:

      1.1 Normal.

      1.2 Failed open when not commanded, with failure rate λ_{OD}.

      1.3 Failed shut when not commanded, with failure rate λ_{SD}.

      1.4 Failed open when commanded, with failure rate λ_{OE}.

      1.5 Failed shut when commanded, with failure rate λ_{SE}.

    (2) What is the reliability of this subsystem when only two modes of function are considered and on command only: normal or failed.

    (3) Which reliability value would be higher, that of Case 1 or 2? Why?

6-3. A subsystem consists of two 2-ohm resistors in parallel arrangement physically. Assume that each resistor has three modes of function: N (normal), O (failed open), and S (failed shorted). Subsystem success is defined as a subsystem resistance of 1 ohm or of 2 ohms. If a resistor fails open, take its resistance to be ∞ ohms, and if it fails shorted, take its resistance to be 0 ohms. What is this subsystem's reliability?

6-4. Four static, normally open valves, whose physical arrangement is shown in Fig. 6.6, are functioning in an equipment. Success for this four-valve arrangement is defined as follows:

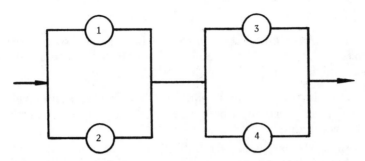

Fig. 6.6 – Schematic of the four static valves of Problem 6-4.

1. Cut off flow when commanded to close.
2. Not to cut off flow when not commanded to close.

The function modes of each valve are to (1) function normally, or successfully, or close without failing when commanded to close, (2) fail closed on command, and (3) fail open on command. The failure rates of failing closed or open when not commanded are assumed to be zero. Do the following:

(1) Determine the reliability of this four-valve arrangement.
(2) Determine the unreliability of this four-valve arrangement directly.

6-5. A subsystem consists of three, unequal, 3-ohm resistors arranged physically in parallel. Each resistor has three modes of function: normal, failed open, and failed short. The subsystem succeeds if the circuit's resistance is 1 ohm, 1.5 ohms, or 3 ohms. Find this subsystem's reliability.

Chapter 7

RELIABILITY OF SYSTEMS OPERATING AT VARIOUS LEVELS OF STRESS DURING A MISSION

7.1   FOR THE EXPONENTIAL CASE

If during a mission the level of stress imposed on the unit changes, then λ changes also. For example, at takeoff the jet engine has to generate a greater torque to get the higher engine thrust required. At cruising altitude and speed, the torque requirements are reduced.

Let's assume the following torque and thereby stress characteristics are involved in an airplane's flight:

No.	Period	Stress level	λ_u
1	t_1	Takeoff	λ_1
2	t_2	Climb	λ_2
3	t_3	Cruise	λ_3
4	t_4	Descent	λ_4
5	t_5	Landing	λ_5

The stress profile may then look as shown in Fig. 7.1. Assuming the engine is essentially in its useful life, then the mission reliability of the engine will be given by

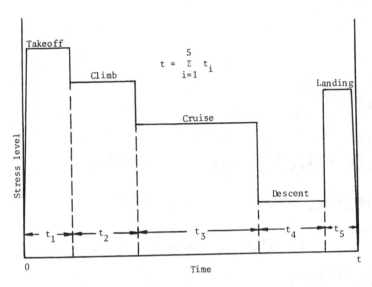

Fig. 7.1 – Stress profile for an airplane engine in a typical flight.

$$R_E(t) = \prod_{i=1}^{5} R_{Ei} = R_1 \times R_2 \times R_3 \times R_4 \times R_5,$$

or

$$R_E(t) = e^{-\lambda_1 t_1} \cdot e^{-\lambda_2 t_2} \cdot e^{-\lambda_3 t_3} \cdot e^{-\lambda_4 t_4} \cdot e^{-\lambda_5 t_5}. \quad (7.1)$$

The various λ_i for the same unit or subsystem at different stress levels of function need to be known. A similar analysis should be carried out for each unit in the system to arrive at the exact mission reliability of the system.

7.2 FOR THE WEIBULL CASE

If the engines' time-to-failure distribution is Weibullian, then one approach to evaluating $R_E(t)$ is the following: It may be assumed that the engine's reliability, when subjected to different stress levels of operation during a mission, would be given by the probability of surviving t_1 hours of operation at the takeoff stress level; *and* the probability of surviving t_2 hours of operation at the climb stress level *plus* the *equivalent* of t_1 hours of operation at the climb stress level, t_{1e}, given that the engine already survived t_1 hours of operation; *and* the probability of surviving t_3 hours of operation at the cruise stress level *plus* the *equivalent* of $(t_{1e} + t_2)$ at the cruise stress level, $(t_{1e} + t_2)_e$, given that the engine already survived $(t_{1e} + t_2)$ hours of operation; *and* the probability of surviving t_4 hours of operation at the descent stress level *plus* the *equivalent* of $[(t_{1e} + t_2)_e + t_3]$ hours of operation at the descent stress level, $[(t_{1e} + t_2)_e + t_3]_e$, given that the engine already survived $[(t_{1e} + t_2)_e + t_3]$ hours of operation; *and* the probability of surviving t_5 hours of operation at the landing stress level *plus* the equivalent of $\{[(t_{1e} + t_2)_e + t_3]_e + t_4\}$ hours of operation at the landing stress level, $\{[(t_{1e} + t_2)_e + t_3]_e + t_4\}_e$, given that the engine already survived $\{[(t_{1e} + t_2)_e + t_3]_e + t_4\}$ hours of operation.

The previous probabilistically expressed overall engine reliability may be expressed mathematically as follows:

$$R_E(t) = R_1(t_1) \cdot \frac{R_2(t_{1e} + t_2)}{R_1(t_1)} \cdot \frac{R_3[(t_{1e} + t_2)_e + t_3]}{R_2(t_{1e} + t_2)}$$
$$\cdot \frac{R_4\{[(t_{1e} + t_2)_e + t_3]_e + t_4\}}{R_3[(t_{1e} + t_2)_e + t_3]}$$
$$\cdot \frac{R_5\Big\{\{[(t_{1e} + t_2)_e + t_3]_e + t_4\}_e + t_5\Big\}}{R_4\{[(t_{1e} + t_2)_e + t_3]_e + t_4\}}.$$

Simplification yields

$$R_E(t) = R_5\Big\{\{[(t_{1e} + t_2)_e + t_3]_e + t_4\}_e + t_5\Big\}. \qquad (7.2)$$

Now all of the *equivalent* hours of operation must be quantified to enable the quantification of $R_E(t)$.

The equivalent hours of operating t_{1e} will be quantified based on the assumption that

$$R_1(t_1) = R_2(t_{1e}).$$

That is t_{1e} must be the equivalent operating hours at the second, or climb, stress level that give the same reliability as that of operating t_1 hours at the first, or takeoff, stress level; consequently,

$$e^{-[(t_1 - \gamma_1)/\eta_1]^{\beta_1}} = e^{-[(t_{1e} - \gamma_2)/\eta_2]^{\beta_2}},$$
$$\left(\frac{t_1 - \gamma_1}{\eta_1}\right)^{\beta_1} = \left(\frac{t_{1e} - \gamma_2}{\eta_2}\right)^{\beta_2},$$
$$\frac{t_{1e} - \gamma_2}{\eta_2} = \left(\frac{t_1 - \gamma_1}{\eta_1}\right)^{\beta_1/\beta_2},$$

or

$$t_{1e} = \eta_2 \left(\frac{t_1 - \gamma_1}{\eta_1}\right)^{\beta_1/\beta_2} + \gamma_2. \tag{7.3}$$

Similarly,

$$R_2(t_{1e} + t_2) = R_3(t_{1e} + t_2)_e,$$

$$e^{[(t_{1e}+t_2-\gamma_2)/\eta_2]^{\beta_2}} = e^{\{[(t_{1e}+t_2)_e-\gamma_3]/\eta_3\}^{\beta_3}},$$

$$\left(\frac{t_{1e} + t_2 - \gamma_2}{\eta_2}\right)^{\beta_2/\beta_3} = \left[\frac{(t_{1e} + t_2)_e - \gamma_3}{\eta_3}\right],$$

or

$$(t_{1e} + t_2)_e = \eta_3 \left(\frac{t_{1e} + t_2 - \gamma_2}{\eta_2}\right)^{\beta_2/\beta_3} + \gamma_3. \tag{7.4}$$

Also,

$$R_3[(t_{1e} + t_2)_e + t_3] = R_4\{[(t_{1e} + t_2) + t_3]_e\},$$

$$e^{-\{[(t_{1e} + t_2)_e + t_3 - \gamma_3]/\eta_3\}^{\beta_3}} = e^{-(\{[(t_{1e} + t_2)_e + t_3]_e - \gamma_4\}/\eta_4)^{\beta_4}},$$

or

$$[(t_{1e} + t_2)_e + t_3]_e = \eta_4\left[\frac{(t_{1e} + t_2)_e + t_3 - \gamma_3}{\eta_3}\right]^{\beta_3/\beta_4} + \gamma_4. \tag{7.5}$$

Finally,

$$R_4\{[(t_{1e} + t_2)_e + t_3]_e + t_4\} = R_5\left\{\{[(t_{1e} + t_2)_e + t_3]_e + t_4\}_e\right\},$$

$$e^{-\left\{\frac{[(t_{1e}+t_2)_e+t_3]_e+t_4-\gamma_4}{\eta_4}\right\}^{\beta_4}} = e^{-\left\{\frac{\{[(t_{1e}+t_2)_e+t_3]_e+t_4\}_e-\gamma_5}{\eta_5}\right\}^{\beta_5}},$$

or

$$\{[(t_{1e} + t_2)_e + t_3]_e + t_4\}_e = \eta_5\left\{\frac{[(t_{1e} + t_2)_e + t_5]_e + t_4 - \gamma_4}{\eta_4}\right\}^{\beta_4/\beta_5}$$
$$+ \gamma_5. \tag{7.6}$$

Consequently, $R_E(t)$, given by Eq. (7.2), can now be evaluated by substituting Eqs. (7.3) through (7.6) into Eq. (7.2), or in

$$R_E(t) = e^{-\left\{\left\{\{[(t_{1e}+t_2)_e+t_3]_e+t_4\}_e+t_5-\gamma_5\right\}/\eta_5\right\}^{\beta_5}}. \tag{7.7}$$

For an engine with an exponential time-to-failure distribution, the $\beta_i = 1$. If we assume the $\gamma_i = 0$, then Eq. (7.3) yields

$$t_{1e} = \frac{\eta_2}{\eta_1} t_1,$$

Eq. (7.4) yields

$$(t_{1e} + t_2)_e = \frac{\eta_3}{\eta_2}(\frac{\eta_2}{\eta_1}t_1 + t_2) = \frac{\eta_3}{\eta_1}t_1 + \frac{\eta_3}{\eta_2}t_2,$$

Eq. (7.5) yields

$$[(t_{1e} + t_2)_e + t_3]_e = \frac{\eta_4}{\eta_3}(\frac{\eta_3}{\eta_1}t_1 + \frac{\eta_3}{\eta_2}t_2 + t_3),$$

and Eq. (7.6) yields

$$\{[(t_{1e} + t_2)_e + t_3]_e + t_4\}_e = \frac{\eta_5}{\eta_4}[(\frac{\eta_4}{\eta_1}t_1 + \frac{\eta_4}{\eta_2}t_2 + \frac{\eta_4}{\eta_3}t_3) + t_4]$$
$$= \frac{\eta_5}{\eta_1}t_1 + \frac{\eta_5}{\eta_2}t_2 + \frac{\eta_5}{\eta_3}t_3 + \frac{\eta_5}{\eta_4}t_4. \quad (7.8)$$

Substitution of Eq. (7.8) into Eq. (7.7) yields

$$R_E(t) = e^{-\left\{\left[(\eta_5/\eta_1)t_1 + (\eta_5/\eta_2)t_2 + (\eta_5/\eta_3)t_3 + (\eta_5/\eta_4)t_4 + t_5 - \gamma_5\right]/\eta_5\right\}^{\beta_5}}.$$
$$(7.9)$$

But with $\beta_5 = 1$ and $\gamma_5 = 0$, Eq. (7.9) becomes

$$R_E(t) = e^{-[(t_1/\eta_1) + (t_2/\eta_2) + (t_3/\eta_3) + (t_4/\eta_4) + (t_5/\eta_5)]}, \quad (7.10)$$

where $1/\eta_i = \lambda_i$; therefore, Eq. (7.10) finally becomes

$$R_E(t) = e^{-(\lambda_1 t_1 + \lambda_2 t_2 + \lambda_3 t_3 + \lambda_4 t_4 + \lambda_5 t_5)}, \quad (7.11)$$

which is the same as Eq. (7.1), as it should be.

7.3   RELIABILITY OF CYCLICAL OPERATIONS

Define λ_{cy} as the cyclic or switching failure rate expressed in terms of failures per cycle, or failures per switching operation. If λ_{cy} is constant with time, then

$$R_c = e^{-\lambda_{cy}c},$$

where

c = number of cycles of operation during a mission.

If during a mission 100 such operations are involved and

$$\lambda_{cy} = 5 \text{ fr}/10^6 \text{ cycles},$$

then

$$R_c = e^{-5 \times 10^{-6} \times 100} = e^{-5 \times 10^{-4}} = e^{-0.0005},$$

or

$$R_c = 0.9995.$$

If a part is cycled on and off during its normal operation in an equipment and exhibits an operating failure rate of λ_{on} in failures per hour, a nonoperating failure rate of λ_{off} in failures per hour, and a cycling or switching failure rate of λ_{cy} in failures per cycle, then the average failure rate of the part, λ_{av}, under these conditions is given by

$$\lambda_{av} = \frac{1}{t}[\lambda_{cy}c_f t_{cy} + \lambda_{on}t_{on} + \lambda_{off}(t - t_{cy} - t_{on})],$$

where

t = mission time in hours,

c_f = cycling or switching frequency in cycles per hour,

t_{cy} = accumulated time consumed in cycling, or switching, in hours,

and

t_{on} = accumulated time in the operating condition in hours.

The part's reliability, $R(t)$, is then given by

$$R(t) = e^{-\lambda_{av}t}.$$

PROBLEMS

7-1. A system is operating at n stress levels, s_i, $i = 1, 2, ..., n$, during a T-hr mission. The proportion of operating time at stress level s_i is p_i, $i = 1, 2, ..., n$, $\sum_{i=1}^{n} p_i = 1$. Assume that the system's failure rate at stress level s_i is λ_i, $i = 1, 2, ..., n$. Find the system's reliability for this mission.

7-2. A unit is operating at three different stress levels, s_1, s_2, and s_3, during a 500-hr mission. The proportion of operating time at stress level s_1 is 0.25, at stress level s_2 is 0.45, and at stress level s_3 is 0.30. The cumulative times-to-failure distributions at stress levels s_1, s_2, and s_3 are

$$F_1(T) = 1 - e^{-\left(\frac{T}{150}\right)^{1.2}},$$
$$F_2(T) = 1 - e^{-\left(\frac{T}{3,000}\right)^{2.0}},$$

and

$$F_3(T) = 1 - e^{-\left(\frac{T}{500}\right)^{2.3}},$$

respectively. Find the reliability of the unit for this mission for the following stress application sequences:

(1) $s_1 \rightarrow s_2 \rightarrow s_3$.
(2) $s_2 \rightarrow s_1 \rightarrow s_3$.
(3) $s_3 \rightarrow s_2 \rightarrow s_1$.

7-3. During a 100-hr mission, a unit is functioning cyclically. Sixty-five percent of the time the unit is switched on, 34% of the time the unit is switched off, and the switching operation consumes 1% of the mission time. The failure rates are

$$\lambda_{on} = 5 \text{ fr}/10^6 \text{ hr},$$
$$\lambda_{off} = 0.1 \text{ fr}/10^6 \text{hr},$$

and

$$\lambda_{cy} = 6 \text{ fr}/10^4 \text{ cycles}.$$

The switching frequency is 1,800 cycles/hr. Find the unit's reliability.

7-4. A system is operating under two stress levels, s_1 and s_2, alternatively, for n cycles in a mission time of t. The operating times at each stress level are the same. Show that the system's reliability, $R_E(t)$, is given by

$$R_E(t) = e^{-\left\{\left[(\eta_1 + \eta_2)/(2\eta_1\eta_2)\right]t\right\}^{\beta}},$$

if the times-to-failure distribution at each stress level is Weibull with $\beta_1 = \beta_2 = \beta$ and $\gamma_1 = \gamma_2 = 0$.

7-5. A system is operating under three stress levels, s_1, s_2, and s_3, alternatively, for n cycles in a mission time of t. The operating times at each stress level are the same. Find the system's reliability, $R_E(t)$, if the times-to-failure distribution at each stress level is Weibull with $\beta_1 \neq \beta_2 \neq \beta_3$ and $\gamma_1 = \gamma_2 = \gamma_3 = 0$.

Chapter 8

LOAD-SHARING RELIABILITY

8.1 RELIABILITY OF TWO PARALLEL LOAD-SHARING SWITCHES

The quantification of the reliability of parallel units, as determined previously, is based on the assumption that, when a redundant unit fails, the failure rate or the reliability of the surviving units does not change during the mission. There will be situations, however, when this will not prevail and the failure rate, or the reliability, of the surviving units will change. Usually their failure rate will increase and their reliability will decrease, because the surviving units will be sharing the load during the mission; consequently, their share of the load will increase, as indicated in Fig. 8.1. To correctly determine the reliability of such units, the change of the failure rate, or of the reliability, of the surviving units has to be properly taken into account.

The reliability of the two load-sharing exponential units shown in Fig. 8.1 is given by

$R(t)$ = Probability that Units 1 and 2 complete their mission successfully with *pdf*'s $f_1(T)$ and $f_2(T)$, respectively, *or* the probability that Unit 1 fails at $t_1 < t$ with *pdf* $f_1(T)$, and Unit 2 functions till t_1 with *pdf* $f_2(T)$ *and* then functions for the rest of the mission, or in $(t - t_1)$, with *pdf* $f_{2'}(T)$, *or* the probability that Unit 2 fails at $t_2 < t$ with *pdf* $f_2(T)$ *and* Unit 1 functions till t_2 with *pdf* $f_1(T)$ *and* then functions with *pdf* $f_{1'}(T)$ in $(t - t_2)$.

This situation is shown in tabular form in Table 8.1 and plotted in *pdf* form in Fig. 8.2 when the units are exponential. The previous

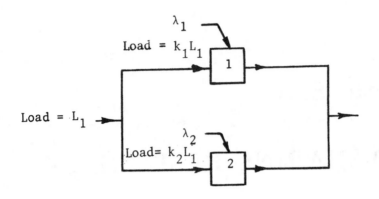

(a) Both units are functioning and k_i is less than 1; thus the load L_1 is being shared throughout the mission.

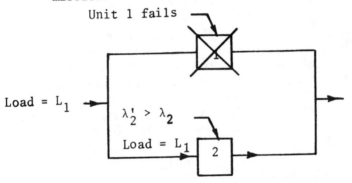

(b) Unit 1 fails; hence Unit 2 has to carry the full load L_1 for the remainder of the mission.

Fig. 8.1 – Load sharing with two units in parallel.

TABLE 8.1 – Matrix of units, function modes, pdf's, and time domains for all system success function modes for a system of two load-sharing parallel units.

System success function mode number	Units, function modes, pdf's, and time domains	
	Unit 1	Unit 2
1	G^* $f_1(T);\ t$	G $f_2(T);\ t$
2	B^{**} $f_1(T);\ t_1 < t$	G $f_2(T);\ t_1$ $f_{2'}(T);\ t - t_1$
3	G $f_1(T);\ t_2$ $f_{1'}(T);\ t - t_2$	B $f_2(T);\ t_2 < t$

$^*\ G$ = unit is good throughout the designated period.
$^{**}B$ = unit goes bad at the designated time.

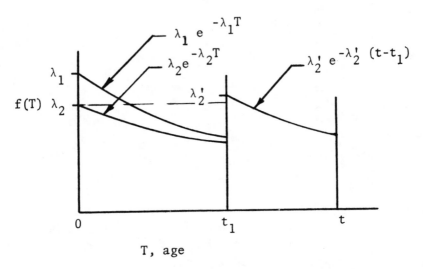

(a) Load-sharing pdf's of two units when Unit 1 fails.

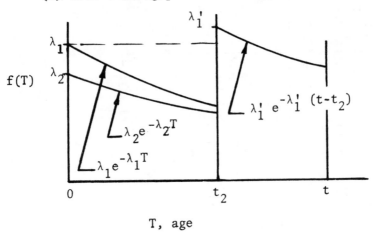

(b) Load-sharing pdf's of two units when Unit 2 fails.

Fig. 8.2 — The *pdf*'s of two load-sharing units when one unit
fails before the mission is completed and the units
are exponential.

reliability statement may be quantified by

$$R(t) = R_1(t)R_2(t) + Q_1(t_1)R_2(t_1)R_{2'}[t_1, (t - t_1)]$$
$$+ Q_2(t_2)R_1(t_2)R_{1'}[t_2, (t - t_2)],$$

or

$$R(t) = R_1(t)R_2(t) + \int_0^t f_1(t_1)R_2(t_1)\frac{R_{2'}(t)}{R_{2'}(t_1)} \, dt_1$$
$$+ \int_0^t f_2(t_2)R_1(t_2)\frac{R_{1'}(t)}{R_{1'}(t_2)} \, dt_2. \tag{8.1}$$

If the units are exponential, then in Eq. (8.1)

$$\frac{R_{2'}(t)}{R_{2'}(t_1)} = \frac{e^{-\lambda_2' t}}{e^{-\lambda_2' t_1}} = e^{-\lambda_2'(t-t_1)} = R_{2'}(t - t_1), \tag{8.2}$$

and

$$\frac{R_{1'}(t)}{R_{1'}(t_2)} = \frac{e^{-\lambda_1' t}}{e^{-\lambda_1' t_2}} = e^{-\lambda_1'(t-t_2)} = R_{1'}(t - t_2). \tag{8.3}$$

It may be seen from the distribution in Section 3.2 page 45 of this volume, that for the *exponential* case the equivalent time t_{1e} at the stress level of λ_i drops out yielding the correct Eqs. (8.2) and (8.3). The final results in Eqs. (8.2) and (8.3) explain why the time domains given in Table 8.1 have been used when a unit fails in System Success Function Modes 2 and 3.

When the units are exponential, Eq. (8.1) becomes

$$R(t) = e^{-\lambda_1 t}e^{-\lambda_2 t} + \int_0^t \lambda_1 e^{-\lambda_1 t_1} e^{-\lambda_2 t_1} e^{-\lambda_2'(t-t_1)} \, dt_1$$
$$+ \int_0^t \lambda_2 e^{-\lambda_2 t_2} e^{-\lambda_1 t_2} e^{-\lambda_1'(t-t_2)} dt_2. \tag{8.4}$$

Equation (8.4) integrated out yields

$$R(t) = e^{-(\lambda_1+\lambda_2)t}\left(1 - \frac{\lambda_1}{\lambda_1 + \lambda_2 - \lambda_2'} - \frac{\lambda_2}{\lambda_1 - \lambda_1' + \lambda_2}\right)$$
$$+ \frac{\lambda_1 e^{-\lambda_2' t}}{\lambda_1 + \lambda_2 - \lambda_2'} + \frac{\lambda_2 e^{-\lambda_1' t}}{\lambda_1 - \lambda_1' + \lambda_2}. \tag{8.4'}$$

If the two units are equal, i.e., they have the same *pdf*, then $\lambda_1 = \lambda_2 = \lambda$ and $\lambda_1' = \lambda_2' = \lambda'$, and Eq. (8.4') becomes

$$R(t) = e^{-2\lambda t} + \frac{2\lambda}{\lambda' - 2\lambda}\left(e^{-2\lambda t} - e^{-\lambda' t}\right). \tag{8.5}$$

8.2   RELIABILITY OF THREE LOAD-SHARING CYCLIC SWITCHES ARRANGED PHYSICALLY IN PARALLEL

8.2.1   THREE UNEQUAL CYCLIC SWITCHES

The arrangement of three, physically in parallel, cyclic switches is shown in Fig. 8.3. If the switches function normally, their failure rate is λ; when one fails, the remaining two have a failure rate of λ'; and when two fail, the remaining one has a failure rate of λ''. Also,

$$\lambda = \lambda_o + \lambda_c, \quad \lambda' = \lambda'_o + \lambda'_c, \quad \text{and} \quad \lambda'' = \lambda''_o + \lambda''_c. \tag{8.6}$$

(For details on the reliability of cyclic switches, see Chapter 10.) Then the system's success function modes matrix is as given in Table 8.2, and the system's reliability is given by

$$R_S(t) = \sum_{i=1}^{10} R_i(t), \tag{8.7}$$

where $R_i(t)$ is the reliability for system success function mode i. These $R_i(t)$ are determined as follows based on the time domains given in Fig. 8.4:

$$R_1(t) = e^{-(\lambda_1 + \lambda_2 + \lambda_3)t}.$$

Considering that

$$\lambda_{3o} e^{-\lambda_{3o} t_1} e^{-\lambda_{3c} t_1} = \lambda_{3o} e^{-(\lambda_{3o} + \lambda_{3c}) t_1} = \lambda_{3o} e^{-\lambda_3 t_1},$$

$$R_2(t) = \int_{t_1=0}^{t} \lambda_{3o} e^{-\lambda_3 t_1} e^{-\lambda_2 t_1} e^{-\lambda'_2 (t-t_1)} e^{-\lambda_1 t_1} e^{-\lambda'_1 (t-t_1)} \, dt_1,$$

$$R_2(t) = \lambda_{3o} e^{-(\lambda'_1 + \lambda'_2)t} \int_{t_1=0}^{t} e^{-(\lambda_1 - \lambda'_1 + \lambda_2 - \lambda'_2 + \lambda_3) t_1} \, dt_1,$$

or

$$R_2(t) = \lambda_{3o} e^{-(\lambda'_1 + \lambda'_2)t} \frac{1 - e^{-(\lambda_1 - \lambda'_1 + \lambda_2 - \lambda'_2 + \lambda_3)t}}{\lambda_1 - \lambda'_1 + \lambda_2 - \lambda'_2 + \lambda_3}.$$

$$R_3(t) = \int_{t_2=0}^{t} \lambda_{2o} e^{-\lambda_2 t_2} e^{-\lambda_3 t_2} e^{-\lambda'_3 (t-t_2)} e^{-\lambda_1 t_2} e^{-\lambda'_1 (t-t_2)} \, dt_2,$$

$$R_3(t) = \lambda_{2o} e^{-(\lambda'_1 + \lambda'_3)t} \int_{t_2=0}^{t} e^{-(\lambda_1 - \lambda'_1 + \lambda_2 + \lambda_3 - \lambda'_3) t_2} \, dt_2,$$

or

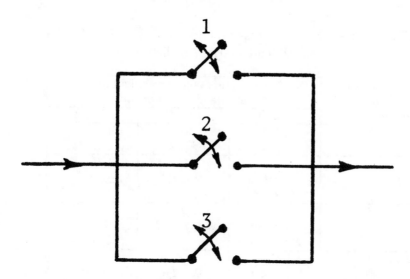

Fig. 8.3 – Schematic of three cyclic switches arranged physically in parallel.

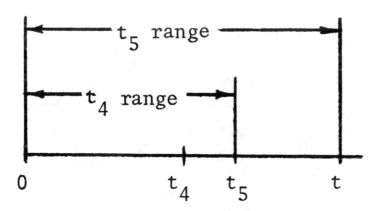

Fig. 8.4 – Time domains used in Function Mode 5 of Table 8.2.

TABLE 8.2 – System success function modes matrix for three unequal, load-sharing, cyclic switches.

System success function mode number	Units, function modes, failure rates, and time domains		
	Unit 1	Unit 2	Unit 3
1	G $\lambda_1; t$	G $\lambda_2; t$	G $\lambda_3; t$
2	G $\lambda_1; t_1$ $\lambda_1'; t - t_1$	G $\lambda_2; t_1$ $\lambda_2'; t - t_1$	$B_o{}^*$ $\lambda_{3o}; t_1 < t$
3	G $\lambda_1; t_2$ $\lambda_1'; t - t_2$	B_o $\lambda_2; t_2 < t$	G $\lambda_3; t_2$ $\lambda_3'; t - t_2$
4	B_o $\lambda_{1o}; t < t_3$	G $\lambda_2; t_3$ $\lambda_2'; t - t_3$	G $\lambda_3; t_3$ $\lambda_3'; t - t_3$
5	G $\lambda_1; t_4$ $\lambda_1'; t_5 - t_4$ $\lambda_1''; t - t_5$	$G; B_o$ $\lambda_2; t_4$ $\lambda_{2o}'; t_5 - t_4$	B_o $\lambda_{3o}; t_4 < t$
6	G $\lambda_1; t_6$ $\lambda_1'; t_7 - t_6$ $\lambda_1''; t - t_7$	B_o $\lambda_{2o}; t_6 < t$	$G; B_o$ $\lambda_3; t_6$ $\lambda_{3o}'; t_7 - t_6$
7	$G; B_o$ $\lambda_1; t_8$ $\lambda_{1o}'; t_9 - t_8$	B_o $\lambda_{2o}; t_8 < t$	G $\lambda_3; t_8$ $\lambda_3'; t_9 - t_8$ $\lambda_3''; t - t_9$
8	B_o $\lambda_{1o}; t_{10} < t$	$G; B_o$ $\lambda_2; t_{10}$ $\lambda_{2o}'; t_{11} - t_{10}$	G $\lambda_3; t_{10}$ $\lambda_3'; t_{11} - t_{10}$ $\lambda_3''; t - t_{11}$
9	$G; B_o$ $\lambda_1; t_{12}$ $\lambda_{1o}'; t_{13} - t_{12}$	G $\lambda_2; t_{12}$ $\lambda_2'; t_{13} - t_{12}$ $\lambda_2''; t - t_{13}$	B_o $\lambda_{3o}; t_{12} < t$
10	B_o $\lambda_{1o}; t_{14} < t$	G $\lambda_2; t_{14}$ $\lambda_2'; t_{15} - t_{14}$ $\lambda_2''; t - t_{15}$	$G; B_o$ $\lambda_3; t_{14}$ $\lambda_{3o}'; t_{15} - t_{14}$

$*B_o$ = failed open.

$$R_3(t) = \lambda_{2_0} e^{-(\lambda_1' + \lambda_3')t} \frac{1 - e^{-(\lambda_1 - \lambda_1' + \lambda_2 + \lambda_3 - \lambda_3')t}}{\lambda_1 - \lambda_1' + \lambda_2 + \lambda_3 - \lambda_3'}.$$

$$R_4(t) = \int_{t_3=0}^{t} \lambda_{1_0} e^{-\lambda_1 t_3} e^{-\lambda_2 t_3} e^{-\lambda_2'(t-t_3)} e^{-\lambda_3 t_3} e^{-\lambda_3'(t-t_3)} \, dt_3,$$

$$R_4(t) = \lambda_{1_0} e^{-(\lambda_2' + \lambda_3')t} \int_{t_3=0}^{t} e^{-(\lambda_1 + \lambda_2 - \lambda_2' + \lambda_3 - \lambda_3')t_3} \, dt_3,$$

or

$$R_4(t) = \lambda_{1_0} e^{-(\lambda_2' + \lambda_3')t} \frac{1 - e^{-(\lambda_1 + \lambda_2 - \lambda_2' + \lambda_3 - \lambda_3')t}}{\lambda_1 + \lambda_2 - \lambda_2' + \lambda_3 - \lambda_3'}.$$

$$R_5(t) = \int_{t_5=0}^{t} \int_{t_4=0}^{t_5} \lambda_{3_0} e^{\lambda_3 t_4} e^{-\lambda_2 t_4} \lambda_{2_0}' e^{-\lambda_2'(t_5-t_4)}$$
$$\cdot \, e^{-\lambda_1 t_4} e^{-\lambda_1'(t_5-t_4)} e^{-\lambda_1''(t-t_5)} \, dt_4 \, dt_5,$$

$$R_5(t) = \lambda_{2_0}' \lambda_{3_0} e^{-\lambda_1'' t} \int_{t_5=0}^{t} e^{-(\lambda_1' - \lambda_1'' + \lambda_2')t_5}$$
$$\cdot \int_{t_4=0}^{t_5} e^{-(\lambda_1 - \lambda_1' + \lambda_2 - \lambda_2' + \lambda_3)t_4} \, dt_4 \, dt_5,$$

$$R_5(t) = \lambda_{2_0}' \lambda_{3_0} e^{-\lambda_1'' t} \int_{t_5=0}^{t} e^{-(\lambda_1' - \lambda_1'' + \lambda_2')t_5}$$
$$\cdot \frac{1 - e^{-(\lambda_1 - \lambda_1' + \lambda_2 - \lambda_2' + \lambda_3)t_5}}{\lambda_1 - \lambda_1' + \lambda_2 - \lambda_2' + \lambda_3} \, dt_5,$$

$$R_5(t) = \lambda_{2_0}' \lambda_{3_0} \frac{e^{-\lambda_1'' t}}{\lambda_1 - \lambda_1' + \lambda_2 - \lambda_2' + \lambda_3}$$
$$\cdot \int_{t_5=0}^{t} \left[e^{-(\lambda_1' - \lambda_1'' + \lambda_2')t_5} - e^{-(\lambda_1 - \lambda_1'' + \lambda_2 + \lambda_3)t_5} \right] \, dt_5,$$

or

$$R_5(t) = \lambda_{2_0}' \lambda_{3_0} \frac{e^{-\lambda_1'' t}}{\lambda_1 - \lambda_1' + \lambda_2 - \lambda_2' + \lambda_3}$$
$$\cdot \left\{ \frac{1 - e^{-(\lambda_1' - \lambda_1'' + \lambda_2')t}}{\lambda_1' - \lambda_1'' + \lambda_2'} - \frac{1 - e^{-(\lambda_1 - \lambda_1'' + \lambda_2 + \lambda_3)t}}{\lambda_1 - \lambda_1'' + \lambda_2 + \lambda_3} \right\}.$$

$$R_6(t) = \int_{t_7=0}^{t} \int_{t_6=0}^{t_7} \lambda_{2_0} e^{\lambda_2 t_6} e^{-\lambda_3 t_6} \lambda_{3_0}' e^{-\lambda_3'(t_7-t_6)}$$
$$\cdot \, e^{-\lambda_1 t_6} e^{-\lambda_1'(t_7-t_6)} e^{-\lambda_1''(t-t_7)} \, dt_6 \, dt_7,$$

$$R_6(t) = \lambda_{2o}\lambda_{3o}' e^{-\lambda_1'' t} \int_{t_7=0}^{t} e^{-(\lambda_1'-\lambda_1''+\lambda_3')t_7}$$

$$\cdot \int_{t_6=0}^{t_7} e^{-(\lambda_1-\lambda_1'+\lambda_2+\lambda_3-\lambda_3')t_6}\, dt_6\, dt_7,$$

$$R_6(t) = \lambda_{2o}\lambda_{3o}' e^{-\lambda_1'' t} \int_{t_7=0}^{t} e^{-(\lambda_1'-\lambda_1''+\lambda_3')t_7}$$

$$\cdot \frac{1 - e^{-(\lambda_1-\lambda_1'+\lambda_2+\lambda_3-\lambda_3')t_7}}{\lambda_1 - \lambda_1' + \lambda_2 + \lambda_3 - \lambda_3'}\, dt_7,$$

$$R_6(t) = \lambda_{2o}\lambda_{3o}' \frac{e^{-\lambda_1'' t}}{\lambda_1 - \lambda_1' + \lambda_2 + \lambda_3 - \lambda_3'}$$

$$\cdot \int_{t_7=0}^{t} \left[e^{-(\lambda_1'-\lambda_1''+\lambda_3')t_7} - e^{-(\lambda_1-\lambda_1''+\lambda_2+\lambda_3)t_7} \right] dt_7,$$

or

$$R_6(t) = \lambda_{2o}\lambda_{3o}' \frac{e^{-\lambda_1'' t}}{\lambda_1 - \lambda_1' + \lambda_2 + \lambda_3 - \lambda_3'}$$

$$\cdot \left\{ \frac{1 - e^{-(\lambda_1'-\lambda_1''+\lambda_3')t}}{\lambda_1' - \lambda_1'' + \lambda_3'} - \frac{1 - e^{-(\lambda_1-\lambda_1''+\lambda_2+\lambda_3)t}}{\lambda_1 - \lambda_1'' + \lambda_2 + \lambda_3} \right\}.$$

$$R_7(t) = \int_{t_9=0}^{t} \int_{t_8=0}^{t_9} \lambda_{2o} e^{\lambda_2 t_8} e^{-\lambda_1 t_8} \lambda_{1o}' e^{-\lambda_1'(t_9-t_8)}$$

$$\cdot e^{-\lambda_3 t_8} e^{-\lambda_3'(t_9-t_8)} e^{-\lambda_3''(t-t_9)}\, dt_8\, dt_9,$$

$$R_7(t) = \lambda_{1o}'\lambda_{2o} e^{-\lambda_3'' t} \int_{t_9=0}^{t} e^{-(\lambda_1'+\lambda_3'-\lambda_3'')t_9}$$

$$\cdot \int_{t_8=0}^{t_9} e^{-(\lambda_1-\lambda_1'+\lambda_2+\lambda_3-\lambda_3')t_8}\, dt_8 dt_9,$$

$$R_7(t) = \lambda_{1o}'\lambda_{2o} e^{-\lambda_3'' t} \int_{t_9=0}^{t} e^{-(\lambda_1'+\lambda_3'-\lambda_3'')t_9}$$

$$\cdot \frac{1 - e^{-(\lambda_1-\lambda_1'+\lambda_2+\lambda_3-\lambda_3')t_9}}{\lambda_1 - \lambda_1' + \lambda_2 + \lambda_3 - \lambda_3'}\, dt_9,$$

$$R_7(t) = \lambda_{1o}'\lambda_{2o} \frac{e^{-\lambda_3'' t}}{\lambda_1 - \lambda_1' + \lambda_2 + \lambda_3 - \lambda_3'}$$

$$\cdot \int_{t_9=0}^{t} \left[e^{-(\lambda_1'+\lambda_3'-\lambda_3'')t_9} - e^{-(\lambda_1+\lambda_2+\lambda_3-\lambda_3'')t_9} \right] dt_9,$$

or

$$R_7(t) = \lambda_{1o}'\lambda_{2o} \frac{e^{-\lambda_3'' t}}{\lambda_1 - \lambda_1' + \lambda_2 + \lambda_3 - \lambda_3'}$$

$$\cdot \left\{ \frac{1 - e^{-(\lambda_1' + \lambda_3' - \lambda_3'')t}}{\lambda_1' + \lambda_3' - \lambda_3''} - \frac{1 - e^{-(\lambda_1 + \lambda_2 + \lambda_3 - \lambda_3'')t}}{\lambda_1 + \lambda_2 + \lambda_3 - \lambda_3''} \right\}.$$

$$R_8(t) = \int_{t_{11}=0}^{t} \int_{t_{10}=0}^{t_{11}} \lambda_{1o} e^{\lambda_1 t_{10}} e^{-\lambda_2 t_{10}} \lambda_{2o}' e^{-\lambda_2'(t_{11}-t_{10})}$$
$$\cdot e^{-\lambda_3 t_{10}} e^{-\lambda_3'(t_{11}-t_{10})} e^{-\lambda_3''(t-t_{11})} \, dt_{10} \, dt_{11},$$

$$R_8(t) = \lambda_{1o} \lambda_{2o}' e^{-\lambda_3'' t} \int_{t_{11}=0}^{t} e^{-(\lambda_2' + \lambda_3' - \lambda_3'')t_{11}}$$
$$\cdot \int_{t_{10}=0}^{t_{11}} e^{-(\lambda_1 + \lambda_2 - \lambda_2' + \lambda_3 - \lambda_3')t_{10}} \, dt_{10} \, dt_{11},$$

$$R_8(t) = \lambda_{1o} \lambda_{2o}' e^{-\lambda_3'' t} \int_{t_{11}=0}^{t} e^{-(\lambda_2' + \lambda_3' - \lambda_3'')t_{11}}$$
$$\cdot \frac{1 - e^{-(\lambda_1 + \lambda_2 - \lambda_2' + \lambda_3 - \lambda_3')t_{11}}}{\lambda_1 + \lambda_2 - \lambda_2' + \lambda_3 - \lambda_3'} \, dt_{11},$$

$$R_8(t) = \lambda_{1o} \lambda_{2o}' \frac{e^{-\lambda_3'' t}}{\lambda_1 + \lambda_2 - \lambda_2' + \lambda_3 - \lambda_3'}$$
$$\cdot \int_{t_{11}=0}^{t} \left[e^{-(\lambda_2' + \lambda_3' - \lambda_3'')t_{11}} - e^{-(\lambda_1 + \lambda_2 + \lambda_3 - \lambda_3'')t_{11}} \right] \, dt_{11},$$

or

$$R_8(t) = \lambda_{1o} \lambda_{2o}' \frac{e^{-\lambda_3'' t}}{\lambda_1 + \lambda_2 - \lambda_2' + \lambda_3 - \lambda_3'}$$
$$\cdot \left\{ \frac{1 - e^{-(\lambda_2' + \lambda_3' - \lambda_3'')t}}{\lambda_2' + \lambda_3' - \lambda_3''} - \frac{1 - e^{-(\lambda_1 + \lambda_2 + \lambda_3 - \lambda_3'')t}}{\lambda_1 + \lambda_2 + \lambda_3 - \lambda_3''} \right\}.$$

$$R_9(t) = \int_{t_{13}=0}^{t} \int_{t_{12}=0}^{t_{13}} \lambda_{3o} e^{\lambda_3 t_{12}} e^{-\lambda_1 t_{12}} \lambda_{1o}' e^{-\lambda_1'(t_{13}-t_{12})}$$
$$\cdot e^{-\lambda_2 t_{12}} e^{-\lambda_2'(t_{13}-t_{12})} e^{-\lambda_2''(t-t_{13})} \, dt_{12} \, dt_{13},$$

$$R_9(t) = \lambda_{1o}' \lambda_{2o} e^{-\lambda_2'' t} \int_{t_{13}=0}^{t} e^{-(\lambda_1' + \lambda_2' - \lambda_2'')t_{13}}$$
$$\cdot \int_{t_{12}=0}^{t_{13}} e^{-(\lambda_1 - \lambda_1' + \lambda_2 - \lambda_2' + \lambda_3)t_{12}} \, dt_{12} \, dt_{13},$$

$$R_9(t) = \lambda_{1o}' \lambda_{2o} e^{-\lambda_2'' t} \int_{t_{13}=0}^{t} e^{-(\lambda_1' + \lambda_2' - \lambda_2'')t_{13}}$$
$$\cdot \frac{1 - e^{-(\lambda_1 - \lambda_1' + \lambda_2 - \lambda_2' + \lambda_3)t_{13}}}{\lambda_1 - \lambda_1' + \lambda_2 - \lambda_2' + \lambda_3} \, dt_{13},$$

$$R_9(t) = \lambda'_{1o}\lambda_{3o}\frac{e^{-\lambda''_2 t}}{\lambda_1 - \lambda'_1 + \lambda_2 - \lambda'_2 + \lambda_3}$$
$$\cdot \int_{t_{13}=0}^{t} \left[e^{-(\lambda'_1+\lambda'_2-\lambda''_2)t_{13}} - e^{-(\lambda_1+\lambda_2-\lambda''_2+\lambda_3)t_{13}} \right] dt_{13},$$

or

$$R_9(t) = \lambda'_{1o}\lambda_{3o}\frac{e^{-\lambda''_2 t}}{\lambda_1 - \lambda'_1 + \lambda_2 - \lambda'_2 + \lambda_3}$$
$$\cdot \left\{ \frac{1 - e^{-(\lambda'_1+\lambda'_2-\lambda''_2)t}}{\lambda'_1 + \lambda'_2 - \lambda''_2} - \frac{1 - e^{-(\lambda_1+\lambda_2-\lambda''_2+\lambda_3)t}}{\lambda_1 + \lambda_2 - \lambda''_2 + \lambda_3} \right\}.$$

$$R_{10}(t) = \int_{t_{15}=0}^{t}\int_{t_{14}=0}^{t_{15}} \lambda_{1o}e^{\lambda_1 t_{14}}e^{-\lambda_3 t_{14}}\lambda'_{3o}e^{-\lambda'_3(t_{15}-t_{14})}$$
$$\cdot e^{-\lambda_2 t_{14}}e^{-\lambda'_2(t_{15}-t_{14})}e^{-\lambda''_2(t-t_{15})} \, dt_{14} \, dt_{15},$$

$$R_{10}(t) = \lambda_{1o}\lambda'_{3o}e^{-\lambda''_2 t} \int_{t_{15}=0}^{t} e^{-(\lambda'_2-\lambda''_2+\lambda'_3)t_{15}}$$
$$\cdot \int_{t_{14}=0}^{t_{15}} e^{-(\lambda_1+\lambda_2-\lambda'_2+\lambda_3-\lambda'_3)t_{14}} \, dt_{14} \, dt_{15},$$

$$R_{10}(t) = \lambda_{1o}\lambda'_{3o}e^{-\lambda''_2 t} \int_{t_{15}=0}^{t} e^{-(\lambda'_2-\lambda''_2+\lambda'_3)t_{15}}$$
$$\cdot \frac{\left[1 - e^{-(\lambda_1+\lambda_2-\lambda'_2+\lambda_3-\lambda'_3)t_{15}} \right]}{\lambda_1 + \lambda_2 - \lambda'_2 + \lambda_3 - \lambda'_3} \, dt_{15},$$

$$R_{10}(t) = \lambda_{1o}\lambda'_{3o}\frac{e^{-\lambda''_2 t}}{\lambda_1 + \lambda_2 - \lambda'_2 + \lambda_3 - \lambda'_3}$$
$$\cdot \int_{t_{15}=0}^{t} \left[e^{-(\lambda'_2-\lambda''_2+\lambda'_3)t_{15}} - e^{-(\lambda_1+\lambda_2-\lambda''_2+\lambda_3)t_{15}} \right] dt_{15},$$

or

$$R_{10}(t) = \lambda_{1o}\lambda'_{3o}\frac{e^{-\lambda''_2 t}}{\lambda_1 + \lambda_2 - \lambda'_2 + \lambda_3 - \lambda'_3}$$
$$\cdot \left\{ \frac{1 - e^{-(\lambda'_2-\lambda''_2+\lambda'_3)t}}{\lambda'_2 - \lambda''_2 + \lambda'_3} - \frac{1 - e^{-(\lambda_1+\lambda_2-\lambda''_2+\lambda_3)t}}{\lambda_1 + \lambda_2 - \lambda''_2 + \lambda_3} \right\}.$$

The system's reliability is

$$R_S(t) \;=\; e^{-(\lambda_1+\lambda_2+\lambda_3)t}$$

$$+ \; \lambda_{3_o} e^{-(\lambda_1'+\lambda_2')t} \; \frac{1 - e^{-(\lambda_1-\lambda_1'+\lambda_2-\lambda_2'+\lambda_3)t}}{\lambda_1 - \lambda_1' + \lambda_2 - \lambda_2' + \lambda_3}$$

$$+ \; \lambda_{2_o} e^{-(\lambda_1'+\lambda_3')t} \; \frac{1 - e^{-(\lambda_1-\lambda_1'+\lambda_2+\lambda_3-\lambda_3')t}}{\lambda_1 - \lambda_1' + \lambda_2 + \lambda_3 - \lambda_3'}$$

$$+ \; \lambda_{1_o} e^{-(\lambda_2'+\lambda_3')t} \; \frac{1 - e^{-(\lambda_1+\lambda_2-\lambda_2'+\lambda_3-\lambda_3')t}}{\lambda_1 + \lambda_2 - \lambda_2' + \lambda_3 - \lambda_3'}$$

$$+ \; \frac{\lambda_{2_o}'\lambda_{3_o} e^{-\lambda_1''t}}{\lambda_1 - \lambda_1' + \lambda_2 - \lambda_2' + \lambda_3}$$

$$\cdot \left\{ \frac{1 - e^{-(\lambda_1'-\lambda_1''+\lambda_2')t}}{\lambda_1' - \lambda_1'' + \lambda_2'} - \frac{1 - e^{-(\lambda_1-\lambda_1''+\lambda_2+\lambda_3)t}}{\lambda_1 - \lambda_1'' + \lambda_2 + \lambda_3} \right\}$$

$$+ \; \frac{\lambda_{2_o}\lambda_{3_o}' e^{-\lambda_1''t}}{\lambda_1 - \lambda_1' + \lambda_2 + \lambda_3 - \lambda_3'}$$

$$\cdot \left\{ \frac{1 - e^{-(\lambda_1'-\lambda_1''+\lambda_3')t}}{\lambda_1' - \lambda_1'' + \lambda_3'} - \frac{1 - e^{-(\lambda_1-\lambda_1''+\lambda_2+\lambda_3)t}}{\lambda_1 - \lambda_1'' + \lambda_2 + \lambda_3} \right\}$$

$$+ \; \frac{\lambda_{1_o}'\lambda_{2_o} e^{-\lambda_3''t}}{\lambda_1 - \lambda_1' + \lambda_2 + \lambda_3 - \lambda_3'}$$

$$\cdot \left\{ \frac{1 - e^{-(\lambda_1'+\lambda_3'-\lambda_3'')t}}{\lambda_1' + \lambda_3' - \lambda_3''} - \frac{1 - e^{-(\lambda_1+\lambda_2+\lambda_3-\lambda_3'')t}}{\lambda_1 + \lambda_2 + \lambda_3 - \lambda_3''} \right\}$$

$$+ \; \frac{\lambda_{1_o}\lambda_{2_o}' e^{-\lambda_3''t}}{\lambda_1 + \lambda_2 - \lambda_2' + \lambda_3 - \lambda_3'}$$

$$\cdot \left\{ \frac{1 - e^{-(\lambda_2'+\lambda_3'-\lambda_3'')t}}{\lambda_2' + \lambda_3' - \lambda_3''} - \frac{1 - e^{-(\lambda_1+\lambda_2+\lambda_3-\lambda_3'')t}}{\lambda_1 + \lambda_2 + \lambda_3 - \lambda_3''} \right\}$$

$$+ \; \frac{\lambda_{1_o}'\lambda_{3_o} e^{-\lambda_2''t}}{\lambda_1 - \lambda_1' + \lambda_2 - \lambda_2' + \lambda_3}$$

$$\cdot \left\{ \frac{1 - e^{-(\lambda_1'+\lambda_2'-\lambda_2'')t}}{\lambda_1' + \lambda_2' - \lambda_2''} - \frac{1 - e^{-(\lambda_1+\lambda_2-\lambda_2''+\lambda_3)t}}{\lambda_1 + \lambda_2 - \lambda_2'' + \lambda_3} \right\}$$

$$+ \; \frac{\lambda_{1_o}\lambda_{3_o}' e^{-\lambda_2''t}}{\lambda_1 + \lambda_2 - \lambda_2' + \lambda_3 - \lambda_3'}$$

$$\cdot \left\{ \frac{1 - e^{-(\lambda_2'-\lambda_2''+\lambda_3')t}}{\lambda_2' - \lambda_2'' + \lambda_3'} - \frac{1 - e^{-(\lambda_1+\lambda_2-\lambda_2''+\lambda_3)t}}{\lambda_1 + \lambda_2 - \lambda_2'' + \lambda_3} \right\} . \quad (8.8)$$

8.2.2 THREE EQUAL CYCLIC SWITCHES

If all units are equal, i.e., all three cyclic switches have the same failure rate, then this system's total reliability becomes

$$
\begin{aligned}
R_S(t) = \ & e^{-3\lambda t} + 3 \int_0^t \lambda_o e^{-\lambda_o t_1} e^{-\lambda_c t_1} e^{-\lambda t_1} e^{-\lambda'(t-t_1)} e^{-\lambda t_1} e^{-\lambda'(t-t_1)} \, dt_1 \\
& + 6 \int_0^t \int_0^{t_2} \lambda_o e^{-\lambda_o t_1} e^{-\lambda_c t_1} e^{-\lambda t_1} \lambda_o' e^{-\lambda_o'(t_2-t_1)} e^{-\lambda_c'(t_2-t_1)} \\
& \cdot e^{-\lambda t_1} e^{-\lambda'(t_2-t_1)} e^{-\lambda''(t-t_2)} \, dt_1 \, dt_2,
\end{aligned}
\tag{8.9}
$$

$$
\begin{aligned}
R_S(t) = \ & e^{-3\lambda t} + 3\lambda_o e^{-2\lambda' t} \int_0^t e^{-(\lambda_o+\lambda_c+2\lambda-2\lambda')t_1} \, dt_1 \\
& + 6\lambda_o \lambda_o' e^{-\lambda'' t} \int_0^t \int_0^{t_2} e^{-(\lambda_o+\lambda_c-\lambda_o'-\lambda_c'+2\lambda-\lambda')t_1} \\
& \cdot e^{-(\lambda_o'+\lambda_c'+\lambda'-\lambda'')t_2} \, dt_1 \, dt_2,
\end{aligned}
\tag{8.10}
$$

or, integrated out,

$$
\begin{aligned}
R_S(t) = \ & e^{-3\lambda t} + \frac{3\lambda_o e^{-2\lambda' t}}{3\lambda - 2\lambda'}\left[1 - e^{-(3\lambda - 2\lambda')t}\right] + \frac{6\lambda_o \lambda_o' e^{-\lambda'' t}}{3\lambda - 2\lambda'} \\
& \cdot \left\{ \frac{1}{2\lambda' - \lambda''}\left[1 - e^{-(2\lambda' - \lambda'')t}\right] - \frac{1}{3\lambda - \lambda''}\left[1 - e^{-(3\lambda - \lambda'')t}\right]\right\}.
\end{aligned}
\tag{8.11}
$$

8.3 RELIABILITY OF TWO LOAD-SHARING WEIBULLIAN UNITS ARRANGED RELIABILITYWISE IN PARALLEL

The two-unit, load-sharing system with Weibullian units reliabilitywise in parallel is shown in Fig. 8.5. The following definitions apply:

$f_1(T)$ = *pdf* of Weibullian Unit 1 with parameters β_1, γ_1, and η_1, when Unit 1 carries the load $k_1 L$, as shown in Fig. 8.5(a),

$f_2(T)$ = *pdf* of Weibullian Unit 2 with parameters β_2, γ_2, and η_2, when Unit 2 carries the load $k_2 L$, as shown in Fig. 8.5(b),

$f_1'(T)$ = *pdf* of Unit 1 with parameters β_1', γ_1', and η_1', when Unit 1 carries load L,

$f_2'(T)$ = pdf of Unit 2 with parameters β_2', γ_2', and η_2', when Unit 2 carries load L,

$f_1''(T - t_2)$ = conditional pdf of Unit 1 after Unit 2 fails at time t_2 with $t_2 < T$, as shown in Fig. 8.5(c),

and

$f_2''(T - t_1)$ = conditional pdf of Unit 2 after Unit 1 fails at time t_1 with $t_1 < T$, as shown in Fig. 8.5(b).

It may be seen from the previous definitions that $f_2''(T - t_1)$ is the pdf of Unit 2, when Unit 2 carries the full load L, after Unit 1 fails at time t_1. Furthermore, $f_2''(T - t_1)$ depends on both the age of Unit 2 and the time t_1 when Unit 1 fails. For example, $f_2''(T - t_1)$ when $t_1 = 10$ hr is different from $f_2''(T - t_1)$ when $t_1 = 20$ hr, and the unit with pdf $f_2''(T - t_1)$ when $t_1 = 20$ hr is weaker than the unit with pdf $f_2''(T - t_1)$ when $t_1 = 10$ hr.

In practice, it is difficult to obtain the pdf of $f_2''(T-t_1)$ with $t_1 = 10$ hr or of $f_2''(T - t_1)$ with $t_1 = 20$ hr, because the value of t_1 is randomly distributed in the time interval $[0, t]$.

The pdf's $f_1'(T)$ or $f_2'(T)$ can be determined through life tests; e.g., by taking a sample of size n of Unit 1, applying the total load, L, on each unit to get time-to-failure data, and determining the pdf, $f_1'(T)$, by estimating its parameters. The problem, however, is to find the pdf's $f_1''(T - t_2)$ and $f_2''(T - t_1)$. The details of the method to find $f_1''(T - t_2)$ and $f_2''(T - t_1)$, using the equivalent-time technique, are given in Chapter 7, and the method's application is illustrated by the cases covered here.

The system's success modes are listed in Table 8.3. The system's reliability function for its first mission, starting the mission at age 0, at the end of which mission the age T is t is given by

$$R(t) = R_1(t)R_2(t) + Q_1(t_1)R_2(t_1)R_2''(t - t_1)$$
$$+ Q_2(t_2)R_1(t_2)R_1''(t - t_2). \qquad (8.12)$$

For subsequent missions of t duration of the same system, starting the mission at age T, the reliability should be calculated from

$$R(T,t) = \frac{R(T + t)}{R(T)}.$$

The first term of Eq. (8.12) is the probability that Units 1 and 2 complete their mission from 0 to t successfully with pdf's $f_1(T)$ and $f_2(T)$, respectively. It can be written as

$$R_1(t)R_2(t) = e^{-[(t-\gamma_1)/\eta_1]^{\beta_1}} \cdot e^{-[(t-\gamma_2)/\eta_2]^{\beta_2}},$$

or

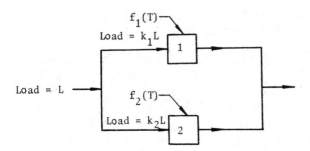

(a) - Both units are functioning and $k_1 + k_2 = 1$,
thus the load L is being shared throughout
the mission.

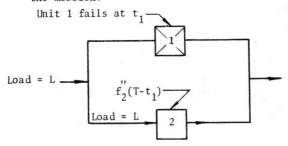

(b) - Unit 1 fails at t_1, hence Unit 2 has to carry
the full load L for the remainder of the
mission.

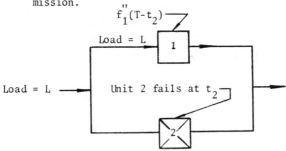

(c) - Unit 2 fails at t_2, hence Unit 1 has to carry
the full load L for the remainder of the
mission.

Fig. 8.5 – Two load-sharing units in parallel.

TABLE 8.3 – System success function modes matrix of two load-sharing Weibullian units reliabilitywise in parallel.

System success function modes number	Unit, function modes, *pdf*'s, and time domains		Reliability function
	Unit 1	Unit 2	
1	G	G	
	$f_1(T)$; $T < t$	$f_2(T)$; $T < t$	$R_1(t)R_2(t)$
2	B	G	
	$f_1(T)$; $T < t_1$	$f_2(T)$; $T < t_1$	$Q_1(t_1)R_2(t_1)$ $\cdot R_2''(t - t_1)$
		$f_2''(T - t_1)$; $t_1 < T < t$	
3	G	B	
	$f_1(T)$; $t_2 < t$	$f_2(T)$; $T < t_2$	$Q_2(t_2)R_1(t_2)$ $\cdot R_1''(t - t_2)$
	$f_1''(T - t_2)$; $t_2 < T < t$		

$$R_1(t)R_2(t) = e^{-\{[(t-\gamma_1)/\eta_1]^{\beta_1} + [(t-\gamma_2)/\eta_2]^{\beta_2}\}}. \tag{8.13}$$

The second term of Eq. (8.12) is the probability that Unit 1 fails at $t_1 < t$ with *pdf* $f_1(T)$, *and* Unit 2 functions till t_1 with *pdf* $f_2(T)$ *and* then functions for the rest of the mission with *pdf* $f_2''(T - t_1)$, as shown in Fig. 8.6(a). To determine the $f_2''(T-t_1)$ *pdf* or the conditional reliability function $R_2''(T - t_1)$, the equivalent-time technique will be used as illustrated in Fig. 8.6(b). The equivalent time, t_{1e}, should be chosen such that, up to time t_1, the probability of failure of Unit 2 when it carries the full load L is the same as the probability of failure of Unit 2 when it carries the load k_2L; i.e.,

$$\int_{\gamma_2}^{t_1} f_2(T) \, dT = \int_{t_1-t_{1e}+\gamma_2'}^{t_1} f_2'[T - (t_1 - t_{1e})] \, dT. \tag{8.14}$$

Let $\xi = T - (t_1 - t_{1e})$; then $d\xi = dT$. And when $T \to t_1$, $\xi \to t_{1e}$, and when $T \to t_1 - t_{1e} + \gamma_2'$, $\xi \to \gamma_2'$. Then Eq. (8.14) can be written as

$$\int_{\gamma_2}^{t_1} f_2(T) \, dT = \int_{\gamma_2'}^{t_{1e}} f_2'(\xi) \, d\xi,$$

or

$$1 - e^{-[(t_1 - \gamma_2)/\eta_2]^{\beta_2}} = 1 - e^{-[(t_{1e}-\gamma_2')/\eta_2']^{\beta_2'}}. \tag{8.15}$$

From Eq. (8.15)

$$t_{1e} = \eta_2'(\frac{t_1 - \gamma_2}{\eta_2})^{\beta_2/\beta_2'} + \gamma_2'. \tag{8.16}$$

Equation (8.16) is for the case when $t_1 \geq \gamma_2$. When $t_1 < \gamma_2$, t_{1e} is not defined by Eq. (8.16) because there is no influence on the value of the reliability of Unit 2 until t_1 starts to exceed γ_2. That means the reliability of Unit 2 is always equal to 1 when $0 < t_1 \leq \gamma_2$.

The value of equivalent time, t_{1e}, should be in between zero and γ_2' when t_1 is in the interval $[0, \gamma_2]$. $t_{1e} = 0$ when $t_1 = 0$, and $t_{1e} = \gamma_2'$ when $t_1 = \gamma_2$. If we take γ_2 and γ_2' as the minimum lives of Unit 2 at lower and higher stress levels, respectively, and assume that the modified Miner's rule [1] applies here, i.e.,

$$\sum_{i=1}^{n} \frac{N_i'}{N_i} = \delta,$$

where

N_i', $i = 1, 2, ..., n$ are the operating times at stress levels S_i, $i = 1, 2, ..., n$, respectively, and

N_i, $i = 1, 2, ..., n$ are the lives of the units at stress levels S_i, $i = 1, 2, ..., n$, respectively,

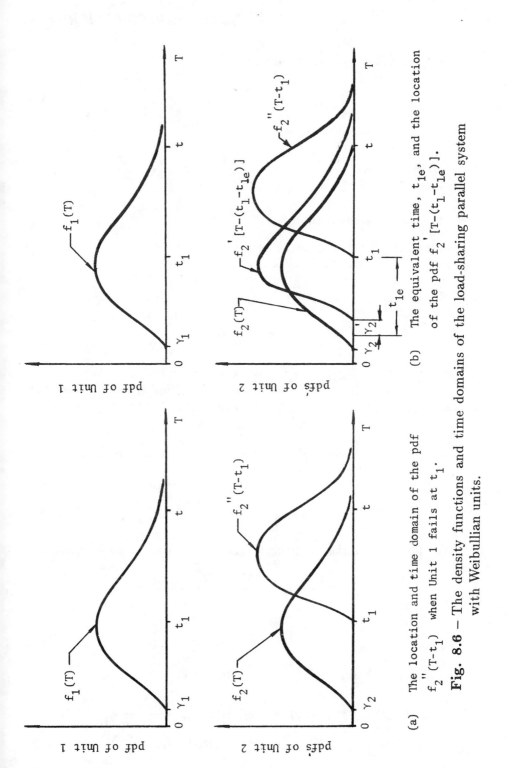

(a) The location and time domain of the pdf $f_2''(T-t_1)$ when Unit 1 fails at t_1.

(b) The equivalent time, t_{1e}, and the location of the pdf $f_2'[T-(t_1-t_{1e})]$.

Fig. 8.6 – The density functions and time domains of the load-sharing parallel system with Weibullian units.

161

and if δ is taken to be 1 in value, then

$$\frac{t_1}{\gamma_2} + \frac{t_1'}{\gamma_2'} = 1, \tag{8.16'}$$

where

$$t_1' = \gamma_2' - t_{1e},$$

and

$$t_{1e} = \text{equivalent time of } t_1.$$

Substituting t_1' into Eq. (8.16') yields

$$\frac{t_1}{\gamma_2} + \frac{\gamma_2' - t_{1e}}{\gamma_2'} = 1,$$

or

$$t_{1e} = t_1 \frac{\gamma_2'}{\gamma_2}.$$

Miner's rule for $\delta = 1$ is derived in Appendix 8A.

Experimentally, the value of δ has been found to be between 0.18 and 23.0, i.e., $0.18 < \delta < 23.0$ [1]. δ varies with materials, the properties of the materials, and so on. Because in most cases δ is greater than 1 when units fail, Miner's rule is conservative in most cases; therefore, it is reasonable to assume that Miner's rule may be applied here conservatively. Now t_{1e} becomes

$$t_{1e} = \begin{cases} \eta_2'[(t_1 - \gamma_2)/\eta_2]^{\beta_2/\beta_2'} + \gamma_2', & t_1 \geq \gamma_2, \\ t_1(\gamma_2'/\gamma_2), & t_1 < \gamma_2. \end{cases} \tag{8.17}$$

From Fig. 8.6(b), the conditional reliability of Unit 2 can be expressed as

$$R_2''(t - t_1) = \frac{R_2'[t - (t_1 - t_{1e})]}{R_2'(t_{1e})},$$

or

$$R_2''(t - t_1) = \frac{R_2'[t - (t_1 - t_{1e})]}{R_2(t_1)}. \tag{8.18}$$

The second term of Eq. (8.12) can now be written as

$$\begin{aligned} Q_1(t_1)R_2(t_1)R_2''(t - t_1) &= Q_1(t_1)R_2(t_1) \cdot \frac{R_2'[t - (t_1 - t_{1e})]}{R_2(t_1)}, \\ &= Q_1(t_1)R_2'[t - (t_1 - t_{1e})]. \end{aligned} \tag{8.19}$$

Substituting the *pdf*'s involved into Eq. (8.19) and considering that time t_1 is an arbitrary value in time interval $[0, t]$ yields

$$Q_1(t_1)R_2(t_1)R_2''(t - t_1) = \int_{\gamma_1}^{t} \frac{\beta_1}{\eta_1} (\frac{t_1 - \gamma_1}{\eta_1})^{\beta_1 - 1}$$

$$\cdot e^{-\left\{[(t_1 - \gamma_1/\eta_1]^{\beta_1} + \{[t - (t_1 - t_{1e}) - \gamma_2']/\eta_2'\}^{\beta_2'}\right\}} dt_1.$$
$$(8.20)$$

The third term of Eq. (8.12) is the probability that Unit 2 fails at $t_2 < t$ with *pdf* $f_2(T)$ *and* Unit 1 functions till t_2 with *pdf* $f_1(T)$ *and* then functions for the rest of the mission with *pdf* $f_1''(T - t_2)$. Using the same approach as before, the equivalent time, t_{2e}, is

$$t_{2e} = \begin{cases} \eta_1'[(t_2 - \gamma_1)/\eta_1]^{\beta_1/\beta_1'} + \gamma_1', & t_2 \geq \gamma_1, \\ t_2(\gamma_1'/\gamma_1), & t_2 < \gamma_1, \end{cases} \quad (8.21)$$

and the conditional reliability for a mission of duration t, when Unit 2 fails at t_2, is

$$R_1''(t - t_2) = \frac{R_1'[t - (t_2 - t_{2e})]}{R_1(t_2)}. \quad (8.22)$$

Then the third term of Eq. (8.12) can be written as

$$Q_2(t_2)R_1(t_2)R_1''(t - t_2) = Q_2(t_2)R_1(t_2) \cdot \frac{R_1'[t - (t_2 - t_{2e})]}{R_1(t_2)},$$
$$= Q_2(t_2)R_1'[t - (t_2 - t_{2e})]. \quad (8.23)$$

Substituting the *pdf*'s involved into Eq. (8.23) and considering that t_2 is an arbitrary value in time interval $[0, t]$ yields

$$Q_2(t_2)R_1(t_2)R_1''(t - t_2) = \int_{\gamma_2}^{t} \frac{\beta_2}{\eta_2} (\frac{t_2 - \gamma_2}{\eta_2})^{\beta_2 - 1}$$

$$\cdot e^{-\left\{[(t_2 - \gamma_2)/\eta_2]^{\beta_2} + \{[t - (t_2 - t_{2e}) - \gamma_1']/\eta_1'\}^{\beta_1'}\right\}} dt_2.$$
$$(8.24)$$

Consequently, the system's reliability for a mission of duration t is given by

$$R(t) \;=\; e^{-\{[(t-\gamma_1)/\eta_1]^{\beta_1}+[(t-\gamma_2)/\eta_2]^{\beta_2}\}}$$

$$+ \int_{\gamma_1}^{t} \frac{\beta_1}{\eta_1}\Big(\frac{t_1-\gamma_1}{\eta_1}\Big)^{\beta_1-1}$$

$$\cdot \, e^{-\Big\{[(t_1-\gamma_1)/\eta_1]^{\beta_1}+\{[t-(t_1-t_{1e})-\gamma_2']/\eta_2'\}^{\beta_2'}\Big\}} \, dt_1$$

$$+ \int_{\gamma_2}^{t} \frac{\beta_2}{\eta_2}\Big(\frac{t_2-\gamma_2}{\eta_2}\Big)^{\beta_2-1}$$

$$\cdot \, e^{-\Big\{[(t_2-\gamma_2)/\eta_2]^{\beta_2}+\{[t-(t_2-t_{2e})-\gamma_1']/\eta_1'\}^{\beta_1'}\Big\}} \, dt_2,$$

$$(8.25)$$

where t_{1e} and t_{2e} are defined by Eqs. (8.17) and (8.21), respectively.

EXAMPLE 8–1

A system with two units reliabilitywise in parallel is to operate in a mission of 20-hr duration. The two units have Weibull *pdf*'s with $\beta_1 = 1.4$, $\gamma_1 = 5$ hr and $\eta_1 = 150$ hr for Unit 1, and with $\beta_2 = 1.8$, $\gamma_2 = 5$ hr and $\eta_2 = 130$ hr for Unit 2 when they are sharing the total load. If the total load is carried by Unit 1, then its Weibull *pdf* parameters become $\beta_1' = 1.7$, $\gamma_1' = 0$ hr and $\eta_1' = 90$ hr. If the total load is carried by Unit 2, then its Weibull *pdf* parameters become $\beta_2' = 2.0$, $\gamma_2' = 0$ hr and $\eta_2' = 70$ hr.

Find the reliability of this system for the following three cases:

1. The units are unequal and load sharing as stated above.

2. The *pdf*'s do not change when a unit fails and each unit carries the $k_i L$ load for the full mission.

3. The units carry the full load L for the full mission.

4. Comparatively discuss the previous results.

SOLUTIONS TO EXAMPLE 8–1

1. The equivalent times are

$$t_{1e} = \eta_2'\Big(\frac{t_1-\gamma_2}{\eta_2}\Big)^{\beta_2/\beta_2'} + \gamma_2' = 70\Big(\frac{t_1-5}{130}\Big)^{1.8/2.0} + 0,$$

and

$$t_{2e} = \eta_1'\Big(\frac{t_2-\gamma_1}{\eta_1}\Big)^{\beta_1/\beta_1'} + \gamma_1' = 90\Big(\frac{t_2-5}{150}\Big)^{1.4/1.7} + 0.$$

Substituting the parameters and the equivalent times t_{1e} and t_{2e} into Eq. (8.25) yields

$$
\begin{aligned}
R(20) ={}& e^{-[(\frac{20-5}{150})^{1.4}+(\frac{20-5}{130})^{1.8}]} \\
&+ \int_5^{20} \frac{1.4}{150}\left(\frac{t_1-5}{150}\right)^{1.4-1} \\
&\cdot e^{-\left\{[(t_1-5)/150]^{1.4}+\{[(20-t_1)/70]+[(t_1-5)/130]^{1.8}/2.0\}^{2.0}\right\}}\, dt_1 \\
&+ \int_5^{20} \frac{1.8}{130}\left(\frac{t_2-5}{130}\right)^{1.8-1} \\
&\cdot e^{-\left\{[(t_2-5)/130]^{1.8}+\{[(20-t_2)/90]+[(t_2-5)/150]^{1.4}/1.7\}^{1.7}\right\}}\, dt_2.
\end{aligned}
\tag{8.26}
$$

Using a numerical integration computer program yields

$$R(20) = 0.94146696 + 0.03781458 + 0.01939125,$$

or

$$R(20) = 0.9986728.$$

2. If it is assumed that the *pdf*'s do not change when a unit fails, then the system's reliability, as a special case, becomes

$$
\begin{aligned}
R(t) ={}& e^{-\{[(t-\gamma_1)/\eta_1]^{\beta_1}+[(t-\gamma_2)/\eta_2]^{\beta_2}\}} \\
&+ \int_{\gamma_1}^{t} \frac{\beta_1}{\eta_1}\left(\frac{t_1-\gamma_1}{\eta_1}\right)^{\beta_1-1} \\
&\cdot e^{-\left\{[(t_1-\gamma_1)/\eta_1]^{\beta_1}+\{[t-(t_1-t_{1e})-\gamma_2]/\eta_2\}^{\beta_2}\right\}}\, dt_1 \\
&+ \int_{\gamma_2}^{t} \frac{\beta_2}{\eta_2}\left(\frac{t_2-\gamma_2}{\eta_2}\right)^{\beta_2-1} \\
&\cdot e^{-\left\{[(t_2-\gamma_2)/\eta_2]^{\beta_2}+\{[t-(t_2-t_{2e})-\gamma_1]/\eta_1\}^{\beta_1}\right\}}\, dt_2,
\end{aligned}
$$

where the equivalent times become

$$t_{1e} = \eta_2\left(\frac{t_1-\gamma_2}{\eta_2}\right)^{\beta_2/\beta_2} + \gamma_2,$$

or

$$t_{1e} = t_1,$$

and

$$t_{2e} = \eta_1\left(\frac{t_2-\gamma_1}{\eta_1}\right)^{\beta_1/\beta_1} + \gamma_1,$$

or

$$t_{2e} = t_2.$$

Substituting t_{1e} and t_{2e} into the reliability equation yields

$$
\begin{aligned}
R(t) &= e^{-\{[(t-\gamma_1)/\eta_1]^{\beta_1}+[(t-\gamma_2)/\eta_2]^{\beta_2}\}} \\
&\quad + e^{-[(t-\gamma_2)/\eta_2]^{\beta_2}} \int_{\gamma_1}^{t} \frac{\beta_1}{\eta_1}\left(\frac{t_1-\gamma_1}{\eta_1}\right)^{\beta_1-1} e^{-[(t_1-\gamma_1)/\eta_1]^{\beta_1}}\, dt_1 \\
&\quad + e^{-[(t-\gamma_1)/\eta_1]^{\beta_1}} \int_{\gamma_2}^{t} \frac{\beta_2}{\eta_2}\left(\frac{t_2-\gamma_2}{\eta_2}\right)^{\beta_2-1} e^{-[(t_2-\gamma_2)/\eta_2]^{\beta_2}}\, dt_2, \\
&= e^{-\{[(t-\gamma_1)/\eta_1]^{\beta_1}+[(t-\gamma_2)/\eta_2]^{\beta_2}\}} \\
&\quad + e^{-[(t-\gamma_2)/\eta_2]^{\beta_2}}\{1 - e^{-[(t-\gamma_1)/\eta_1]^{\beta_1}}\} \\
&\quad + e^{-[(t-\gamma_1)/\eta_1]^{\beta_1}}\{1 - e^{-[(t-\gamma_2)/\eta_2]^{\beta_2}}\},
\end{aligned}
$$

or

$$R(t) = e^{-[(t-\gamma_1)/\eta_1]^{\beta_1}} + e^{-[(t-\gamma_2)/\eta_2]^{\beta_2}} - e^{-\{[(t-\gamma_1)/\eta_1]^{\beta_1}+[(t-\gamma_2)/\eta_2]^{\beta_2}\}}.$$

Substituting the parameters and the mission time into the reliability equation yields

$$
\begin{aligned}
R(20) &= e^{-\left(\frac{20-5}{150}\right)^{1.4}} + e^{-\left(\frac{20-5}{130}\right)^{1.8}} - e^{-\left[\left(\frac{20-5}{150}\right)^{1.4}+\left(\frac{20-5}{130}\right)^{1.8}\right]}, \\
&= 0.9609713 + 0.9797035 - 0.9414669,
\end{aligned}
$$

or

$$R(20) = 0.9992079.$$

3. When both units carry the full load, L, the system's reliability can be obtained using Eq. (8.25), and the equivalent times become

$$t_{1e} = \eta_2'\left(\frac{t_1-\gamma_2'}{\eta_2'}\right)^{\beta_2'/\beta_2'} + \gamma_2',$$

or

$$t_{1e} = t_1,$$

and

$$t_{2e} = \eta_1'\left(\frac{t_2-\gamma_1'}{\eta_1'}\right)^{\beta_1'/\beta_1'} + \gamma_1',$$

or

$$t_{2e} = t_2.$$

The system's reliability now becomes

$$R(t) = e^{-\{[(t-\gamma_1')/\eta_1']^{\beta_1'} + [(t-\gamma_2')/\eta_2']^{\beta_2'}\}}$$

$$+ \int_{\gamma_1'}^{t} \frac{\beta_1'}{\eta_1'} \left(\frac{t_1 - \gamma_1'}{\eta_1'}\right)^{\beta_1'-1}$$

$$\cdot e^{-\left\{[(t_1-\gamma_1')/\eta_1']^{\beta_1'} + \{[t-(t_1-t_{1c})-\gamma_2']/\eta_2'\}^{\beta_2'}\right\}} dt_1$$

$$+ \int_{\gamma_2'}^{t} \frac{\beta_2'}{\eta_2'} \left(\frac{t_2 - \gamma_2'}{\eta_2'}\right)^{\beta_2'-1}$$

$$\cdot e^{-\left\{[(t_2-\gamma_2')/\eta_2']^{\beta_2'} + \{[t-(t_2-t_{2c})-\gamma_1']/\eta_1'\}^{\beta_1'}\right\}} dt_2,$$

$$= e^{-\{[(t-\gamma_1')/\eta_1']^{\beta_1'} + [(t-\gamma_2')/\eta_2']^{\beta_2'}\}}$$

$$+ e^{-[(t-\gamma_2')/\eta_2']^{\beta_2'}} \{1 - e^{-[(t-\gamma_1')/\eta_1']^{\beta_1'}}\}$$

$$+ e^{-[(t-\gamma_1')/\eta_1']^{\beta_1'}} \{1 - e^{-[(t-\gamma_2')/\eta_2']^{\beta_2'}}\},$$

or

$$R(t) = e^{-[(t-\gamma_1')/\eta_1']^{\beta_1'}} + e^{-[(t-\gamma_2')/\eta_2']^{\beta_2'}}$$

$$- e^{-\{[(t-\gamma_1')/\eta_1']^{\beta_1'} + [(t-\gamma_2')/\eta_2']^{\beta_2'}\}}.$$

Substituting the mission time and the parameters of the distributions into the system's reliability equation yields

$$R(20) = e^{-\left(\frac{20-0}{90}\right)^{1.7}} + e^{-\left(\frac{20-0}{70}\right)^{2.0}} - e^{-[(\frac{20-0}{90})^{1.7} + (\frac{20-0}{70})^{2.0}]},$$

$$= 0.92538787 + 0.92161045 - 0.85284713,$$

or

$$R(20) = 0.9941512.$$

4. It is interesting to compare the reliabilities obtained in the three previous cases.

Case 2 gives the *highest reliability* because the *pdf*'s used are those for partial full load, or for $k_i L$, where $0 < k_i < 1$, and these are assumed not to increase during the mission when one of the units fails.

Case 3 gives the *lowest reliability* because it is assumed that the units are operating for the full mission at their highest load.

Case 1 gives the correct load-sharing reliability and, as to be expected, it has a value between those of Cases 2 and 3; i.e.,

$$R_3(20) = 0.9941512 < R_1(20) = 0.9986728 < R_2(20) = 0.9992079.$$

PROBLEMS

8-1. (1) Given are *two unequal cyclic switches* arranged *physically in series*. What is the reliability of this arrangement, whose success is defined as the cyclic passing and interruption of an electrical signal, under load-sharing conditions. Assume that, when both switches are functioning normally, their failure rates are λ_1 and λ_2; when Switch 1 fails, λ_2 becomes λ_2'; and when Switch 2 fails, λ_1 becomes λ_1', $\lambda_1' > \lambda_1$, and $\lambda_2' > \lambda_2$. Note that $\lambda_i = \lambda_{io} + \lambda_{ic}$ and $\lambda_i' = \lambda_{io}' + \lambda_{ic}'$. Work out the solution completely and give the answer explicitly in terms of the failure rates involved.

      (2) Find the reliability of this arrangement with no load sharing.

      (3) If $\lambda_1 = 0.55 \text{ fr}/10^6$ hr, $\lambda_2 = 0.45 \text{ fr}/10^6$ hr, $\lambda_{1o} = 0.35 \text{ fr}/10^6$ hr, $\lambda_{1c} = 0.20 \text{ fr}/10^6$ hr, $\lambda_{2o} = 0.28 \text{ fr}/10^6$ hr, $\lambda_{2c} = 0.17 \text{ fr}/10^6$ hr, $\lambda_1' = 0.65 \text{ fr}/10^6$ hr, and $\lambda_2' = 0.55 \text{ fr}/10^6$ hr, find the numerical answer for the reliability of this arrangement in Case 1 for a mission of 10 hr.

      (4) Find the reliability of this arrangement in Case 2 for a mission of 10 hr.

      (5) Discuss the results in Cases 3 and 4 comparatively.

8-2. Given are *three equal cyclic switches* arranged *physically in series*. What is the reliability of this system, whose success is defined as the cyclic passing and interruption of an electrical signal, under load-sharing conditions? Assume that, when all three switches are functioning normally, their failure rate is λ; when one fails, the failure rate of the surviving switches is λ'; and when two fail, the failure rate of the surviving switch is λ''. Note that $\lambda = \lambda_o + \lambda_c$, $\lambda' = \lambda_o' + \lambda_c'$, and $\lambda'' = \lambda_o'' + \lambda_c''$. Work out the solution completely and provide the answer explicitly in terms of the failure rates involved.

8-3. Given are *three unequal cyclic switches* arranged *physically in series*. What is the reliability of this system, whose success is defined as the cyclic passing and interruption of an electrical signal, under load-sharing conditions? Assume that, when all three switches are functioning normally, their failure rate is λ_i; when one fails, the failure rate of the surviving switches is λ_i'; and when two fail, the failure rate of the surviving switch is λ_i''. Note that $\lambda = \lambda_o + \lambda_c$, $\lambda' = \lambda_o' + \lambda_c'$, and $\lambda'' = \lambda_o'' + \lambda_c''$. Work out the solution completely and provide the answer explicitly in terms of the failure rates involved.

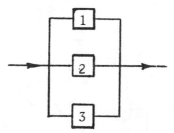

Fig. 8.7 – Physical arrangement of the three, unequal, load-sharing units of Problem 8-4.

8-4. The system shown in Fig. 8.7 consists of *three unequal load-sharing units* exhibiting constant failure rates. The following failure rates prevail during a mission:

$$\lambda_1 = 1.1; \quad \lambda_{1,2} = 1.3; \quad \lambda_{1,3} = 1.5;$$
$$\lambda_2 = 2.0; \quad \lambda_{2,1} = 2.5; \quad \lambda_{2,3} = 3.0;$$

and

$$\lambda_3 = 3.0; \quad \lambda_{3,1} = 3.5; \quad \lambda_{3,2} = 4.0.$$

$\lambda_{1,3}$ gives the failure rate of Unit 1 when Unit 3 has failed, etc. All failure rates are in failures per 1,000 hr. For system success *at least two* satisfactorily functioning units are required. If the mission is of 15-hr duration, determine the reliability of the system under the following situations:

(1) Assuming no load sharing.
(2) Assuming load sharing.

Work out the answer completely numerically.

8-5. In the system in Fig. 8.8 there are *three equal functioning units* with a reliability of R_1 and a standby unit with a reliability of R_2. *Two units are required for system success.* All units have constant failure rates. Consider the following function modes: standby quiescent and energized, sensing energized, switch quiescent, switch energized, switch failing open, and *load sharing*. When one functioning unit fails, λ_1 becomes λ_1'. When two functioning units fail, λ_1 becomes λ_1''. What is the reliability of this system? Work out the answer completely.

8-6. The load-sharing, standby system of Fig. 8.9 is given. Do the following:

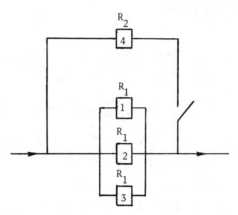

Fig. 8.8 – Physical arrangement of the standby system of Problem 8-5.

(1) Prepare a table of all system success arrangements.

(2) Write out the reliability math model with all respective failure rates substituted.

(3) Integrate out *one* of the multiple integral terms of the solution.

8-7. A system consists of two units which function reliabilitywise in parallel. The two units have Weibull *pdf*'s with $\beta_1 = 1.8$, $\gamma_1 = 5$ hr, and $\eta_1 = 180$ hr for Unit 1, and with $\beta_2 = 1.95$, $\gamma_2 = 0$ hr, and $\eta_2 = 150$ hr for Unit 2 when they are sharing the full load. If the full load is carried by Unit 1, then its Weibull *pdf*'s parameters become $\beta_1' = 1.8$, $\gamma_1' = 0$ hr, and $\eta_1' = 100$ hr. If the full load is carried by Unit 2, then its Weibull *pdf*'s parameters become $\beta_2' = 2.0$, $\gamma_2' = 0$ hr, and $\eta_2' = 90$ hr. Find the reliability of this system for a 25-hr mission for the following three cases:

(1) The units are load sharing as stated above.

(2) The *pdf*'s do not change when a unit fails and each unit carries the partial load, $k_i L$, for the full mission.

(3) Each unit carries the full load for the full mission.

(4) Comparatively discuss the previous results.

8-8. A system consists of three units which are reliabilitywise in parallel. The system is a load-sharing system of three units which have Weibull *pdf*'s. The parameters of each unit are listed in Table 8.4. Find the reliability of this system for a 20-hr mission for the following cases:

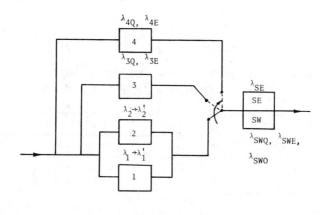

Note: $\lambda_1' > \lambda_1$ and $\lambda_2' > \lambda_2$. The λ_i' are the load-sharing failure rates.

Fig. 8.9 – The load-sharing standby system of Problem 8-6.

(1) The units are load sharing.

(2) The *pdf*'s do not change when a unit fails and each unit carries the partial load, $k_i L$, for the full mission.

(3) Each unit carries the full load for the full mission.

(4) Comparatively discuss the previous results.

REFERENCE

1. Kececioglu, D., Chester, L.B., and Gardner, E.O., "Sequential Cumulative Fatigue Reliability," *Proceedings Reliability and Maintainability Symposium*, Los Angeles, Calif., pp. 533-539, Jan. 29-31, 1972.

TABLE 8.4 – The parameters of each unit in various states of function for Problem 8-8.

State of each unit in the system		Parameters of each unit		
		Unit 1	Unit 2	Unit 3
Complete load sharing where each unit carries the load, $k_i L$, $k_1+k_2+k_3=1$.		$\beta_1 = 1.5$, $\gamma_1 = 5$ hr, $n_1 = 150$ hr.	$\beta_2 = 1.45$, $\gamma_2 = 3$ hr, $n_2 = 135$ hr.	$\beta_3 = 1.6$, $\gamma_3 = 7$ hr, $n_3 = 130$ hr.
Two units carry the full load.	Unit 1 and the corresponding unit in the column for which the parameters are given are carrying the full load.		$\beta'_{21} = 1.5$ $\gamma'_{21} = 0$, $n'_{21} = 100$ hr.	$\beta'_{31} = 1.8$, $\gamma'_{31} = 3$ hr, $n'_{31} = 110$ hr.
	Unit 2 and the corresponding unit in the column for which the parameters are given are carrying the full load.	$\beta'_{12} = 1.55$, $\gamma'_{12} = 2$ hr, $n'_{12} = 120$ hr.		$\beta'_{32} = 1.85$, $\gamma'_{32} = 2$ hr, $n'_{32} = 99$ hr.
	Unit 3 and the corresponding unit in the column for which the parameters are given are carrying the full load.	$\beta'_{13} = 1.55$, $\gamma'_{13} = 2$ hr, $n'_{13} = 105$ hr.	$\beta'_{23} = 1.5$, $\gamma'_{23} = 0$ hr, $n'_{23} = 98$ hr.	
One unit is operating and carries the full load.		$\beta''_1 = 1.8$, $\gamma''_1 = 0$ hr, $n''_1 = 85$ hr.	$\beta''_2 = 1.7$, $\gamma''_2 = 0$ hr, $n''_2 = 76$ hr.	$\beta''_3 = 1.95$, $\gamma''_3 = 0$ hr, $n''_3 = 75$ hr.

APPENDIX 8A
THE DERIVATION OF MINER'S RULE [1]

The following are assumed:

1. The maximum amount of energy that is absorbed by a component before failure is equal at various stress levels of operation.

2. The amount of energy absorbed by a component, at a specific stress level, is in proportion to the total operating time, or cycles, at that stress level.

If N_1, N_2, ..., N_n are the lives of components at stress levels S_1, S_2, ..., S_n, respectively, and N_1', N_2', ..., N_n' are the operating times at stress levels S_1, S_2, ..., S_n, respectively, then according to the previous assumptions at stress level S_1 the amount of energy absorbed by the component would be given by

$$W_1 = W \frac{N_1'}{N_1},$$

where W is the maximum amount of energy absorbed by the component before failure.

At stress level S_2 the amount of energy absorbed by the component would be

$$W_2 = W \frac{N_2'}{N_2},$$

and at stress level S_n it would be

$$W_n = W \frac{N_n'}{N_n}.$$

If the component fails, the total energy absorbed by it would be

$$W = W_1 + W_2 + \cdots + W_n,$$

or

$$W = W \frac{N_1'}{N_1} + W \frac{N_2'}{N_2} + \cdots + W \frac{N_n'}{N_n}.$$

Dividing both sides by W yields

$$\frac{N_1'}{N_1} + \frac{N_2'}{N_2} + \cdots + \frac{N_n'}{N_n} = 1,$$

which is Miner's rule. However, it must be pointed out that the two assumptions on which it is based are valid in relatively few cases. Nevertheless, they lead to conservative life estimates in the majority of cases encountered in practice.

Chapter 9

RELIABILITY OF STATIC SWITCHES

9.1 OBJECTIVES

A static switch is defined as a switch that has either to close or open a circuit during a mission. A static relay, a common single-acting switch, a squib, and a stop valve are examples. Such switches have a multiplicity of modes of function and of failure, such as close normally, fail in the quiescent mode, fail on command, fail open on command, fail closed on command, fail open after closing normally, etc. All such modes need to be considered when determining the reliability of static switches. First, the reliability of a single switch will be determined.

9.2 SINGLE-SWITCH RELIABILITY

9.2.1 NORMALLY OPEN SWITCH WHOSE FUNCTION IS TO CLOSE ON COMMAND

A static switch is considered to have the following function modes: normal (not failed in any mode); failed closed or failed open on command; failed closed or failed open in the quiescent function mode; failed closed after failing open; failed open after failing closed; failed open after closing normally; failed closed after closing normally; failed open after opening normally; and failed closed after opening normally.

The function cycle of a single, normally open, switch is shown in Fig. 9.1. Such a switch's reliability is given by

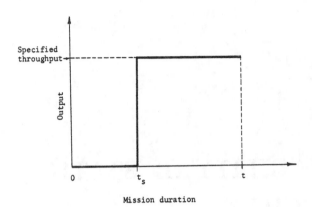

Fig. 9.1 – Duty cycle of a normally open static switch.

$R_{S,SW/o}(t) =$ probability the switch will not fail either open or closed by t_s, when it is called upon to close and thus transmit a signal or provide the flow of a current, *and* the probability it will close normally on command, *and* the probability it will not fail open after it closed normally during the rest of the mission or in $(t-t_s)$, *or* the probability it will not fail either open or closed by t_s when it is called upon to close, *and* the probability it will fail closed on command, *and* the probability it will not fail open thereafter for the rest of the mission or in $(t - t_s)$.

Consequently,

$$R_{S,SW/o}(t) = r_{q/o}(t_s)r_{cn}r_{o/cn}(t - t_s) + r_{q/o}(t_s)q_c r_{o/fc}(t - t_s), \quad (9.1)$$

where

$$r_{q/o}(t_s) = e^{-\lambda_{q/o}t_s},$$

and a good estimate of $\lambda_{q/o}$, or $\hat{\lambda}_{q/o}$, is given by

$$\hat{\lambda}_{q/o} = \frac{N_{fq/o}}{N \times \Delta t},$$

where

N = number of switches being tested,

$N_{fq/o}(\Delta t)$ = number of failures either open or closed in the quiescent mode in Δt, when the switch is normally open,

$\hat{r}_{cn} = (N_{cn}/N_T)$, which is a good estimate of r_{cn},

N_{cn} = number of times the switch closes normally on command,

N_T = total number of commands,

$r_{o/cn}(t - t_s) = e^{-\lambda_{o/cn}(t-t_s)}$,

$\hat{\lambda}_{o/cn} = \frac{N_{fo/cn}(\Delta t)}{N'_{cn}\Delta t}$,

$N_{fo/cn}(\Delta t)$ = number of test times the normally closed switch fails open thereafter in Δt test or function time,

N'_{cn} = number of switches that closed normally on command,

$\hat{q}_c = \frac{N_{fc}}{N_T}$,

N_{fc} = number of times the switch fails closed on command,

$r_{o/fc}(t - t_s) = e^{-\lambda_{o/fc}(t-t_s)}$,

$\hat{\lambda}_{o/fc} = \frac{N_{fo/fc}(\Delta t)}{N'_{fc}\Delta t}$,

$N_{fo/fc}(\Delta t)$ = number of times the failed-closed switch fails open thereafter in Δt,

N'_{fc} = number of switches that failed closed on command,

and

$N_{cn} + N_{fc} + N_{fo} = N_T$,

where

N_{fo} = number of times the switch fails open on command.

Substitution of the above into Eq. (9.1) yields

$$R_{S,SW/o}(t) = e^{-\lambda_{q/o}t_s}\left[\frac{N_{cn}}{N_T}e^{-\lambda_{o/cn}(t-t_s)} + \frac{N_{fc}}{N_T}e^{-\lambda_{o/fc}(t-t_s)}\right]. \quad (9.2)$$

9.2.2   SPECIAL CASES

1. If the switch is activated at the beginning of the mission, then $t_s = 0$, and Eq. (9.2) becomes

$$R_{S,SW/o}(t) = \frac{N_{cn}}{N_T}e^{-\lambda_{o/cn}t} + \frac{N_{fc}}{N_T}e^{-\lambda_{o/fc}t}. \tag{9.3}$$

2. If the switch cannot fail open after it fails closed on command, then $\lambda_{o/fc} = 0$, and Eq. (9.3) becomes

$$R_{S,SW/o}(t) = \frac{N_{cn}}{N_T}e^{-\lambda_{o/cn}t} + \frac{N_{fc}}{N_T}. \tag{9.4}$$

3. If only a single closing action is required and the function of the switch ends immediately thereafter, then $\lambda_{o/cn} = 0$ and $\lambda_{o/fc} = 0$; consequently, Eq. (9.2) becomes

$$R_{S,SW/o}(t) = e^{-\lambda_{q/o}t_s}\frac{N_{cn} + N_{fc}}{N_T}. \tag{9.5}$$

4. If the switch is activated at the beginning of the mission and a single action is required, with the function of the switch ending immediately thereafter, then $t_s = 0$, $\lambda_{o/cn} = 0$, and $\lambda_{o/fc} = 0$; consequently, Eq. (9.2) becomes

$$R_{S,SW/o}(t) = \frac{N_{cn} + N_{fc}}{N_T} = R_{S,SW/o}. \tag{9.6}$$

9.2.3   NORMALLY CLOSED SWITCH WHOSE FUNCTION IS TO OPEN ON COMMAND

The function cycle of a normally closed switch is shown in Fig. 9.2. Such a switch's reliability is given by

$R_{S,SW/c}(t) =$ probability the switch will not fail either open or closed by t_s, thus not interrupting the flow of a signal when not called upon, and opening on command because it did not fail closed already, and the probability it will open on command, and the probability it will stay open during the rest of the mission, i.e., it will not fail closed in $(t - t_s)$, or the probability it will not fail either open or closed by t_s, and the probability it will fail open on command, and the probability it will not fail closed for the rest of the mission, or in $(t - t_s)$.

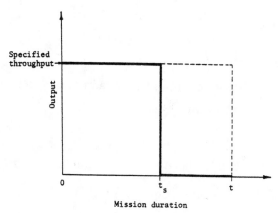

Fig. 9.2 – Duty cycle of a normally closed static switch.

Consequently,

$$R_{S,SW/c}(t) = r_{q/c}(t_s)r_{on}r_{c/on}(t - t_s) + r_{q/c}(t_s)q_o r_{c/fo}(t - t_s), \quad (9.7)$$

where

$r_{q/c}(t_s) = e^{-\lambda_{q/c}t_s}$,

$\lambda_{q/c} = \dfrac{N_{fq/c}(\Delta t)}{N\Delta t}$,

$N_{fq/c}(\Delta t)$ = number of failures either open or closed in the quiescent mode in Δt, when the switch is normally closed,

$r_{on} = N_{on}/N_T$,

N_{on} = number of times the switch opens normally on command,

$r_{c/on} = e^{-\lambda_{c/on}(t-t_s)}$,

$\lambda_{c/on} = N_{fc/on}(\Delta t)/N'_{on}\Delta t$,

$N_{fc/on}(\Delta t)$ = number of times the normally opened switch fails
closed in Δt,

N'_{on} = number of switches that opened normally on command,

$q_o = N_{fo}/N_T$,

N_{fo} = number of times the switch fails open on command,

$$r_{c/fo}(t - t_s) = e^{-\lambda_{c/fo}(t-t_s)},$$

$$\lambda_{c/fo} = N_{fc/fo}(\Delta t)/N'_{fo}\Delta t,$$

$N_{fc/fo}(\Delta t)$ = number of times the failed open switch closes thereafter in Δt,

and

N'_{fo} = number of switches that failed open on command.

Substitution of the above into Eq. (9.7) yields

$$R_{S,SW/c}(t) = e^{-\lambda_{q/c}t_s}\left[\frac{N_{on}}{N_T}e^{-\lambda_{c/on}(t-t_s)} + \frac{N_{fo}}{N_T}e^{-\lambda_{c/fo}(t-t_s)}\right]. \quad (9.8)$$

9.2.4   SPECIAL CASES

1. If the switch is activated at the beginning of the mission, then $t_s = 0$, and Eq. (9.8) becomes

$$R_{S,SW/c}(t) = \frac{N_{on}}{N_T}e^{-\lambda_{c/on}t} + \frac{N_{fo}}{N_T}e^{-\lambda_{c/fo}t}. \quad (9.9)$$

2. If the switch cannot fail closed after having failed open, then $\lambda_{c/fo} = 0$, and Eq. (9.9) becomes

$$R_{S,SW/c}(t) = \frac{N_{on}}{N_T}e^{-\lambda_{c/on}t} + \frac{N_{fo}}{N_T}. \quad (9.10)$$

3. If only a single opening action is required and the function of the switch ends immediately thereafter, then $\lambda_{c/on} = 0$ and $\lambda_{c/fo} = 0$; consequently, Eq. (9.8) becomes

$$R_{S,SW/c}(t) = e^{-\lambda_{q/c}t_s}\frac{N_{on} + N_{fo}}{N_T}. \quad (9.11)$$

4. If the switch is activated at the beginning of the mission and a single action is required, with the function of the switch ending immediately thereafter, then $t_s = 0$, $\lambda_{c/on} = 0$, and $\lambda_{c/fo} = 0$; consequently, Eq. (9.11) becomes

$$R_{S,SW/c}(t) = \frac{N_{on} + N_{fo}}{N_T}. \quad (9.12)$$

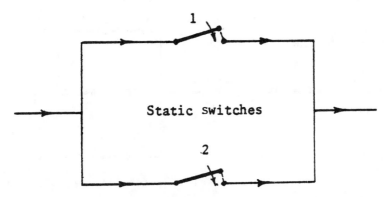

Fig. 9.3 – Schematic and reliability block diagram of two static switches which are normally open and their function is to close the circuit and keep it closed and thus allow a signal or a current to go through.

9.3 STATIC SWITCHES RELIABILITYWISE IN PARALLEL

9.3.1 NORMALLY OPEN SWITCHES

If the reliability of switching needs to be increased, then parallel redundancy may be considered. If *two switches are normally open* and their function is to close the circuit and keep it closed, then they should be arranged physically in parallel, as shown in Fig. 9.3, because with this configuration the system of switches will succeed when either one or both close and stay closed. Consequently, they are also reliabilitywise in parallel. The reliability of the arrangement shown in Fig. 9.3 is then given by

$$R_{2PS,SW/o}(t) = 1 - [1 - R_{1S,SW/o}(t)][1 - R_{2S,SW/o}(t)], \qquad (9.13)$$

where $R_{1S,SW/o}$ and $R_{2S,SW/o}$ are obtained from Eqs. (9.1) through (9.6), depending on the specific case.

Equation (9.13) may be expanded to yield

$$R_{2PS,SW/o}(t) = R_{1S,SW/o}(t) + R_{2S,SW/o}(t) - R_{1S,SW/o}(t)R_{2S,SW/o}(t). \qquad (9.13')$$

If the two switches are equal, then

$$R_{2PS,SW/o}(t) = 1 - [1 - R_{S,SW/o}(t)]^2, \qquad (9.14)$$

or

$$R_{2PS,SW/o}(t) = 2R_{S,SW/o}(t) - R_{S,SW/o}^2(t). \qquad (9.14')$$

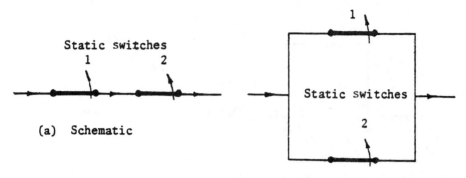

(b) Reliability block diagram

Fig. 9.4 − Schematic and reliability block diagram of two
static switches which are normally closed and
their function is to open the circuit and keep it
open, thus interrupting the flow of a signal or a
current.

9.3.2 NORMALLY CLOSED SWITCHES

If *two switches are normally closed* and their function is to open the
circuit and keep it open, then they should be arranged physically in
series as shown in Fig. 9.4(a), because this system of switches will
succeed when either one switch succeeds or both switches succeed, i.e.,
they open and stay open. Consequently, they are reliabilitywise in
parallel, as shown in Fig. 9.4(b). The reliability of the arrangement
shown in Fig. 9.4 is given by

$$R_{2SS,SW/c}(t) = 1 - [1 - R_{1S,SW/c}(t)][1 - R_{2S,SW/c}(t)], \qquad (9.15)$$

where $R_{1S,SW/c}$ and $R_{2S,SW/c}$ are obtained from Eqs. (9.7) through
(9.12), depending on the specific case.

Equation (9.15) may be expanded to yield

$$R_{2SS,SW/c}(t) = R_{1S,SW/c}(t) + R_{2S,SW/c}(t) - R_{1S,SW/c}(t)R_{2S,SW/c}(t).$$
$$(9.15')$$

If the two switches are equal, then

$$R_{2SS,SW/c}(t) = 1 - [1 - R_{S,SW/c}(t)]^2, \qquad (9.16)$$

or

$$R_{2SS,SW/c}(t) = 2R_{S,SW/c}(t) - R_{S,SW/c}^2(t). \qquad (9.16')$$

If even higher reliabilities are required, then, to close and keep closed,
arrange $N > 2$ switches physically in parallel. Then the circuit relia-

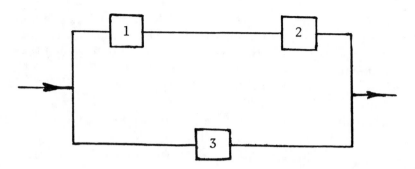

Fig. 9.5 — Physical arrangement of the three switches of Problem 9-1.

bility is given by

$$R_{NPS,SW/o}(t) = 1 - [1 - R_{S,SW/o}(t)]^N. \qquad (9.17)$$

To open and keep open, arrange $N > 2$ switches physically in series. Then the circuit reliability is given by

$$R_{NSS,SW/c}(t) = 1 - [1 - R_{S,SW/c}(t)]^N. \qquad (9.18)$$

PROBLEMS

9-1. Given the three-unit switching circuit of Fig. 9.5, determine the circuit's reliability using (1) the generalized *Bayes theorem* approach and (2) the *tabular multimode function* approach, for the following situations:

The three units are *static switches* and they are *normally open*. The circuit *succeeds* when a *signal passes on command*.

Consider the following function modes: *Normal (n)*; *failed open (o)*; and *failed shut (s)*.

Use the following probabilities:

$$r_i(t) = e^{-\lambda_i t_s} \times \frac{N_{isn}}{N_{iT}} \times e^{-\lambda_{io/sn}(t - t_s)},$$

$$q_i(t) = e^{-\lambda_i t_s} \times \frac{N_{ifo}}{N_{iT}} \times e^{-\lambda_{is/fo}(t - t_s)},$$

and

$$q_{is}(t) = e^{-\lambda_i t_s} \times \frac{N_{ifs}}{N_{iT}} \times e^{-\lambda_{io/fs}(t - t_s)}.$$

9-2.   Given three physically in parallel static switches which are normally open, determine the reliability of this arrangement when success requires that a signal be allowed to go through when they are commanded to close at $t_s = 2$ hr after the start of the mission, and it is required that the surviving switches stay closed for the rest of the mission. The mission duration is $t = 10$ hr.

λ = failure rate of the switch when normally open, 0.05 fr/10^6 hr,

q_c = probability of failing closed on command, given it is normally open, 0.01,

q_o = probability of failing open on command, given it is normally open, 0.005,

λ_{on} = failing open failure rate, given it closed normally after command, 0.008 fr/10^6 hr,

and

λ_{oc} = failing open failure rate, given it failed closed after command, 0.002 fr/10^6 hr.

9-3.   Given three physically in series static switches which are normally closed, determine the reliability of this arrangement when success requires that a signal be stopped, when the switches are commanded to open at $t_s = 2$ hr after the start of the mission, and it is required that no signal be allowed to go through during the rest of the mission. The mission duration is $t = 10$ hr.

λ = failure rate of the switch when it closes normally, 0.05 fr/10^6 hr,

q_c = probability of failing closed on command, given it closed normally, 0.01,

q_o = probability of failing open on command, given it closed normally, 0.005,

λ_{cn} = failing closed failure rate, given it opened normally after command, 0.008 fr/10^6 hr,

and

λ_{oc} = failing open failure rate, given it failed open after command, 0.002 fr/10^6 hr.

9-4.   Repeat Problem 9–2 for a similar arrangement, but of *two* parallel static swithches.

9-5.   Repeat Problem 9–3 for a similar arrangement, but of *two* series static switches.

Chapter 10

RELIABILITY OF CYCLIC SWITCHES

10.1 OBJECTIVES

A cyclic switch is defined as a switch that performs a continuous, cyclic switching on-and-off operation during a mission. Semiconductor rectifying diodes which perform a switching operation every half-cycle, choppers, cyclic relays, flip-flops, and control valves are examples. Such switches have a multiplicity of modes of function and of failure, such as neither failed open nor failed closed, failed open, failed closed, and either in quiescent or energized mode. The quantification of the reliability of such cyclically functioning switches is covered next.

10.2 SINGLE CYCLIC SWITCH RELIABILITY

The function cycle of a single cyclic switch is shown in Fig. 10.1. A single cyclic switch succeeds when it neither fails open nor closed, because if it fails open it will not pass the desired signal or flow, and if it fails closed it will not cycle. Consequently, the reliability of a single cyclic switch is given by

$R_{C,SW}(t)$ = Probability the cyclic switch does not fail open and it does not fail closed by t_s, when it is called upon to cycle, *and* the probability it will not fail open and it will not fail closed during the rest of the mission or in $(t - t_s)$.

Consequently,

$$R_{C,SW}(t) = r_{oq}(t_s)r_{cq}(t_s)r_{oe}(t - t_s)r_{ce}(t - t_s), \qquad (10.1)$$

185

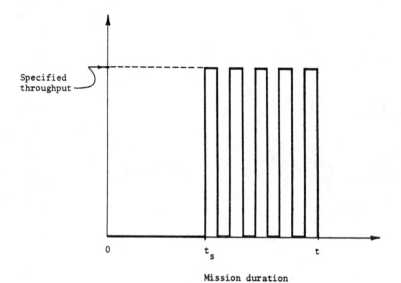

Mission duration

Fig. 10.1 – Duty cycle of a single cyclic switch.

where, for the exponential case,

$$r_{oq}(t_s) = e^{-\lambda_{oq}t_s},$$

$$\lambda_{oq} = N_{foq}(\Delta T)/(N\Delta T),$$

$N_{foq}(\Delta T)$ = number of quiescent failures in the failed open mode in ΔT,

$$r_{cq}(t_s) = e^{-\lambda_{cq}t_s},$$

$$\lambda_{cq} = N_{fcq}(\Delta T)/(N\Delta T),$$

$N_{fcq}(\Delta T)$ = number of quiescent failures in the failed closed mode in ΔT,

$$r_{oe}(t - t_s) = e^{-\lambda_{oe}(t - t_s)},$$

$$\lambda_{oe} = N_{foe}(\Delta T)/(N\Delta T),$$

$N_{foe}(\Delta T)$ = number of energized failures in the failed open mode in ΔT,

$$r_{ce}(t - t_s) = e^{-\lambda_{ce}(t - t_s)},$$

$$\lambda_{ce} = N_{fce}(\Delta T)/N\Delta T,$$

and

$N_{fce}(\Delta T)$ = number of energized failures in the failed closed mode in ΔT.

Substitution of the above into Eq. (10.1) yields

$$R_{C,SW}(t) = e^{-\lambda_{oq}t_s}e^{-\lambda_{cq}t_s}e^{-\lambda_{oe}(t - t_s)}e^{-\lambda_{ce}(t - t_s)}. \qquad (10.2)$$

It must be pointed out that the total failure rate of a cyclic switch in the quiescent failure mode is given by

$$\lambda_q = \lambda_{oq} + \lambda_{cq}. \qquad (10.3)$$

Consequently, the reliability of a cyclic switch in the quiescent function mode is given by

$$r_q(t_s) = r_{oq}(t_s)r_{cq}(t_s) = e^{-\lambda_{oq}t_s}e^{-\lambda_{cq}t_s} = e^{-(\lambda_{oq} + \lambda_{cq})t_s},$$

or

$$r_q(t_s) = e^{-\lambda_q t_s}. \qquad (10.4)$$

Similarly, for the energized function mode

$$\lambda_e = \lambda_{oe} + \lambda_{ce}, \qquad (10.5)$$

and

$$
\begin{aligned}
r_e(t - t_s) &= r_{oe}(t - t_s)r_{ce}(t - t_s), \\
&= e^{-\lambda_{oe}(t - t_s)}e^{-\lambda_{ce}(t - t_s)}, \\
&= e^{-(\lambda_{oe} + \lambda_{ce})(t - t_s)},
\end{aligned}
\qquad (10.6)
$$

or

$$r_e(t - t_s) = e^{-\lambda_e(t - t_s)}. \qquad (10.7)$$

In view of Eqs. (10.4) and (10.7), Eq. (10.2), which gives the total cyclic switch reliability, becomes

$$R_{C,SW}(t) = e^{-\lambda_q t_s}e^{-\lambda_e(t - t_s)}. \qquad (10.8)$$

As a special case, if the cyclic switch has to start functioning at the beginning of the mission, then $t_s = 0$ and Eq. (10.8) becomes

$$R_{C,SW}(t) = e^{-\lambda_e t}.$$ (10.9)

It must be kept in mind that for cyclic switches the λ_{ei} are in failures per cycle of operation and t is in cycles of operation during cyclic function, but in fr/hr during the quiescent function mode and t_s is in hr.

10.3   UNRELIABILITY OF CYCLIC SWITCHES IN FAILING OPEN OR FAILING CLOSED MODE

The unreliability of a cyclic switch in the failing open mode is given by

$q_o(t)$ = probability the cyclic switch fails opengiven that it has not already failed closed.

Consequently,

$$q_o(t) = \int_{t_1=0}^{t_1=t} f_o(t_1)e^{-\lambda_c t_1} \, dt_1,$$ (10.10)

where

$f_o(t_1)$ = probability density function of the times, t_1, to failing open of the cyclic switch,

λ_c = failure rate of the cyclic switch in the failing closed mode,

and

t_1 = variable time to failing open.

The limits of integration in Eq. (10.10) are from 0 to t because the switch may fail open from a split second after the switch is energized up to a split second before the mission is completed, thereby failing open practically any time during the mission. Equation (10.10), therefore, gives all probabilities the cyclic switch fails open during a mission, given that it has not already failed closed.

If $f_o(t)$ is exponential, then

$$f_o(t) = \lambda_o e^{-\lambda_o t}.$$ (10.11)

Substitution of Eq. (10.11) into Eq. (10.10), and integration of Eq. (10.10) thereafter yields

$$q_o(t) = \int_0^t \lambda_o e^{-\lambda_o t_1} e^{-\lambda_c t_1} dt_1,$$

$$= \frac{\lambda_o}{\lambda_c + \lambda_o}[1 - e^{-(\lambda_o + \lambda_c)t}],$$

$$= \frac{\lambda_o}{\lambda}(1 - e^{-\lambda t}), \tag{10.12}$$

or

$$q_o(t) = \frac{\lambda_o}{\lambda} q(t). \tag{10.13}$$

Equation (10.13) says that the probability a cyclic switch fails open is the proportion of its failing open failure rate to its total failure rate times the total unreliability of the cyclic switch.

Similarly, the unreliability of a cyclic switch in the failing closed mode is given by

$q_c(t)$ = probability the cyclic switch fails closed given that it has not already failed open.

Consequently,

$$q_c(t) = \int_0^t f_c(t_1) e^{-\lambda_o t_1} dt_1, \tag{10.14}$$

where

$f_c(t_1)$ = probability density function of times, t_1, to failing closed of cyclic switches,

and

λ_o = failure rate of cyclic switch in the failing open mode.

If $f_c(t)$ is exponential, then

$$f_c(t) = \lambda_c e^{-\lambda_c t}. \tag{10.15}$$

Substitution of Eq. (10.15) into Eq. (10.14) and integration of Eq. (10.14) thereafter yields

$$q_c(t) = \int_0^t \lambda_c e^{-\lambda_c(t_1)} e^{-\lambda_o t_1)} dt_1,$$

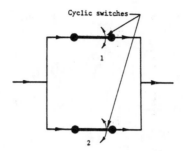

Fig. 10.2 – Two cyclic switches arranged physically in parallel.

$$q_c(t) = \frac{\lambda_c}{\lambda_c + \lambda_o}[1 - e^{-(\lambda_c + \lambda_o)t}],$$

$$q_c(t) = \frac{\lambda_c}{\lambda}(1 - e^{-\lambda t}), \tag{10.16}$$

or

$$q_c(t) = \frac{\lambda_c}{\lambda}q(t). \tag{10.17}$$

Equation (10.17) says that the probability a cyclic switch fails closed is equal to the proportion of its failing closed failure rate to its total failure rate times the total unreliability of the cyclic switch.

The total unreliability of the cyclic switch, as failing open and failing closed are mutually exclusive events, is given by

$$q(t) = q_o(t) + q_c(t), \tag{10.18}$$

and

$$r(t) + q(t) = 1. \tag{10.19}$$

10.4   CYCLIC SWITCHES PHYSICALLY IN PARALLEL

If the reliability of cyclic switching needs to be increased, then parallel redundancy may be used. If two cyclic switches are arranged physically in parallel, as shown in Fig. 10.2, the reliability of this arrangement is given by

$R_{2PC,SW}(t) =$ probability both switches function normally, *or* the first switch functions normally and the second one fails open, *or* the second switch functions normally and the first one fails open.

It must be noted that if any one or both switches fail closed, then the arrangement cannot cycle; consequently, it fails. Quantitatively, the reliability of this arrangement is given by

$$R_{2PC,SW}(t) = r_1(t)r_2(t) + r_1(t)q_{2o}(t) + q_{1o}(t)r_2(t). \tag{10.20}$$

Substitution of Eq. (10.12) into Eq. (10.20) yields

$$R_{2PC,SW}(t) = e^{-\lambda_1 t}e^{-\lambda_2 t} + \frac{\lambda_{2o}}{\lambda_2}q_2(t)e^{-\lambda_1 t} + \frac{\lambda_{1o}}{\lambda_1}q_1(t)e^{-\lambda_2 t}, \tag{10.21}$$

where

$$q_i(t) = 1 - r_i(t) = 1 - e^{-\lambda_i t},$$

$$\tag{10.22}$$

and $i = 1$ and 2.

10.4.1 SPECIAL CASES

The following special cases may be considered: 1. If the switches are equal, then

$$\lambda_{1o} = \lambda_{2o} = \lambda_o \text{ and } \lambda_1 = \lambda_2 = \lambda.$$

Consequently, Eq. (10.21) becomes

$$R_{2PC,SW}(t) = e^{-2\lambda t} + \frac{2\lambda_o}{\lambda_2}e^{-\lambda t}(1 - e^{-\lambda t}), \tag{10.23}$$

or after rearrangement

$$R_{2PC,SW}(t) = e^{-2\lambda t} + \frac{2\lambda_o}{\lambda}(e^{-\lambda t} - e^{-2\lambda t}). \tag{10.24}$$

2. If the failure rate of failing open is equal to the failure rate of failing closed, then $\lambda_o = \lambda_c$, and

$$\lambda = \lambda_o + \lambda_c = \lambda_o + \lambda_o = 2\lambda_o. \tag{10.25}$$

Consequently, Eq. (10.24) becomes

$$R_{2PC,SW}(t) = e^{-\lambda t}. \tag{10.26}$$

Equation (10.26) says that *nothing is gained by physically paralleling two equal cyclic switches if* $\lambda_o = \lambda_c$, i.e., when the failing open failure rate is equal to the failing closed failure rate. This is so because the reliability of this arrangement is equal to that of a single cyclic switch; consequently, money and spares are wasted, and weight and volume are added needlessly. Therefore, such switches should not be used physically in parallel!

3. If $\lambda_o = 0$, or the cyclic switch does not fail open but *only fails closed*, then Eq. (10.24) becomes

$$R_{2PC,SW}(t) = e^{-2\lambda t}. \tag{10.27}$$

Equation (10.27) says that the *two cyclic switches which are arranged physically in parallel are acting as if they were reliabilitywise in series*, if $\lambda_o = 0$; hence this arrangement gives a reliability that is worse than a single cyclic switch!

4. If $\lambda_c = 0$, or the cyclic switch does not fail closed but only fails open, then $\lambda = \lambda_o$ and Eq. (10.24) becomes

$$R_{2PC,SW}(t) = 2e^{-\lambda t} - e^{-2\lambda t}. \tag{10.28}$$

Equation (10.28) says that the two cyclic switches which are arranged physically in parallel are acting as if they were reliabilitywise in parallel also, hence providing the highest reliability.

It may be concluded from the preceding analysis that for higher reliability of a circuit with *two switches physically in parallel the failing closed failure rate of cyclic switches should be minimized.* Here is where reliability engineering helps the design engineer by telling the design engineer to design the cyclic switches so that they do not fail closed.

10.5   CYCLIC SWITCHES PHYSICALLY IN SERIES

If two cyclic switches are arranged physically in series, as shown in Fig. 10.3, the reliability of this arrangement is given by

$R_{2SC,SW}(t)$ = probability both switches function normally, *or* the first switch functions normally and the second one fails closed, *or* the second switch functions normally and the first one fails closed.

If any one or both switches fail open, then the arrangement cannot transmit a signal or a flow; consequently, it fails. Quantitatively, the reliability of this arrangement is given by

$$R_{2SC,SW}(t) = r_1(t)r_2(t) + r_1(t)q_{2c}(t) + q_{1c}(t)r_2(t). \tag{10.29}$$

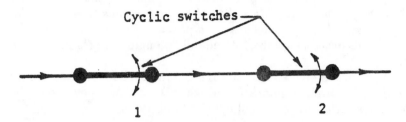

Fig. 10.3 – Two cyclic switches arranged physically in series.

Substitution of Eq. (10.16) into Eq. (10.29) yields

$$R_{2SC,SW}(t) = e^{-\lambda_1 t}e^{-\lambda_2 t} + \frac{\lambda_{2c}}{\lambda_2}e^{-\lambda_1 t}(1 - e^{-\lambda_2 t})$$
$$+ \frac{\lambda_{1c}}{\lambda_1}(1 - e^{-\lambda_1 t})e^{-\lambda_2 t}. \tag{10.30}$$

10.5.1 SPECIAL CASES

The following special cases may be considered:

1. If the switches are equal, then Eq. (10.30) becomes

$$R_{2SC,SW}(t) = e^{-2\lambda t} + \frac{2\lambda_c}{\lambda}e^{-\lambda t}(1 - e^{-\lambda t}). \tag{10.31}$$

2. If $\lambda_o = \lambda_c$, then $\lambda = 2\lambda_c$ and Eq. (10.31) becomes

$$R_{2SC,SW}(t) = e^{-\lambda t}. \tag{10.32}$$

Equation (10.32) says that nothing is gained by using two such cyclic switches physically in series.

3. If $\lambda_c = 0$, then Eq. (10.31) becomes

$$R_{2SC,SW}(t) = e^{-2\lambda t}. \tag{10.33}$$

Equation (10.33) says that the two cyclic switches, physically in series, are acting as if they were reliabilitywise in series also; hence it is a worse arrangement than a single switch!

4. If $\lambda_o = 0$, then $\lambda = \lambda_c$ and Eq. (10.31) becomes

$$R_{2SC,SW}(t) = 2e^{-\lambda t} - e^{-2\lambda t}. \tag{10.34}$$

Equation (10.34) says that the two cyclic switches, physically in series, are acting as if they were reliabilitywise in parallel, hence providing the highest reliability.

It may be concluded from the preceding analysis that for higher reliability the *failing open failure rate of cyclic switches should be minimized if the switches are to be arranged physically* in series. Again the benefits to the design engineer of reliability engineering analysis are obvious.

As a rule, when $\lambda_o \gg \lambda_c$, the redundant cyclic switches should be *arranged physically in parallel,* and when $\lambda_o \ll \lambda_c$, they should be *arranged physically in series* for higher reliability.

10.6   COMPLEX SYSTEMS WITH CYCLICALLY FUNCTIONING UNITS

For the successful function of complex systems containing cyclically functioning units, it should be determined which failed function mode of a unit in the system leads to the system's success. For example, in Case 1 of Table 10.1, for system success it is required that a cyclic unit fail closed, not open, whereas in Case 2 it is required that a cyclic unit fail open for system success. Using the concepts developed earlier, the reliability and the mean life of 14 such complex systems are presented in Table 10.1. In this table the following nomenclature is used:

   r = probability of functioning normally or successfully for the function period intended, $r(t)$, or the reliability of each unit,

   q = total probability of failure , $q(t)$, $q = q_o + q_c$,

   q_o = probability of failing open, $q_o(t)$, as given by Eq. (10.13), or $q_o = (\lambda_o/\lambda)q(t) = (\lambda_o/\lambda)q$,

   q_c = probability of failing closed, $q_c(t)$, as given by Eq. (10.17), or $q_c = (\lambda_c/\lambda)q(t) = (\lambda_c/\lambda)q$,

and

   m = mean life of each unit in the system where all units are assumed to be equal and have a constant failure rate.

Two examples follow where Cases 5 and 9 of Table 10.1 are derived to provide an idea of how the reliability and mean life of the remaining cases have been determined.

EXAMPLE 10–1

For Case 5 of Table 10.1, do the following:

Case number	Circuit	Reliability of circuit	Mean life of circuit
1		$r^2 + 2\dfrac{\lambda_c}{\lambda} rq$	$\left(1.5 - \dfrac{\lambda_c}{\lambda}\right) m$
2		$r^2 + 2\dfrac{\lambda_o}{\lambda} rq$	$\left(1.5 - \dfrac{\lambda_o}{\lambda}\right) m$
3		$r^3 + 3\dfrac{\lambda_o}{\lambda} r^2 q + 3\left(\dfrac{\lambda_o}{\lambda}\right)^2 rq^2$	$\left[2 + 3\dfrac{\lambda_o}{\lambda} + 6\left(\dfrac{\lambda_o}{\lambda}\right)^2\right]\dfrac{m}{6}$
4		$r^3 + 3\dfrac{\lambda_c}{\lambda} r^2 q + 3\left(\dfrac{\lambda_c}{\lambda}\right)^2 rq^2$	$\left[2 + 3\dfrac{\lambda_c}{\lambda} + 6\left(\dfrac{\lambda_c}{\lambda}\right)^2\right]\dfrac{m}{6}$
5		$r^3 + \left(2 + \dfrac{\lambda_o}{\lambda}\right)r^2 q + \left[4\dfrac{\lambda_o}{\lambda} - 3\left(\dfrac{\lambda_o}{\lambda}\right)^2\right] rq^2$	$\left[\dfrac{2}{3} + \dfrac{3}{2}\left(\dfrac{\lambda_o}{\lambda}\right) - \left(\dfrac{\lambda_o}{\lambda}\right)^2\right] m$
6		$r^3 + \left(2 + \dfrac{\lambda_c}{\lambda}\right)r^2 q + \left[4\dfrac{\lambda_c}{\lambda} - 3\left(\dfrac{\lambda_c}{\lambda}\right)^2\right] rq^2$	$\left[\dfrac{2}{3} + \dfrac{3}{2}\left(\dfrac{\lambda_c}{\lambda}\right) - \left(\dfrac{\lambda_c}{\lambda}\right)^2\right] m$

TABLE 10.1–(Continued)

Case number	Circuit	Reliability of circuit	Mean life of circuit
7		$r^4 + 4\frac{\lambda_o}{\lambda}r^3 q + 6\left(\frac{\lambda_o}{\lambda}\right)^2 r^2 q^2 + 4\left(\frac{\lambda_o}{\lambda}\right)^3 rq^3$	$\left[\frac{1}{4} + \frac{1}{3}\left(\frac{\lambda_o}{\lambda}\right) + \frac{1}{2}\left(\frac{\lambda_o}{\lambda}\right)^2 + \left(\frac{\lambda_o}{\lambda}\right)^3\right]m$
8		$r^4 + 4\frac{\lambda_c}{\lambda}r^3 q + 6\left(\frac{\lambda_c}{\lambda}\right)^2 r^2 q^2 + 4\left(\frac{\lambda_c}{\lambda}\right)^3 rq^3$	$\left[\frac{1}{4} + \frac{1}{3}\left(\frac{\lambda_c}{\lambda}\right) + \frac{1}{2}\left(\frac{\lambda_c}{\lambda}\right)^2 + \left(\frac{\lambda_c}{\lambda}\right)^3\right]m$
9		$r^4 + 4r^3 q + \left[2+8\frac{\lambda_o}{\lambda} - 6\left(\frac{\lambda_o}{\lambda}\right)^2\right]r^2 q^2 + 4\left[\frac{\lambda_o}{\lambda} - \left(\frac{\lambda_o}{\lambda}\right)^3\right]rq^3$	$\left[\frac{3}{4} + \frac{5}{3}\left(\frac{\lambda_o}{\lambda}\right) - \frac{1}{2}\left(\frac{\lambda_o}{\lambda}\right)^2 - \left(\frac{\lambda_o}{\lambda}\right)^3\right]m$
10		$r^4 + 4r^3 q + \left[2+8\frac{\lambda_c}{\lambda} - 6\left(\frac{\lambda_c}{\lambda}\right)^2\right]r^2 q^2 + 4\left[\frac{\lambda_c}{\lambda} - \left(\frac{\lambda_c}{\lambda}\right)^3\right]rq^3$	$\left[\frac{3}{4} + \frac{5}{3}\left(\frac{\lambda_c}{\lambda}\right) - \frac{1}{2}\left(\frac{\lambda_c}{\lambda}\right)^2 - \left(\frac{\lambda_c}{\lambda}\right)^3\right]m$
11		$r^4 + \left(3+\frac{\lambda_o}{\lambda}\right)r^3 q + 3\left[1+\frac{\lambda_o}{\lambda} - \left(\frac{\lambda_o}{\lambda}\right)^2\right]r^2 q^2 + \left[6\frac{\lambda_o}{\lambda} - 9\left(\frac{\lambda_o}{\lambda}\right)^2 + 4\left(\frac{\lambda_o}{\lambda}\right)^3\right]rq^3$	$\left[\frac{3}{4} + \frac{11}{6}\left(\frac{\lambda_o}{\lambda}\right) - \frac{5}{2}\left(\frac{\lambda_o}{\lambda}\right)^2 + \left(\frac{\lambda_o}{\lambda}\right)^3\right]m$

TABLE 10.1– (Continued)

Case number	Circuit	Reliability of circuit	Mean life of circuit
12		$r^4 + \left(3+\frac{\lambda_c}{\lambda}\right)r^3 q + 3\left[1+\frac{\lambda_c}{\lambda}-\left(\frac{\lambda_c}{\lambda}\right)^2\right]r^2 q^2 + \left[6\frac{\lambda_c}{\lambda}-9\left(\frac{\lambda_c}{\lambda}\right)^2+4\left(\frac{\lambda_c}{\lambda}\right)^3\right]rq^3$	$\left[\frac{3}{4}+\frac{11}{6}\left(\frac{\lambda_c}{\lambda}\right)-\frac{5}{2}\left(\frac{\lambda_c}{\lambda}\right)^2+\left(\frac{\lambda_c}{\lambda}\right)^3\right]m$
13		$r^4 + \left(2+2\frac{\lambda_o}{\lambda}\right)r^3 q + 6\frac{\lambda_o}{\lambda}r^2 q^2 + \left[6\left(\frac{\lambda_o}{\lambda}\right)^2-4\left(\frac{\lambda_o}{\lambda}\right)^3\right]rq^3$	$\left[\frac{5}{12}+\frac{2}{3}\left(\frac{\lambda_o}{\lambda}\right)+\frac{3}{2}\left(\frac{\lambda_o}{\lambda}\right)^2-\left(\frac{\lambda_o}{\lambda}\right)^3\right]m$
14		$r^4 + \left(2+2\frac{\lambda_c}{\lambda}\right)r^3 q + 6\frac{\lambda_c}{\lambda}r^2 q^2 + \left[6\left(\frac{\lambda_c}{\lambda}\right)^2-4\left(\frac{\lambda_c}{\lambda}\right)^3\right]rq^3$	$\left[\frac{5}{12}+\frac{2}{3}\left(\frac{\lambda_c}{\lambda}\right)+\frac{3}{2}\left(\frac{\lambda_c}{\lambda}\right)^2-\left(\frac{\lambda_c}{\lambda}\right)^3\right]m$

r = reliability of each unit,

$m = \int_0^\infty r\, dt = \frac{1}{\lambda}$, mean life of each unit,

$\lambda = \lambda_o + \lambda_c$, the total failure rate of each unit,

λ_o = failure rate in the failing open mode,

λ_c = failure rate in the failing closed mode.

197

1. Derive the equation for the reliability of the system.

2. Derive the equation for the mean life of the system.

SOLUTIONS TO EXAMPLE 10–1

1. The derivation of the equation for the reliability of the system in Case 5 of Table 10.1 is as follows: The system may be divided into two subsystems:

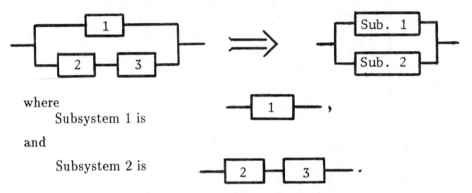

where
  Subsystem 1 is

and
  Subsystem 2 is

Then the reliability of the system can be expressed as

$$
\begin{aligned}
R_S \;=\;\; & P(\text{Both Sub. 1 and Sub. 2 succeed}) \\
& + P(\text{Sub. 1 succeeds and Sub. 2 fails open}) \\
& + P(\text{Sub. 2 succeeds and Sub. 1 fails open}),
\end{aligned}
$$

or

$$
\begin{aligned}
R_S \;=\;\; & P(\text{Sub. 1 succeeds}) \cdot P(\text{Sub. 2 succeeds}) \\
& + P(\text{Sub. 1 succeeds}) \cdot P(\text{Sub. 2 fails open}) \\
& + P(\text{Sub. 1 fails open}) \cdot P(\text{Sub. 2 succeeds}). \quad (10.35)
\end{aligned}
$$

Individually,

$$
P(\text{Sub. 2 succeeds}) \;=\; r^2 + 2\frac{\lambda_c}{\lambda} qr \text{ (from Case 1)}, \qquad (10.36)
$$

$$
\begin{aligned}
P(\text{Sub. 2 fails open}) \;=\;\; & 2P(\text{one succeeds and the other fails open}) \\
& + P(\text{both fail open}) \\
& + 2P(\text{one fails closed and the other fails} \\
& \quad\ \text{open}),
\end{aligned}
$$

or

$$P(\text{Sub. 2 fails open}) \;=\; 2(r\frac{\lambda_o}{\lambda}q) + (\frac{\lambda_o}{\lambda}q)^2 + 2(\frac{\lambda_c\lambda_o}{\lambda^2}q^2). \tag{10.37}$$

$$P(\text{Sub. 1 succeeds}) \;=\; r, \tag{10.38}$$

and

$$P(\text{Sub. 1 fails open}) \;=\; \frac{\lambda_o}{\lambda}q. \tag{10.39}$$

Substituting Eqs. (10.36) through (10.39), into Eq. (10.35) yields

$$
\begin{aligned}
R_S \;=\;& r(r^2 + 2\frac{\lambda_c}{\lambda}qr) + r[2(r\frac{\lambda_o}{\lambda}q) + (\frac{\lambda_o}{\lambda})^2q^2 + 2(\frac{\lambda_c\lambda_o}{\lambda^2}q)^2] \\
& + (\frac{\lambda_o}{\lambda}q)(r^2 + 2\frac{\lambda_c}{\lambda}qr), \\
\;=\;& r^3 + (2\frac{\lambda_c}{\lambda} + 3\frac{\lambda_o}{\lambda})r^2q + [2\frac{\lambda_c\lambda_o}{\lambda_2} + (\frac{\lambda_o}{\lambda})^2 + 2\frac{\lambda_c\lambda_o}{\lambda^2}]rq^2.
\end{aligned}
$$

Considering $\lambda_c + \lambda_o = \lambda$, then

$$R_S \;=\; r^3 + (2\frac{\lambda - \lambda_o}{\lambda} + 3\frac{\lambda_o}{\lambda})r^2q + [4\frac{(\lambda - \lambda_o)\lambda_o}{\lambda^2} + (\frac{\lambda_o}{\lambda})^2]rq^2,$$

or

$$R_S \;=\; r^3 + (2 + \frac{\lambda_o}{\lambda})r^2q + [4\frac{\lambda_o}{\lambda} - 3(\frac{\lambda_o}{\lambda})^2]rq^2. \tag{10.40}$$

2. The derivation of the equation for the mean life of the system in Case 5 of Table 10.1 is as follows:

$$
\begin{aligned}
m_S \;=\;& \int_0^\infty R_S\,dt, \\
\;=\;& \int_0^\infty \left\{ r^3 + (2 + \frac{\lambda_o}{\lambda})r^2q + \left[4\frac{\lambda_o}{\lambda} - 3(\frac{\lambda_o}{\lambda})^2\right]rq^2 \right\}dt,
\end{aligned}
$$

or

$$m_S = \int_0^\infty r^3\,dt + (2 + \frac{\lambda_o}{\lambda})\int_0^\infty r^2q\,dt + [4\frac{\lambda_o}{\lambda} - 3(\frac{\lambda_o}{\lambda})^2]\int_0^\infty rq^2\,dt, \tag{10.41}$$

where

$$r = r(t) = e^{-\lambda t},$$

and

$$q = q(t) = 1 - e^{-\lambda t}.$$

Then, for the first term of Eq. (10.41)

$$\int_0^\infty r^3\,dt = \int_0^\infty e^{-3\lambda t}\,dt = \frac{1}{3\lambda} = \frac{m}{3}, \tag{10.42}$$

where

$m = \frac{1}{\lambda}$, mean life of each unit.

For the second term of Eq. (10.41),

$$\int_0^\infty r^2 q\, dt = \int_0^\infty e^{-2\lambda t}(1 - e^{-\lambda t})\, dt = \int_0^\infty e^{-2\lambda t}\, dt - \int_0^\infty e^{-3\lambda t}\, dt,$$

$$= \frac{1}{2\lambda} - \frac{1}{3\lambda} = \frac{m}{2} - \frac{m}{3},$$

or

$$\int_0^\infty r^2 q\, dt = \frac{m}{6}. \qquad (10.43)$$

For the third term of Eq. (10.41),

$$\int_0^\infty r q^2\, dt = \int_0^\infty e^{-\lambda t}(1 - e^{-\lambda t})^2\, dt,$$

$$= \int_0^\infty (e^{-\lambda t} - 2e^{-2\lambda t} + e^{-3\lambda t})\, dt,$$

$$= \int_0^\infty e^{-\lambda t}\, dt - \int_0^\infty 2e^{-2\lambda t}\, dt + \int_0^\infty e^{-3\lambda t}\, dt,$$

$$= \frac{1}{\lambda} - \frac{2}{2\lambda} + \frac{1}{3\lambda} = \frac{1}{3\lambda},$$

or

$$\int_0^\infty r q^2\, dt = \frac{m}{3}. \qquad (10.44)$$

Substituting Eqs. (10.42), (10.43) and (10.44) into Eq. (10.41) yields

$$m_S = \frac{m}{3} + (2 + \frac{\lambda_o}{\lambda})\frac{m}{6} + [4\frac{\lambda_o}{\lambda} - 3(\frac{\lambda_o}{\lambda})^2]\frac{m}{3},$$

or

$$m_S = [\frac{2}{3} + \frac{3}{2}(\frac{\lambda_o}{\lambda}) - (\frac{\lambda_o}{\lambda})^2]m. \qquad (10.45)$$

EXAMPLE 10-2

For Case 9 of Table 10.1, do the following:

1. Derive the equation for the reliability of the system.

2. Derive the equation for the mean life of the system.

SOLUTIONS TO EXAMPLE 10-2

1. The derivation of the equation for the reliability of the system in Case 9 of Table 10.1 is as follows: The system may be divided into two subsystems:

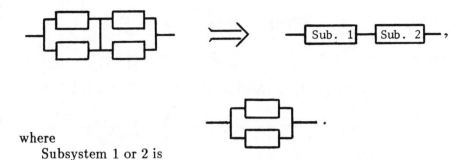

where
  Subsystem 1 or 2 is

Then the reliability of the system can be expressed as

$$R_S = P(\text{Both Sub. 1 and Sub. 2 succeed}) + P(\text{any one of them succeeds and the other one fails closed}),$$

or

$$R_S = P(\text{Sub. 1 succeeds}) \cdot P(\text{Sub. 2 succeeds}) + 2P(\text{one succeeds}) \cdot P(\text{other fails closed}). \quad (10.46)$$

Individually,

$$P(\text{Sub. 1 succeeds}) = P(\text{Sub. 2 succeeds}),$$
$$= r^2 + 2\frac{\lambda_o}{\lambda}rq \text{ (from Case 2)}, \quad (10.47)$$

$$P(\text{Sub. 1 fails closed}) = P(\text{Sub. 2 fails closed}),$$
$$= 2P(\text{one succeeds and the other fails closed})$$
$$+ P(\text{both fail closed})$$
$$+ 2P(\text{one fails closed and the other fails open}),$$

or

$$P(\text{Sub. 1 fails closed}) = 2r(\frac{\lambda_o}{\lambda})q + (\frac{\lambda_c}{\lambda})^2 q^2 + 2(\frac{\lambda_c\lambda_o}{\lambda^2})q^2. \quad (10.48)$$

Substituting Eqs. (10.47) and (10.48) into Eq. (10.46) yields

$$R_S = (r^2 + 2\frac{\lambda_o}{\lambda}rq)^2$$

$$+ 2(r^2 + 2\frac{\lambda_o}{\lambda}rq)[2r(\frac{\lambda_c}{\lambda})q + (\frac{\lambda_c}{\lambda})^2 q^2 + 2(\frac{\lambda_c \lambda_o}{\lambda^2})q^2],$$

$$= r^4 + 4\frac{\lambda_o}{\lambda}r^3 q + 4(\frac{\lambda_o}{\lambda})^2 r^2 q^2 + 2(\frac{\lambda_c}{\lambda})^2 r^2 q^2 + 4(\frac{\lambda_o \lambda_c}{\lambda^2})r^2 q^2$$

$$+ 4(\frac{\lambda_o}{\lambda})r^3 q + 4(\frac{\lambda_o}{\lambda})(\frac{\lambda_c}{\lambda})^2 rq^3 + 8(\frac{\lambda_c}{\lambda})(\frac{\lambda_o}{\lambda})^2 rq^3 + 8(\frac{\lambda_o \lambda_c}{\lambda})r^2 q^2,$$

$$= r^4 + (4\frac{\lambda_o}{\lambda} + 4\frac{\lambda_c}{\lambda})r^3 q + [4(\frac{\lambda_o}{\lambda})^2 + 2(\frac{\lambda_c}{\lambda})^2 + 12(\frac{\lambda_c \lambda_o}{\lambda^2})]r^2 q^2$$

$$+ [4\frac{\lambda_o}{\lambda}(\frac{\lambda_c}{\lambda})^2 + 8\frac{\lambda_c}{\lambda}(\frac{\lambda_o}{\lambda})^2]rq^3,$$

or

$$R_S = r^4 + 4r^3 q + [2 + 8\frac{\lambda_o}{\lambda} - 6(\frac{\lambda_o}{\lambda})^2]r^2 q^2 + 4[\frac{\lambda_o}{\lambda} - (\frac{\lambda_o}{\lambda})^3]rq^3. \quad (10.49)$$

2. The derivation of the equation for the mean life of the system in Case 9 of Table 10.1 is as follows:

$$m_S = \int_0^\infty R_S \, dt,$$

$$= \int_0^\infty \{r^4 + 4r^3 q + [2 + 8\frac{\lambda_o}{\lambda} - 6(\frac{\lambda_o}{\lambda})^2]r^2 q^2$$

$$+ 4[\frac{\lambda_o}{\lambda} - (\frac{\lambda_o}{\lambda})^3]rq^3\} \, dt,$$

or

$$m_S = \int_0^\infty r^4 \, dt + 4\int_0^\infty r^3 q \, dt + [2 + 8\frac{\lambda_o}{\lambda} - 6(\frac{\lambda_o}{\lambda})^2]\int_0^\infty r^2 q^2 \, dt$$

$$+ 4[\frac{\lambda_o}{\lambda} - (\frac{\lambda_o}{\lambda})^3]\int_0^\infty rq^3 \, dt \quad (10.50)$$

For the first term of Eq. (10.50),

$$\int_0^\infty r^4 \, dt = \int_0^\infty e^{-4\lambda t} \, dt = \frac{1}{4\lambda} = \frac{m}{4}.$$

$$(10.51)$$

For the second term of Eq. (10.50),

$$\int_0^\infty r^3 q \, dt = \int_0^\infty e^{-3\lambda t}(1 - e^{-\lambda t}) \, dt = \int_0^\infty e^{-3\lambda t} \, dt - \int_0^\infty e^{-4\lambda t} \, dt,$$

$$= \frac{1}{3\lambda} - \frac{1}{4\lambda} = \frac{1}{12\lambda},$$

or

$$\int_0^\infty r^3 q \, dt \;=\; \frac{m}{12}. \tag{10.52}$$

For the third term of Eq. (10.50),

$$
\begin{aligned}
\int_0^\infty r^2 q^2 \, dt &= \int_0^\infty e^{-2\lambda t}(1 - e^{-\lambda t})^2 \, dt, \\
&= \int_0^\infty e^{-2\lambda t} \, dt - 2\int_0^\infty e^{-3\lambda t} \, dt + \int_0^\infty e^{-4\lambda t} \, dt, \\
&= \frac{1}{2\lambda} - \frac{2}{3\lambda} + \frac{1}{4\lambda} = \frac{1}{12\lambda},
\end{aligned}
$$

or

$$\int_0^\infty r^2 q^2 \, dt \;=\; \frac{m}{12}. \tag{10.53}$$

For the fourth term of Eq. (10.50),

$$
\begin{aligned}
\int_0^\infty r q^3 \, dt &= \int_0^\infty e^{-\lambda t}(1 - e^{-\lambda t})^3 \, dt, \\
&= \int_0^\infty (e^{-\lambda t} - 3e^{-2\lambda t} + 3e^{-3\lambda t} - e^{-4\lambda t}) \, dt, \\
&= \frac{1}{\lambda} - \frac{3}{2\lambda} + \frac{3}{3\lambda} - \frac{1}{4\lambda} = \frac{1}{4\lambda},
\end{aligned}
$$

or

$$\int_0^\infty r q^3 \, dt \;=\; \frac{m}{4}. \tag{10.54}$$

Substituting Eqs. (10.51) through (10.54) into Eq. (10.50) yields

$$m_S \;=\; \frac{m}{4} + \frac{4m}{12} + [2 + 8\frac{\lambda_o}{\lambda} - 6(\frac{\lambda_o}{\lambda})^2]\frac{m}{12} + 4[\frac{\lambda_o}{\lambda} - (\frac{\lambda_o}{\lambda})^3]\frac{m}{4},$$

or

$$m_S \;=\; [\frac{3}{4} + \frac{5}{3}(\frac{\lambda_o}{\lambda}) - \frac{1}{2}(\frac{\lambda_o}{\lambda})^2 - \frac{\lambda_o}{\lambda}^3]m. \tag{10.55}$$

PROBLEMS

10-1. Given is a subsystem of two different cyclic switches normally closed and arranged physically in series. Consider the following failure modes: fail open de-energized, FOD; fail short de-energized, FSD; fail open energized, FOE; and fail short energized, FSE. This subsystem succeeds when a signal flow is interrupted regularly on command. Find this subsystem's reliability using Bayes' theorem.

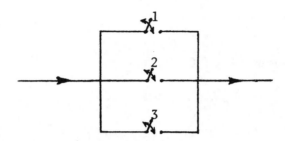

Fig. 10.4 — Schematic of the three cyclic switches of Problem 10-2.

10-2. Given the subsystem of three *equal cyclic switches* which are arranged *physically in parallel,* as shown in Fig. 10.4, find its reliability. Consider three modes of function for each switch: *normal, failed open, and failed closed.* The reliability of this system is defined as the regular and continuous passing through and interruption of a signal. Write out the subsystem's reliability in terms of the respective r'_1s, $q'_{io}s$, and $q'_{ic}s$ of all units involved, assuming the switches are exponential. Preferably use the tabular, multimode function approach to obtain the solution.

10-3. Given are *three different cyclic switches* arranged *physically in series.* Consider two failure modes for each switch: open (O) and closed (C) with the corresponding failure rates of λ_{io} and λ_{ic}. Success is defined as the regular passing and interruption of an electrical signal for a period of 0 to t.

    (1) What is the reliability of this arrangement?

    (2) What is the reliability when all three switches are identical?

    (3) What is the reliability when $\lambda_{io} = 0$?

    (4) What is the reliability when $\lambda_{ic} = 0$?

    (5) What is the reliability when $\lambda_{io} = \lambda_{ic}$?

10-4. (1) Given the three-unit switching circuit of Fig. 10.5, determine the circuit reliability using the generalized *Bayes theorem* approach and also the *tabular multimode function* approach. The three units are *cyclic switches.* The circuit *succeeds* when a signal passes and is then interrupted at a specific frequency throughout the mission.

    (2) Determine the circuit's mean time between failures.

10-5. A semiconductor switch has a failing open failure rate of $\lambda_o = 13$ failures per 10^6 cycles and a failing short failure rate of $\lambda_s = 104$ failures per 10^6 cycles. The switch averages 100 cycles per mission.

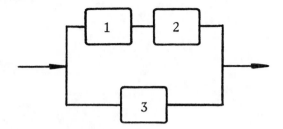

Fig. 10.5 – Physical arrangement of the three switches of
     Problem 10-4.

   (1) What arrangement of switches should be used for a reliabil-
        ity requirement of 0.980?

   (2) What arrangement do you recommend if the reliability re-
        quirement is 0.990? Calculate the reliability of the recom-
        mended arrangement.

10-6. Derive the equations in Table 10.1 for Cases 3, 4, and 6.

10-7. Derive the equations in Table 10.1 for Cases 7, 8, and 10.

10-8. Derive the equations in Table 10.1 for Cases 11 and 12.

10-9. Derive the equations in Table 10.1 for Cases 13 and 14.

Chapter 11

FAULT TREE ANALYSIS

11.1 INTRODUCTION

Fault tree analysis is a technique of reliability and safety analysis and is generally applicable to complex dynamic systems. Fault tree analysis provides an objective basis for analyzing system design, performing trade-off studies, analyzing common mode failures, demonstrating compliance with safety requirements, and justifying system changes and additions.

The concept of fault tree analysis was developed by Bell Telephone Laboratories as a technique with which to perform a safety evaluation of the Minuteman Launch Control System in 1961. Later the Boeing Company modified the concept for computer utilization. It is now widely used in many fields, such as in the nuclear reactor, chemical, and aviation industries.

A fault tree is a model that graphically and logically represents the various combinations of possible events, both faulty and normal, occurring in a system that lead to the top undesired event. Its application can include a complete plant, as well as a system or a subsystem.

Fault tree analysis is of major value in accomplishing the following [1; 2]:

1. Directing the analysis to ferret out failures deductively.

2. Pointing out the aspects of the system involved in the failure of interest.

3. Providing a graphical aid for those in system management who are removed from the system's design changes.

4. Providing options for qualitative, as well as quantitative, system reliability analysis.

5. Allowing the analyst to concentrate on one particular system failure at a time.

6. Providing the analyst with insight into system behavior.

There are three phases in the fault tree analysis process, and these and the support for these are essential to any fault tree analysis:

1. The construction of the logic block diagram, i.e., the fault tree itself. Obviously, this initially requires the definition of the system and its bounds and the definition of the undesired event.

2. The application of Boolean algebra to this logic diagram to produce definite algebraic relationships between events and series of events, i.e., relationships among the fault producing conditions. Within certain limitations the algebraic statements so obtained may then be simplified using Boolean algebra.

3. The application of probabilistic mathematics (based on set theory) and real data to these Boolean equations to determine the probability of each intermediate event and ultimately of the top event. This requires the substitution of the probability of occurrence of each event into the equations. In the case of reliability analysis, the reliability of each component or subsystem for every possible failure mode that would ultimately affect the system must be considered.

11.2   CONSTRUCTION OF THE FAULT TREE

11.2.1   THE ELEMENTS OF THE FAULT TREE

The graphical symbols in a fault tree fall basically into two categories: gate symbols and event symbols.

11.2.1.1– GATE SYMBOLS

The gate symbols are used to interconnect the events according to their causal relations. The gate symbols are given in Table 11.1.

The two most frequently used gates for developing a fault tree are the AND and OR gates. The AND gate provides an output event only if all input events occur simultaneously. On the other hand, the OR gate provides an output event if at least one of the input events occurs. Figure 11.1 gives an example for these two gates. The event "house out of lights" occurs when both events "ordinary lamps stop working" and "emergency light fails" occur simultaneously. The former event

TABLE 11.1– Fault tree gate symbols.

Name of gate	Symbol of gate	Input – output relationship
AND gate	Output • Input 12...n	The output event occurs if all of the n input events occur.
OR gate	Output + Input 12...n	The output event occurs if at least one of the n input events occurs.
m-out-of-n voting gate	Output m Input 12...n	The output event occurs if m or more out of n input events occur.
Priority AND gate	Output Input 12...n	The output event occurs if all input events occur in a certain order.
Exclusive OR gate	Output Input	The output event occurs if only one of the input events occurs.
Inhibit gate	Output Conditional event Input	The input event causes the output event only if the conditional event occurs.

may occur when at least one of the following two events occurs: "no electric power" and "lamps are broken."

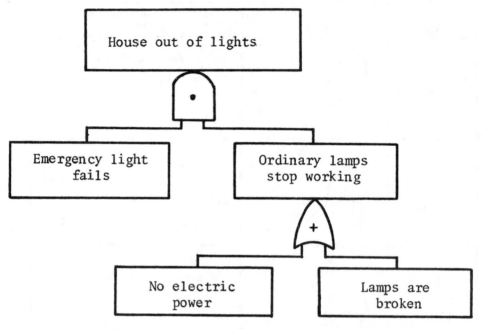

Fig. 11.1–Example of the AND and OR gates.

An *m-out-of-n voting gate* has n input events and the output event occurs if at least m out of n input events occur. This would be the case when a four-engine airplane is allowed to take off for its mission only when at least two engines work normally. If the output event is that "an airplane cannot take off" and there are four input events, i.e., "engine i does not work normally" with $i = 1, 2, 3, 4$, then a 3-out-of-4 voting gate (or 3 out of 4 engines not working) may be used to connect these events, as shown in Fig.11.2.

The *priority AND gate* is a special case of the AND gate in which the output event will occur if the n input events occur in the order from left to right. An example of the priority AND gate is the two-unit standby system. The system will fail if (a) both the active and the standby units fail, or (b) the switch fails open before the failure of the active unit. The order of function of the switch and of the active unit determines whether or not the system fails, because if the active unit fails before the failure of the switch, the system would not fail as long as the standby unit works well. Figure 11.3 shows the use of the priority AND gate.

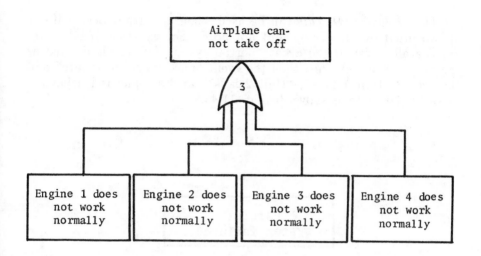

Fig. 11.2–Example of *m*-out-of-*n* gate.

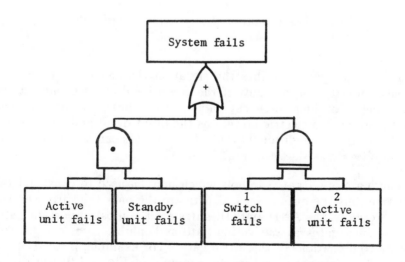

Fig. 11.3–Example of the priority AND gate.

The *exclusive OR gate* can be used when an output occurs if one of two input events occurs, but not both. For example, if in a two-unit parallel system the event "partial system failure" is defined as the failure of either unit, then the event "partial system failure" and the events "Unit 1 fails", "Unit 2 fails" can be connected using the exclusive OR gate as shown in Fig. 11.4.

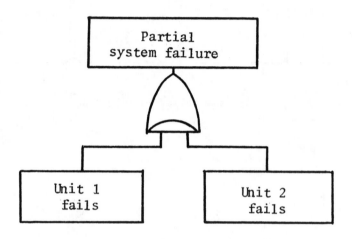

Fig. 11.4–Example of the exclusive OR gate.

The *inhibit gate* describes the situation that an input event will produce an output event only if the conditional event occurs at the same time. Consider the event "house out of lights" again; one can take the event "no electric power" as the input event and "emergency light fails" as the conditional event. These events can be connected by the inhibit gate as shown in Fig. 11.5.

There are other gate symbols which may be used to construct the fault tree, but the ones described here are the ones that are used the most. Actually, the AND and OR gates are the most important gates because such gates as the voting gate and inhibit gate, can be created by combinations of these two gates. For example, instead of the 3-out-of-4 voting gate used in the example of Fig. 11.2, the causal relationship may also be expressed as shown in Fig. 11.6. But it may be seen that Fig. 11.2 is much simpler than Fig. 11.6.

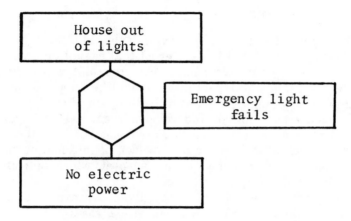

Fig. 11.5–Example of the inhibit gate.

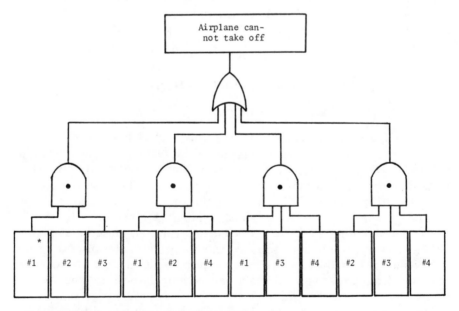

* #1 means Engine #1 does not work normally.

Fig. 11.6–The equivalent fault tree of Fig. 11.2.

11.2.1.2– EVENT SYMBOLS

The fault tree event symbols are given in Table 11.2. A circle stands for the basic fault event which is the event that cannot be decomposed further. The event parameters such as the probabilities of occurrence, the failure rates, and the repair rates, should be known for all basic events in order to obtain the quantitative solution of the fault tree.

A diamond stands for an undeveloped event, which means that at the present time the cause for this event has not been determined either due to lack of interest or due to lack of information or data. The conditional event is denoted with an oval, which indicates any condition or restriction to a logic gate.

A house symbol stands for a fault event which is expected to occur. Sometimes we want to examine various special cases of fault trees by forcing some events to occur and other events not to occur. In such instances the house event symbol may be used.

A rectangle represents a resultant event which results from the combination of fault events through the input of a logic gate.

A transfer-in triangle has a line from its top to a gate, whereas the transfer-out triangle has a line to its side from another gate. These two triangles are used to simplify the representation of a fault tree when there are some identical parts and relations in the fault tree. For example, as shown in Fig. 11.7(a), if the causal relations under gate C are identical to the causal relations under gate B, then this part of the fault tree can be simplified to that of Fig. 11.7(b).

11.2.2 FAULT TREE CONSTRUCTION

11.2.2.1– SYSTEM AND TOP EVENT DEFINITIONS

A correct fault tree analysis requires a thorough definition of the system. First, a functional layout diagram of the system of interest should be drawn to show the functional interconnections and to identify each system component. Physical system bounds are then established to focus the attention of the analyst on the precise area of interest. A common error is failure to establish realistic system bounds, which leads to a diverging analysis. Sufficient information must be available for each system component to allow the analyst to determine the necessary modes of failure of the components. This information can be obtained from experience or from the technical specifications of the components.

For any analysis to be meaningful, the boundary conditions of the system must be determined. These boundary conditions should not be confused with the physical bounds of the system. System boundary conditions define the situation for which the fault tree is to be drawn.

TABLE 11.2– Fault tree event symbols.

Symbol of event	Meaning of symbol
Circle	Basic event.
Diamond	Undeveloped event.
Oval	Conditional event.
House	Trigger event.
Rectangle	Resultant event.
Triangle (in, out)	Transfer-in and transfer-out events.

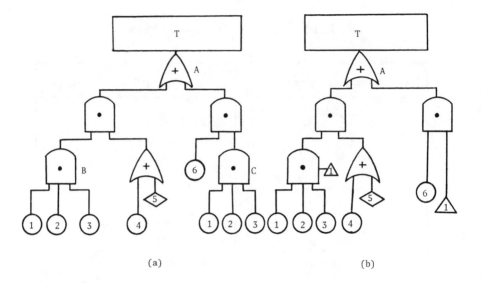

(a)   (b)

Fig. 11.7– An example of transfer-in and transfer-out event symbols.

The system's initial configuration is described by additional system boundary conditions. This configuration must represent the system in the unfailed state. Consequently, these system boundary conditions depend on the definition of the top event. Initial conditions are the system boundary conditions that define the component configurations for which the top event is applicable. All components that have more than one operating state generate initial conditions. For example, the initial quantity of fluid in a tank can be specified. The event "tank is full" becomes one initial condition, while "tank is empty" is another. The time domain must also be specified. For example, a specific electronic circuit may be considered for a specified period of time, such as the warranty period. System boundary conditions also include any fault event declared to exist, or not to be allowed for the duration of the fault tree construction. These events are called *existing* system boundary conditions. An existing system boundary condition is treated as certain to occur, and a *not allowed* system boundary condition is treated as an event with no possibility to occur. Neither existing nor not allowed system boundary condition appear as events in the final system fault tree.

The *top event* is the most important system boundary condition. It is defined as the major system failure. For any given system, a multitude of possibilities for the top event exists, so the correct choice of the top event is sometimes a difficult task. In general, the top event should be chosen as an event (1) whose occurrence must have a definite definition, and the possibility of its occurrence can be quantified, and (2) it can be further decomposed to find its cause.

11.2.2.2– CONSTRUCTION OF THE FAULT TREE

To fully model the system, the analyst must have achieved a thorough understanding of the system or should be able to work with a member of the design team. Every possible cause and effect of each failure condition should be investigated and related to the top event.

EXAMPLE 11–1

Consider the automatic air pumping system shown in Fig. 11.8 [2, p.72]. In this system the tank is filled for 10 minutes and then empties for 50 minutes. After the switch is closed, the timer is set to open the contacts in 10 minutes, and then to close the contacts in another 50 minutes, and so on. If in the process of filling the tank the contacts cannot be opened in 10 minutes the alarm horn sounds and the operator opens the switch to prevent a tank rupture due to overfilling. Prepare the fault tree for this air pumping system.

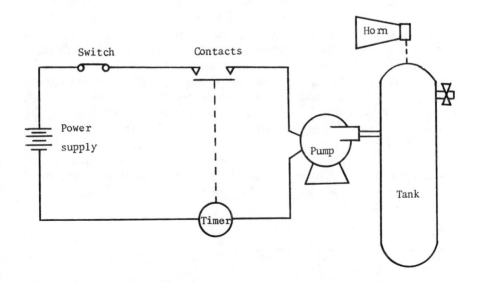

Fig. 11.8–Air pumping system for Example 11–1.

SOLUTION TO EXAMPLE 11-1

In this case, the top event is chosen to be the event of *pressure tank rupture*. Its occurrence is due to the following two causes: (1) failure of the tank itself due to improper design or manufacture; (2) over-pressure in the tank. Since at least one of these two events can cause the top event, the OR gate should be applied to connect them. But overpressure may be due to "pump operates too long," or, equivalently, "current fed to motor for too long." Now consider the event "current fed to motor for too long." It will occur if the following events occur simultaneously: "contacts closed too long" and "switch closed too long." Therefore, an AND gate should be used here. The input events of this gate are "contacts closed too long" and "switch closed too long." The output event is " current fed to motor for too long." Further analysis shows that "contacts closed too long" results from the occurrence of events " basic contacts failure" or "no command to open contacts," and "switch closed too long" results from the occurrence of events "basic switch failure" or " no command to open switch." Using similar analysis, the remainder of the fault tree may be developed as shown in Fig. 11.9.

EXAMPLE 11-2

Consider the pressure tank system shown in Fig. 11.10 [3, pp. 259 -260; 2, pp. 62-64]. In the operational mode, to start the system pumping the reset switch S_1 is closed and then opened immediately. This allows current to flow in the control branch circuit, activating the relay coils K_1 and K_2. Relay K_1 contacts are closed and latched, while relay K_2 contacts close and start the pump motor. In the shutdown mode, after approximately 20 seconds the pressure switch contacts should open (since excess pressure should be detected by the pressure switch), deactivating the control circuit, de-energizing the K_2 coil, opening the K_2 contacts, and thereby shutting the motor off. If there is a pressure switch hang up (emergency shutdown mode), the timer relay contacts should open after 60 seconds, de-energizing the K_1 coil, which in turn de-energizes the K_2 coil and shuts off the pump. The timer resets itself automatically after each cycle. Prepare the fault tree of this pressure tank system.

SOLUTION TO EXAMPLE 11-2

The top event is chosen to be the event *pressure tank rupture*, and the fault tree for this system is constructed as shown in Fig. 11.11.

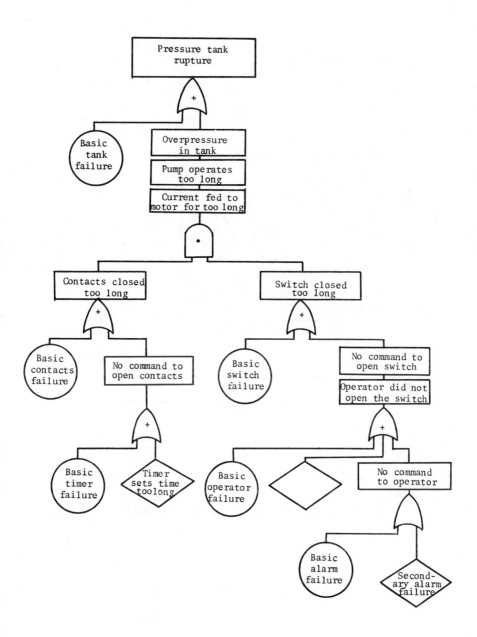

Fig. 11.9– Fault tree for the automatic air pumping system
of Example 11–1.

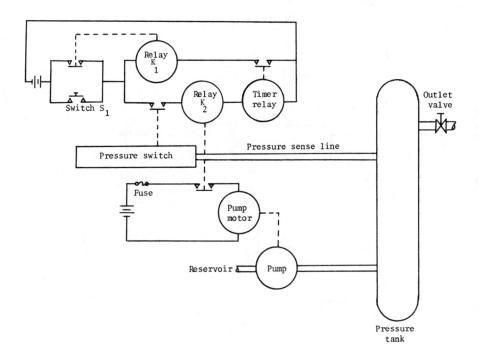

Fig. **11.10**–Pressure tank system of Example 11–2.

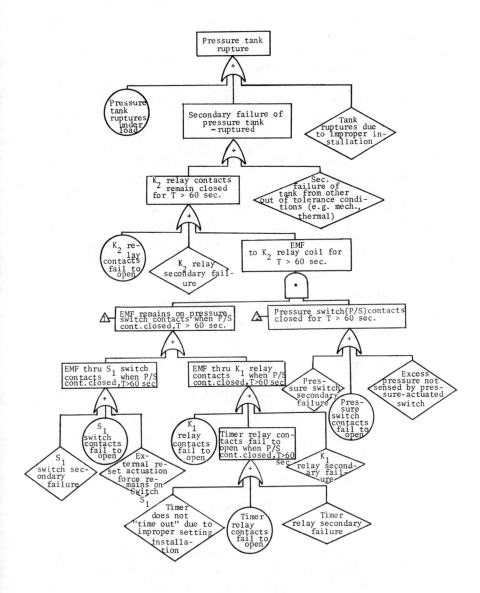

Fig. 11.11– The fault tree for the pressure tank system of
Example 11–2.

11.3 QUALITATIVE EVALUATION OF THE FAULT TREE

11.3.1 MINIMAL CUT SETS AND MINIMAL PATH SETS

A *cut set* for a fault tree is a set of basic events whose occurrence causes the top event to occur. A cut set is said to be a minimal cut set if, when any basic event is removed from the set, the remaining events collectively are no longer a cut set.

A *path set* is a dual set of the cut set. It is also a set of basic events of the fault tree, and if none of the events in this set occurs, the top event is guaranteed not to occur. A path set is minimal if, when any basic event is removed from this set, the remaining events are no longer a path set.

Consider the fault tree shown in Fig. 11.12. By the definitions given above, it can be seen that the set $\{1,2,3,4\}$ is a cut set, since, if all of the four basic events occur the top event occurs. But it is not a minimal cut set, since, if the event 4 or event 3 is removed from this set, the remaining events $\{1,2,3\}$ and $\{1,2,4\}$ are still cut sets. Actually, these two sets are minimal cut sets. The minimal path sets can be simply determined to be the sets $\{3,4\},\{1\}$, and $\{2\}$.

11.3.2 MINIMAL CUT SET ALGORITHMS

11.3.2.1– ALGORITHM 1: MOCUS [4; 2, pp. 111-113]

The MOCUS algorithm generates the minimal cut set for a fault tree in which only AND and OR gates exist. This algorithm is based on the fact that OR gates increase the number of cut sets and, on the other hand, AND gates increase the size of a cut set. The MOCUS algorithm can be described as follows:

1. Name each gate.

2. Number each basic event.

3. Locate the uppermost gate in the first row and column of a matrix.

4. Iterate either one of the fundamental permutations (a) or (b) below in a top-down fashion (when events appear in rectangles, replace them by equivalent gates or basic events).

   (4.1) Replace OR gates by a vertical arrangement of the inputs to the gates, and increase the cut sets.

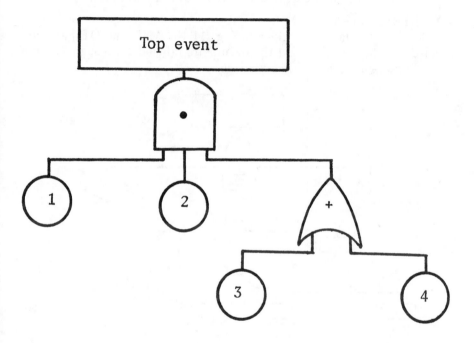

Fig. 11.12– An example for a minimal cut set and a minimal path set.

(4.2) Replace AND gates by a horizontal (separate column) arrangement of the inputs to the gates, and enlarge the size of the cut sets.

5. When all gates are replaced by basic events, obtain the minimal cut set by removing supersets, which are cut sets including some other cut sets.

The procedure is illustrated by the following examples.

EXAMPLE 11-3

Consider the fault tree given in Fig. 11.13. AND and OR gates are labeled G0 through G6, and the events are labeled 1 through 7. Find the minimal cut sets of this fault tree.

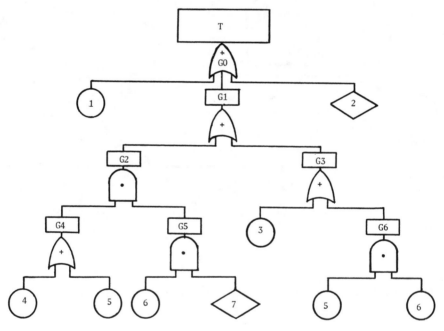

Fig. 11.13–The fault tree for Example 11–3.

SOLUTION TO EXAMPLE 11-3

The uppermost gate G0 is located in the first row and column:

$$G0.$$

Since G0 is an OR gate, it is replaced by its input events in separate rows as follows:

1

2

G1.

G1 is also an OR gate. Replacing it by its input events in separate rows yields

1

2

G2

G3.

G2 is an AND gate. It should be replaced by its input events in separate columns as follows:

1

2

G4, G5

G3.

Replacing G3 by its input events leads to

1

2

G4, G5

3

G6.

Continuing in this fashion, the total cut sets are obtained to be

$$\{1\}, \{2\}, \{4, 6, 7\}, \{5, 6, 7\}, \{3\}, \{5, 6\}$$

and are given in Column 6 of Table 11.3. The individual steps for obtaining the cut sets are shown in Table 11.3.

It may be seen that the set $\{5, 6, 7\}$ includes $\{5, 6\}$. It is a superset and should be removed, since it is not a minimal cut set. Therefore, the minimal cut sets are

$$\{1\}, \{2\}, \{4, 6, 7\}, \{3\}, \{5, 6\}.$$

TABLE 11.3 – The steps for obtaining the cut sets for the fault tree given in Fig. 11.12.

Steps	1	2	3	4	5	6
	1	1	1	1	1	1
	2	2	2	2	2	2
	G1	G2	G4,G5	G4,G5	4,G5	4,6,7
		G3	G3	3	5,G5	5,6,7
				G6	3	3
					5,6	5,6

EXAMPLE 11–4

As the second example, consider the fault tree given in Fig. 11.14, which is a relabeling of the basic events and gates in the pressure tank system fault tree given in Fig. 11.11. Find the minimal cut sets.

SOLUTION TO EXAMPLE 11–4

The AND and OR gates are labeled G0 through G8, and the events are labeled 1 through 16. Locate the uppermost gate G0 in the first row and column,

G0.

G0 is an OR gate. Replacing it by its input events in a separate row yields

1

G1

2.

Since G1 is an OR gate, then it is again replaced by its input events in separate rows as follows:

1

G2

3

2.

As G2 is also an OR gate, it is replaced by its input events as follows:

1

4

5

G3

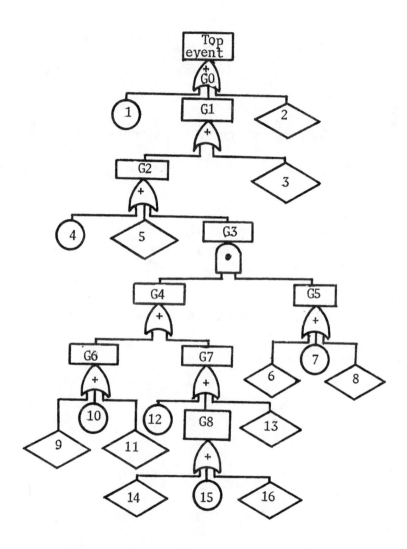

Fig. 11.14–Pressure tank system's fault tree of Example 11–4.

3

2.

Now G3 is an AND gate, so it should be replaced by its input events
in a separate column as follows:

1

4

5

G4, G5

3

2.

Since G4 is an OR gate, replacing G4 by its input events in a separate
row yields

1

4

5

G6, G5

G7, G5

3

2.

G5 is also an OR gate. Replacing it by its input events gives

1

4

5

G6, 6

G6, 7

G6, 8

G7, 6

G7, 7

G7, 8

3

2.

Continuing this way yields Table 11.4. The minimal cut sets are given
in the last column.

TABLE 11.4– The steps for obtaining the cut sets for the fault tree given in Fig. 11.14 of Example 11–4.

Step	1	2	3	4	5	6	7
	1	1	1	1	1	1	1
	G1	G2	4	4	4	4	4
	2	3	5	5	5	5	5
		2	G3	G6,G5	G6,6	9,6	9,6
			3	G7,G5	G6,7	10,6	10,6
			2	3	G6,8	11,6	11,6
				2	G7,6	9,7	9,7
					G7,7	10,7	10,7
					G7,8	11,7	11,7
					3	9,8	9,8
					2	10,8	10,8
						11,8	11,8
						12,6	12,6
						13,6	13,6
						G8,6	14,6
						12,7	15,6
						13,7	16,6
						G8,7	12,7
						12,8	13,7
						13,8	14,7
						G8,8	15,7
						3	16,7
						2	12,8
							13,8
							14,8
							15,8
							16,8
							3
							2

11.3.2.2– ALGORITHM 2

The MOCUS algorithm starts from the uppermost gate and goes to the lowermost gates to find the minimal cut sets. Here Algorithm 2 acts in an opposite way, i.e., it acts in a bottom-up fashion.

This algorithm starts from the lowermost gates: If a gate is an OR gate, set it equal to the union (+) of its input events. If a gate is an AND gate, set it to the intersection (·) of its input events. After all expressions for the lowermost gates are obtained, go to the gates one step up from the lowermost gates, and express these gates by their input events in the same way until the expression for the top event is obtained. At each step the following rules are used to simplify the expressions:

1. $A + A = A.$
2. $A \cdot A = A.$
3. $A + B = B + A.$
4. $A \cdot B = B \cdot A.$
5. $A \cdot (B + C) = A \cdot B + A \cdot C.$
6. $A + A \cdot B = A.$
7. $A + B \cdot C = (A + B) \cdot (A + C).$

EXAMPLE 11–5

Consider Example 11–3, and redetermine the minimal cut sets using Algorithm 2.

SOLUTION TO EXAMPLE 11–5

From Fig. 11.13 it can be seen that the lowermost gates are G4, G5, and G6. G4 is an OR gate, and G5 and G6 are AND gates. Then

$$G4 = X_4 + X_5,$$
$$G5 = X_6 \cdot X_7,$$

and

$$G6 = X_5 \cdot X_6.$$

The one step up gates for G4, G5 and G6 are G2 and G3. Then

$$G2 = G4 \cdot G5,$$
$$= (X_4 + X_5) \cdot (X_6 \cdot X_7),$$

which by Rule 5 gives

$$G2 \; = \; X_4 \cdot X_6 \cdot X_7 + X_5 \cdot X_6 \cdot X_7,$$

and

$$
\begin{aligned}
G3 \; &= \; X_3 + G6, \\
&= \; X_3 + X_5 \cdot X_6.
\end{aligned}
$$

In the same way

$$
\begin{aligned}
G1 \; &= \; G2 + G3, \\
&= \; X_4 \cdot X_6 \cdot X_7 + X_5 \cdot X_6 \cdot X_7 + X_3 + X_5 \cdot X_6,
\end{aligned}
$$

and by Rule 6

$$X_5 \cdot X_6 + X_5 \cdot X_6 \cdot X_7 = X_5 \cdot X_6.$$

Therefore,

$$G1 \; = \; X_3 + X_5 \cdot X_6 + X_4 \cdot X_6 \cdot X_7.$$

Finally,

$$G0 \; = \; X_1 + X_2 + X_3 + X_5 \cdot X_6 + X_4 \cdot X_6 \cdot X_7,$$

where each term gives a minimal cut set for the fault tree, and five minimal cut sets are obtained as

$$\{X_1\}, \{X_2\}, \{X_3\}, \{X_5, X_6\}, \{X_4, X_6, X_7\},$$

which is identical with the result given in Example 11-3 for the minimal cut sets.

11.3.3   DUAL TREES AND THE MINIMAL PATH SETS

The dual tree of a fault tree is the fault tree with the OR gates replaced by AND gates the AND gates replaced by OR gates in the original tree, and the events replaced by their corresponding dual events[3, pp. 266-268].

The dual tree is an important concept since the cut sets in the dual tree are the path sets in the original tree. Therefore, to find the path sets of a fault tree, first obtain its dual tree and then perform the analysis described in the previous section to find the cut sets of the dual tree.

EXAMPLE 11-6

Consider the fault tree given in Fig. 11.13. Prepare its dual tree and then find its minimal cut sets and thereby the minimal path sets of the original fault tree.

SOLUTION TO EXAMPLE 11-6

The dual tree of the fault tree given in Fig. 11.13 is given in Fig. 11.15. The letters or numbers with a prime in Fig. 11.15 denote the dual events. In general, dual events correspond to the nonoccurrence of the original events. For example, if the original top event, T, is "pressure tank rupture," as given in Fig. 11.9, then T' is the event "no tank rupture."

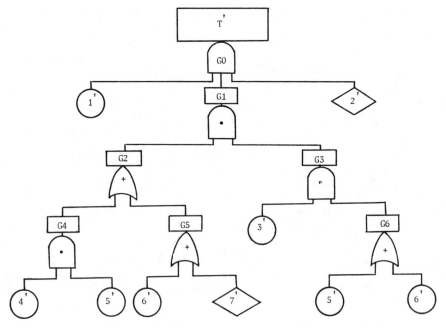

Fig. 11.15– Dual tree of the fault tree given in Fig. 11.13 for Example 11–6.

Performing an analysis of this dual tree using the MOCUS algorithm yields the following minimal cut sets:

$$\{1', 2', 3', 4', 5'\},$$
$$\{1', 2', 3', 6'\},$$

and

$$\{1', 2', 3', 5', 7'\}.$$

The stepwise procedure is given in Table 11.5. Removing the supersets

TABLE 11.5– The steps for obtaining the cut sets for the fault tree of Example 11-6 given in Fig. 11.15 for Example 11–6.

Steps 1	2	3	4	5
$1',G1,2'$	$1',G2,G3,2'$	$1',G4,3',G6,2'$	$1',4',5',G6,2'$	$1',2',3',4',5'$
				$1',2',3',4',5',6'$
			$1',6',3',G6,2'$	$1',2',3',5',6'$
				$1',2',3',6'$
		$1',G5,3',G6,2'$	$1',7',3',G6,2'$	$1',2',3',5',7'$
				$1',2',3',6',7'$

in Column 5 leads to the results given above.

Or, by Algorithm 2, since G4 is an AND gate and both G5 and G6 are OR gates,

$$G4 = X_{4'} \cdot X_{5'},$$
$$G5 = X_{6'} + X_{7'},$$
$$G6 = X_{5'} + X_{6'}.$$

G2 is an OR gate; then

$$G2 = G4 + G5 = X_{4'} \cdot X_{5'} + X_{6'} + X_{7'},$$

G3 is an AND gate; then

$$G3 = X_{3'} \cdot G6 = X_{3'} \cdot X_{5'} + X_{3'} \cdot X_{6'},$$

G1 is an AND gate; then

$$
\begin{aligned}
G1 &= G2 \cdot G3, \\
&= (X_{4'} \cdot X_{5'} + X_{6'} + X_{7'}) \cdot (X_{3'} \cdot X_{5'} + X_{3'} \cdot X_{6'}), \\
&= X_{3'}X_{4'}X_{5'} + X_{3'}X_{5'}X_{6'} + X_{3'}X_{5'}X_{7'}, \\
&\quad + X_{3'}X_{4'}X_{5'}X_{6'} + X_{3'}X_{6'} + X_{3'}X_{6'}X_{7'},
\end{aligned}
$$

or, by the rules of simplification,

$$G1 = X_{3'}X_{4'}X_{5'} + X_{3'}X_{5'}X_{7'} + X_{3'}X_{6'}.$$

G0 is also an AND gate; therefore,

$$
\begin{aligned}
T' &= G1 \cdot X_{1'} \cdot X_{2'}, \\
&= X_{1'}X_{2'}X_{3'}X_{4'}X_{5'} \\
&\quad + X_{1'}X_{2'}X_{3'}X_{5'}X_{7'} \\
&\quad + X_{1'}X_{2'}X_{3'}X_{6'},
\end{aligned}
$$

which gives the same results as the MOCUS algorithm.
Thus the minimal path sets for the original tree are

$$\{1,2,3,4,5\},$$
$$\{1,2,3,6\},$$

and

$$\{1,2,3,5,7\}.$$

11.4 QUANTITATIVE EVALUATION
OF THE FAULT TREE

The quantitative evaluation of the fault tree yields the probability of occurrence, or equivalently of nonoccurrence, of the top event given the probability of the basic events.

11.4.1 PROBABILITY EVALUATION BY
THE INCLUSION-EXCLUSION PRINCIPLE

11.4.1.1– EVALUATION FROM THE MINIMAL CUT SETS

If the minimal cut sets of a fault tree are found to be

$$C_1 = \{c_{1,1}, c_{1,2}, ..., c_{1,n_1}\} = \{\bigcap_{j=1}^{n_1} c_{1,j}\},$$

$$C_2 = \{c_{2,1}, c_{2,2}, ..., c_{2,n_2}\} = \{\bigcap_{j=1}^{n_2} c_{2,j}\},$$

$$\vdots$$

$$C_m = \{c_{m,1}, c_{m,2}, ..., c_{m,n_m}\} = \{\bigcap_{j=1}^{n_m} c_{m,j}\},$$

and $n_1, n_2, ..., n_m$ may or may not be the same, then the fault tree can be expressed equivalently as shown in Fig. 11.16. Here $A \cap B$ reads "A intersects B". An event, say C, is equal to $A \cap B$, when C occurs if only if both events A and B occur at the same time. Another notation used later on in the text is the "union" sign, $\cup$. $A \cup B$ reads "A union B". An event, say E, is equal to $A \cup B$, when E occurs if either A or B occurs or if A and B occur at the same time. Then, in general, the probabilities of the occurrence of events C, and of E, are given by

$$P(C) = P(A \cap B),$$

and

$$P(E) = P(A \cup B),$$
$$= P(A) + P(B) - P(A \cap B),$$

respectively.

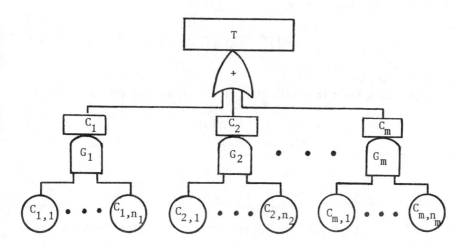

Fig. 11.16–The cut set expressions of a fault tree.

It is clear that the top event occurs if and only if at least all of the basic events in one minimal cut set occur simultaneously, i.e.,

$$\{TE\} = \{\bigcup_{i=1}^{m} C_i\},$$

or

$$\{TE\} = \{\bigcup_{i=1}^{m} (\bigcap_{j=1}^{n_i} c_{i,j})\}.$$

Therefore, the probability of occurrence of the top event is given by

$$P\{TE\} = P\{\bigcup_{i=1}^{m} C_i\},$$

or

$$P\{TE\} = P\{\bigcup_{i=1}^{m} (\bigcap_{j=1}^{n_i} c_{i,j})\}.$$

By the inclusion-exclusion principle, it can be further expanded in terms of the C'_i, $i = 1, 2, ..., m$, as

$$
\begin{aligned}
P\{TE\} &= \sum_{i=1}^{m} P(C_i) - \sum_{i=1}^{m-1} \sum_{j=i+1}^{m} P(C_i \cap C_j) \\
&\quad + \sum_{i=1}^{m_2} \sum_{j=i+1}^{m_1} \sum_{k=j+1}^{m} P(C_i \cap C_j \cap C_k) \\
&\quad + \cdots \\
&\quad + (-1)^{m-1} P(C_1 \cap C_2 \cap C_3 \cdots \cap C_m).
\end{aligned}
\tag{11.1}
$$

EXAMPLE 11–7
Consider a fault tree with the three minimal cut sets of

$$
\begin{aligned}
C_1 &= \{X_1, X_2\}, \\
C_2 &= \{X_3\},
\end{aligned}
$$

and

$$
C_3 = \{X_2, X_4, X_5\},
$$

where X_1, X_2, X_3, X_4, X_5 are independent basic events with the probabilities of occurrence of

$$
\begin{aligned}
P(X_1) &= 0.001, \quad P(X_2) = 0.002, \\
P(X_3) &= 0.050, \quad P(X_4) = 0.010,
\end{aligned}
$$

and

$$
P(X_5) = 0.001.
$$

Find the probability of occurrence of the top event.

SOLUTION TO EXAMPLE 11–7
By Eq. (11.1)

$$
\begin{aligned}
P\{TE\} &= P(C_1) + P(C_2) + P(C_3) - P(C_1 C_2) - P(C_1 C_3) \\
&\quad - P(C_2 C_3) + P(C_1 C_2 C_3,) \\
&= P(X_1 X_2) + P(X_3) + P(X_2 X_4 X_5) - P(X_1 X_2 X_3) \\
&\quad - P(X_1 X_2 X_4 X_5) - P(X_2 X_3 X_4 X_5) + P(X_1 X_2 X_3 X_4 X_5).
\end{aligned}
$$

By the independence of the basic events

$$
P\{TE\} = P(X_1)P(X_2) + P(X_3) + P(X_2)P(X_4)P(X_5)
$$

$$- P(X_1)P(X_2)P(X_3) - P(X_1)P(X_2)P(X_4)P(X_5)$$
$$- P(X_2)P(X_3)P(X_4)P(X_5)$$
$$+ P(X_1)P(X_2)P(X_3)P(X_4)P(X_5),$$
$$= (0.001)(0.002) + 0.050 + (0.002)(0.010)(0.001)$$
$$- (0.001)(0.002)(0.050) - (0.001)(0.002)(0.010)(0.001)$$
$$- (0.002)(0.050)(0.010)(0.001) + (0.001)(0.002)(0.050)$$
$$(0.010)(0.001),$$

or

$$P\{TE\} = 0.0500019.$$

If the top event is the event of failure of the system, then $P\{TE\}$ is the unreliability of the system, and the reliability of the system is given by

$$R_s = 1 - P\{TE\}.$$

Then, for Example 11-7 the system's reliability is

$$R_s = 0.9499981.$$

11.4.1.2– EVALUATION FROM THE MINIMAL PATH SETS

If the minimal path sets of a fault tree are found to be

$$D_1 = \{d_{1,1}, d_{1,2}, ..., d_{1,n_1}\} = \{\bigcup_{j=1}^{n_1} d_{2,j}\},$$

$$D_2 = \{d_{2,1}, d_{2,2}, ..., d_{2,n_2}\} = \{\bigcup_{j=1}^{n_2} d_{2,j}\},$$

$$\vdots$$

$$D_k = \{d_{k,1}, d_{k,2}, ..., d_{k,n_k}\} = \{\bigcup_{j=1}^{n_k} d_{k,j}\},$$

and $n_1, n_2, \cdots, n_k$ may or may not be the same, then the fault tree can be expressed equivalently as shown in Fig. 11.17. Similarly, it can be seen that the top event occurs if at least one basic event in each path set occurs, i.e.,

$$\{TE\} = \{\bigcap_{i=1}^{k} D_i\},$$

or

$$\{TE\} = \{\bigcap_{i=1}^{k}(\bigcup_{j=1}^{n_i} d_{i,j}\},$$

therefore, the probability of the occurrence of the top event can be written as

$$P\{TE\} = P\{\bigcap_{i=1}^{k} D_i\},$$

or

$$P\{TE\} = P\{\bigcap_{i=1}^{k}(\bigcup_{j=1}^{ni} d_{i,j})\}. \tag{11.2}$$

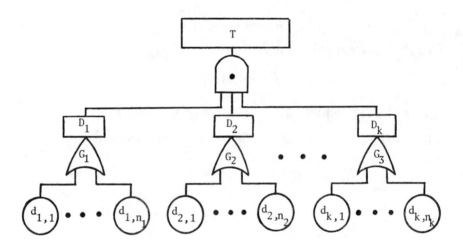

Fig. 11.17–The path set expressions of a fault tree.

EXAMPLE 11–8

A fault tree has the following two path sets:

$$D_1 = \{X_1, X_2\},$$

and

$$D_2 = \{X_3\},$$

where X_1, X_2, X_3 are independent basic events with the probabilities of occurrence

$$P(X_1) = 0.001,$$

$$P(X_2) = 0.003,$$

and

$$P(X_3) = 0.002.$$

Find the probabilty of occurrence of the top event.

SOLUTION TO EXAMPLE 11-8
 By Eq. (11.2)

$$\begin{aligned} P\{TE\} &= P\{D_1 D_2\}, \\ &= P\{(X_1 + X_2)X_3\}, \\ &= P\{X_1 X_3 + X_2 X_3\}, \end{aligned}$$

or

$$\begin{aligned} P\{TE\} &= P(X_1 X_3) + P(X_2 X_3) - P(X_1 X_2 X_3), \\ &= P(X_1)P(X_3) + P(X_2)P(X_3) - P(X_1)P(X_2)P(X_3), \\ &= (0.001)(0.002) + (0.003)(0.002) - (0.001)(0.003)(0.002), \end{aligned}$$

therefore,

$$P\{TE\} = 0.00000799.$$

Then

$$R_s = 1 - P\{TE\},$$

or

$$R_s = 0.99999201.$$

11.4.2 PROBABILITY EVALUATION USING THE STRUCTURE FUNCTION

11.4.2.1– THE STRUCTURE FUNCTION

If the top event and each of the basic events have only two states, i.e., occurring or not occurring, to indicate the state of the ith basic event, a binary indicator variable, say z_i, may be assigned to basic event i:

$$z_i = \begin{cases} 1, & \text{if the basic event is occurring,} \\ 0, & \text{if the basic event is not occurring,} \end{cases}$$

for $i = 1, 2, ..., n$, where n is the number of components in the system.

Similarly, the top event may also be associated with a binary indicator variable $\phi(Z)$ related to the state of the system by

$$\phi(Z) = \begin{cases} 1, & \text{if the top event is occuring,} \\ 0, & \text{if the top event is not occuring,} \end{cases}$$

where

$$Z = (z_1, z_2, ..., z_n).$$

Then the function $\phi(Z)$ is called the structure function of the fault tree [3, pp.1-3].

11.4.2.2– THE STRUCTURE FUNCTION FOR SIMPLE FAULT TREES

Consider the fault tree with an AND gate connected with the basic events. The top event occurs if and only if the input events occur simultaneously. In terms of the structure function

$$\phi(Z) = \prod_{i=1}^{n} z_i = z_1 z_2 ... z_n, \tag{11.3}$$

where z_i is the indicator variable of the ith basic event. It may be seen that $\phi(Z)$ takes on the value 1 if and only if all the z_i's have the value 1.

If the top event is connected with its event by an OR gate, the top event occurs if at least one of the input events occurs; therefore, its structure function is given by

$$\phi(Z) = 1 - (1 - z_1)(1 - z_2)...(1 - z_n), \tag{11.4}$$

and, for $n = 2$,

$$\begin{aligned} \phi(Z) &= 1 - (1 - z_1)(1 - z_2), \\ &= z_1 + z_2 - z_1 z_2. \end{aligned}$$

For the case that the top event is connected with the basic events by a k-out-of-n gate, its structure function is given by

$$\phi(Z) = \begin{cases} 1, & \text{if } \sum_{i=1}^{n} z_i \geq k; \\ 0, & \text{if } \sum_{i=1}^{n} z_i < k. \end{cases} \tag{11.5}$$

Consider a 2-out-of-3 gated fault tree. Its structure function is given by

$$\phi(Z) = 1 - (1 - z_1 z_2)(1 - z_1 z_3)(1 - z_2 z_3),$$

or

$$\phi(Z) = z_1 z_2 + z_1 z_3 + z_2 z_3 - 2z_1 z_2 z_3.$$

The structure function of a fault tree can be obtained in a stepwise way. The structure functions given in Eqs. (11.3), (11.4), and (11.5) are the structure functions of AND, OR, and k-out-of-n gates, respectively. Therefore, the structure function of a fault tree can be obtained by writting out the structure functions for each gate in a bottom-up fashion.

EXAMPLE 11–9

Write out the structure function for the fault tree given in Fig. 11.14.

SOLUTION TO EXAMPLE 11–9

Denote the structure function for gate Gi by $\phi_i(Z), i = 1, 2, 3, 4, 5, 6$, then

$$
\begin{aligned}
\phi_6(Z) &= z_5 + z_6 - z_5 z_6, \\
\phi_5(Z) &= z_6 + z_7 - z_6 z_7, \\
\phi_4(Z) &= z_4 z_5, \\
\phi_3(Z) &= z_3 \phi_6(Z), \\
&= z_3(z_5 + z_6 - z_5 z_6), \\
&= z_3 z_5 + z_3 z_6 - z_3 z_5 z_6, \\
\phi_2(Z) &= \phi_4(Z) + \phi_5(Z) - \phi_4(Z)\phi_5(Z), \\
&= z_4 z_5 + (z_6 + z_7 - z_6 z_7) - z_4 z_5(z_6 + z_7 - z_6 z_7), \\
&= z_4 z_5 + z_6 + z_7 - z_6 z_7 - z_4 z_5 z_6 - z_4 z_5 z_7 + z_4 z_5 z_6 z_7, \\
\phi_1(Z) &= \phi_2(Z)\phi_3(Z), \\
&= (z_3 z_5 + z_3 z_6 - z_3 z_5 z_6)(z_4 z_5 + z_6 + z_7 - z_6 z_7 \\
&\quad - z_4 z_5 z_6 - z_4 z_5 z_7 + z_4 z_5 z_6 z_7).
\end{aligned}
$$

Expanding and then simplifying, the expression $\phi_1(Z)$ yields

$$
\begin{aligned}
\phi_1(Z) &= z_3 z_6 + z_3 z_4 z_5 + z_3 z_4 z_6 + z_3 z_5 z_7 + z_3 z_4 z_5 z_6 z_7 \\
&\quad - z_3 z_5 z_6 z_7 - z_3 z_4 z_5 z_7 - 2z_3 z_4 z_5 z_6.
\end{aligned}
$$

Finally, the system's structure function, $\phi(Z)$, is given by

$$\phi(Z) = z_1 z_2 \phi_1(Z),$$

or

$$
\begin{aligned}
\phi(Z) &= z_1 z_2 z_3 z_6 + z_1 z_2 z_3 z_4 z_5 + z_1 z_2 z_3 z_4 z_6 + z_1 z_2 z_3 z_5 z_7 \\
&\quad + z_1 z_2 z_3 z_4 z_5 z_6 z_7 - z_1 z_2 z_3 z_5 z_6 z_7 - z_1 z_2 z_3 z_4 z_5 z_7 \\
&\quad - 2z_1 z_2 z_3 z_4 z_5 z_6.
\end{aligned}
$$

11.4.2.3 – PROBABILITY EVALUATION USING THE STRUCTURE FUNCTION

Consider the fault tree with two basic events connected with the top event by an AND gate. Its structure function is given by

$$\phi(Z) = z_1 z_2,$$

where z_1, z_2 are the indicator variables of basic events 1 and 2, respectively. If these two events are independent, then the expectation of $\phi(Z)$ is given by

$$E[\phi(Z)] = E(z_1)E(z_2),$$

and

$$E(z_1) \;=\; 1 \cdot P(z_1 = 1) + 0 \cdot P(z_1 = 0),$$

or

$$E(z_1) \;=\; P(z_1 = 1),$$

which is the probability of occurrence of basic event 1. Similarly,

$$E(z_2) = P(z_2 = 1).$$

Thus

$$E[\phi(Z)] = P(z_1 = 1)P(z_2 = 1),$$

which is equal to the probability of occurrence of the top event. Finally,

$$
\begin{aligned}
E[\phi(Z)] \;&=\; 1 \cdot P\{\phi(Z) = 1\} + 0 \cdot P\{\phi(Z) = 0\}, \\
&=\; P\{\phi(Z) = 1\}.
\end{aligned}
$$

Therefore,

$$P(\text{TE}) = E[\phi(Z)], \qquad\qquad (11.6)$$

which says that the probability of the occurrence of the top event is equal to the expectation of its structure function.

EXAMPLE 11–10

Find the probability of occurrence of the top event of the fault tree given in Fig. 11.13 using the structure function, given that

$$P(z_i) = 0.010, \quad i = 1, 2, 3, 4, 5, 6, 7.$$

SOLUTION TO EXAMPLE 11–10

First, the structure function of the fault tree should be found. Denote the structure functions of the gates, Gi, $i = 1, ..., 6$ by $\phi_i(Z)$; then

$$\phi_6(Z) = z_5 z_6,$$
$$\phi_5(Z) = z_6 z_7,$$
$$\phi_4(Z) = z_4 + z_5 - z_4 z_5,$$
$$\phi_3(Z) = z_3 + z_5 z_6 - z_3 z_5 z_6,$$
$$\phi_2(Z) = \phi_4(Z)\phi_5(Z),$$
$$= z_6 z_7(z_4 + z_5 - z_4 z_5),$$
$$= z_4 z_6 z_7 + z_5 z_6 z_7 - z_4 z_5 z_6 z_7,$$
$$\phi_1(Z) = \phi_2(Z) + \phi_3(Z) - \phi_2(Z)\phi_3(Z),$$
$$= (z_4 z_6 z_7 + z_5 z_6 z_7 - z_4 z_5 z_6 z_7) + (z_3 + z_5 z_6 - z_3 z_5 z_6)$$
$$- (z_4 z_6 z_7 + z_5 z_6 z_7 - z_4 z_5 z_6 z_7)(z_3 + z_5 z_6 - z_3 z_5 z_6).$$

Expanding and then simplifying the expression of $\phi_1(Z)$ yields

$$\phi_1(Z) = z_3 + z_5 z_6 + z_4 z_6 z_7 + z_3 z_4 z_5 z_6 z_7 - z_3 z_5 z_6$$
$$- z_3 z_4 z_6 z_7 - z_4 z_5 z_6 z_7.$$

Then the system's structure function is given by

$$\phi(Z) = 1 - (1 - z_1)(1 - z_2)[1 - \phi_1(Z)],$$
$$= \phi_1(Z) + (z_1 + z_2 - z_1 z_2) - (z_1 + z_2 - z_1 z_2)\phi_1(Z),$$
$$= z_1 + z_2 + z_3 - z_1 z_2 - z_1 z_3 - z_2 z_3 + z_5 z_6 + z_1 z_2 z_3$$
$$+ z_4 z_6 z_7 - z_1 z_5 z_6 - z_2 z_5 z_6 - z_3 z_5 z_6 + z_1 z_2 z_5 z_6$$
$$+ z_1 z_3 z_5 z_6 + z_2 z_3 z_5 z_6 - z_3 z_4 z_6 z_7 - z_4 z_5 z_6 z_7$$
$$- z_1 z_4 z_6 z_7 - z_2 z_4 z_6 z_7 + z_1 z_2 z_4 z_6 z_7 + z_3 z_4 z_5 z_6 z_7$$
$$+ z_1 z_3 z_4 z_6 z_7 + z_2 z_3 z_4 z_6 z_7 + z_1 z_4 z_5 z_6 z_7 - z_1 z_2 z_3 z_5 z_6$$
$$+ z_2 z_4 z_5 z_6 z_7 - z_1 z_3 z_4 z_5 z_6 z_7 - z_2 z_3 z_4 z_5 z_6 z_7$$
$$- z_1 z_2 z_3 z_4 z_6 z_7 - z_1 z_2 z_4 z_5 z_6 z_7 + z_1 z_2 z_3 z_4 z_5 z_6 z_7.$$

Therefore,

$$P\{\text{TE}\} = E[\phi(Z)],$$
$$= 3(0.010) - 2(0.010)^2 - (0.010)^3 - (0.010)^4 + 5(0.010)^5$$
$$- 4(0.010)^6 + (0.010)^7,$$

or

$$P\{\text{TE}\} = 0.0297989905.$$

11.4.2.4 – THE STRUCTURE FUNCTION EXPRESSION IN TERMS OF THE MINIMAL CUT SETS OR PATH SETS

The system structure function expression in terms of the minimal cut sets is given by

$$\phi(Z) = 1 - \prod_{i=1}^{m}[1 - \phi_i(Z)], \qquad (11.7)$$

where $\phi_i(Z)$ is the structure function of the ith minimal cut set, $i = 1, 2, ..., m$, where m is the number of the minimal cut sets in the fault tree. Denote the indicator variable of the jth basic event in the ith minimal cut set by z_{ij}; then

$$\phi_i(Z) = \prod_{j=1}^{n_i} z_{ij},$$

and Eq. (11.7) can be written as

$$\phi(Z) = 1 - \prod_{i=1}^{m}[1 - \prod_{j=1}^{n_i} z_{ij}], \qquad (11.8)$$

where n_i is the number of basic events in the ith minimal cut set.

The system's structure function expression in terms of the minimal path sets is given by

$$\phi(Z) = 1 - \prod_{i=1}^{k} \phi_i(Z), \qquad (11.9)$$

where $\phi_i(Z)$ is the structure function of the ith minimal path set, $i = 1, 2, ..., k$, where k is the number of the minimal path sets in the fault tree. Similarly, $\phi(Z)$ can be written in terms of the indicator variable of the basic events in the minimal path set as

$$\phi(Z) = 1 - \prod_{i=1}^{k}[1 - \prod_{j=1}^{n_i}(1 - z_{ij})], \qquad (11.10)$$

where n_i is the number of basic events in the ith minimal path set.

EXAMPLE 11–11

Rework Example 11–7 using the structure function.

SOLUTION TO EXAMPLE 11–11

By Eq. (11.7), the system's structure function is given by

$$\phi(Z) = 1 - [1 - \phi_1(Z)][1 - \phi_2(Z)][1 - \phi_3(Z)],$$

and

$$\phi_1(Z) = z_1 z_2,$$
$$\phi_2(Z) = z_3,$$
$$\phi_3(Z) = z_2 z_4 z_5.$$

Therefore,

$$\begin{aligned}
\phi(Z) &= 1 - (1 - z_1 z_2)(1 - z_3)(1 - z_2 z_4 z_5), \\
&= z_3 + z_1 z_2 - z_1 z_2 z_3 + z_2 z_4 z_5 - z_2 z_3 z_4 z_5 \\
&\quad - z_1 z_2 z_4 z_5 + z_1 z_2 z_3 z_4 z_5.
\end{aligned}$$

Thus

$$\begin{aligned}
P(\text{TE}) &= E[\phi(Z)], \\
&= 0.050 + (0.001)(0.002) - (0.001)(0.002)(0.050) \\
&\quad + (0.002)(0.010)(0.001) - (0.002)(0.050)(0.010)(0.001) \\
&\quad - (0.001)(0.002)(0.010)(0.001) \\
&\quad + (0.001)(0.002)(0.050)(0.010)(0.001),
\end{aligned}$$

or

$$P(\text{TE}) = 0.0500019,$$

which is the same result as obtained in Example 11–7.

PROBLEMS

11-1. Illustrate the use of the gates discussed in this chapter by your own example, in which the input events and the output event are connected by the desired gate.

11-2. Consider the fault tree given in Fig. 11.18. AND and OR gates are labled G0 through G4 and the events are labeled 1 through 10. Using Algorithm 1, find the minimal cut sets of this fault tree.

11-3. Using Algorithm 2, find the minimal cut sets of the fault tree given in Fig. 11.18. Assume that the top event is the breakdown of this system, and each unit has a failure probability of 0.01. Calculate the system's reliability.

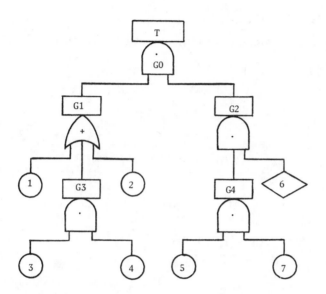

Fig. 11.18 – Fault tree for Problems 11-2 and 11-3.

11-4. Draw the dual tree of the fault tree given in Fig. 11.14, and therefrom find the path sets of the original tree.

11-5. Find the structure function of the system in Fig. 11.19.

11-6. Write out the structure function for the fault tree given in Fig. 11.20, and calculate the system's failure probability assuming that

$$
\begin{aligned}
P(X_1) &= P(X_2) = P(X_6) = 0.01, \\
P(X_3) &= P(X_4) = P(X_5) = 0.05, \\
P(X_7) &= 0.07,
\end{aligned}
$$

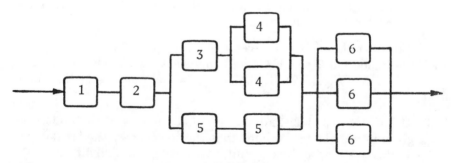

Fig. 11.19 – The system of Problem 11-5.

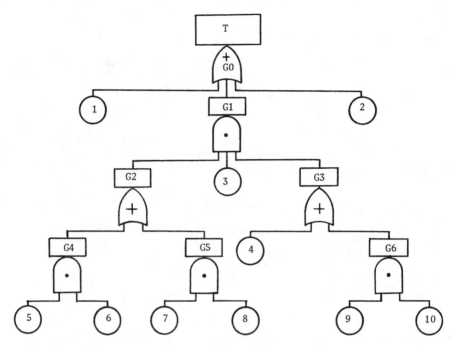

Fig. 11.20 – The system of Problem 11-6.

and

$$P(X_8) = P(X_9) = P(X_{10}) = 0.03.$$

REFERENCES

1. Fussell, J.B., *Fault Tree Analysis–Concepts and Techniques*, Proceedings NATO Advanced Study Institute on Generic Techniques of System Reliability Assessment, Liverpool, England, pp. 133–162, 1974.

2. Henley, E.J., and Kumamoto, H., *Reliability Engineering and Risk Assessment*, Prentice Hall, Englewood Cliffs, N.J., 568 pp., 1981.

3. Barlow, R., and Proschan, F., *Statistical Theory of Reliability and Life Testing–Probability Model*, Holt,Rinehart and Winston, Inc., New York, 290 pp., 1975.

4. Fussell, J.B., Henry, E.B., and Marshall, N.H., "MOCUS–A Computer Program to Obtain Minimal Cut Sets from Fault Tree," ANCR–1156, 1974.

Chapter 12

SYSTEM RELIABILITY PREDICTION AND TARGET RELIABILITY

12.1 TARGET RELIABILITY

Three possible methods of determining the target reliability that a system should be designed to are the following:

1. Calculate the reliability of several different configurations of the system, and some with identical configurations but utilizing different state of the art components. These should be those configurations which will perform the specified functions and will attain the desired mission objectives. Next, calculate the total cost to the user of owning, operating, and maintaining this system for its designed-to life or its life cycle cost in its different configurations. Now find the optimum level of system reliability at which the total cost is a minimum. This optimum reliability may be selected to be the target reliability, which is that of configuration C_4 shown in Fig. 12.1.

2. Obtain a multivalued prediction of the chosen system configuration using such failure rates as published by GIDEP, MIL-HDBK-217, RADC, and other sources. By multivalued is meant that for the same component use the lowest, the most likely, and the highest expected failure rates that component will demonstrate in its actual use environment. Based on the state of the art of the components and subsystems to be actually used in the system, properly modify these failure rates, determine the system's overall failure rate (three of them based on the lowest, the most likely, and the highest failure rates), and select the target reliability, based on the three predicted failure rates, somewhere within the predicted range, and in only exceptional cases above

249

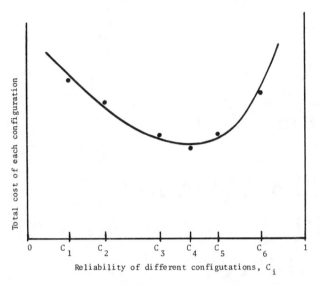

Fig. 12.1 – Determination of the optimum configuration
yielding the minimum total cost reliability. C_i
stands for the ith configuration. Configuration
C_4 is the optimum configuration.

the upper limit of the predicted reliability.

    Figure 16.3 shows the relative placement of the multivalued relia-
bility predictions, the target reliability, and the reliability growth curve
for a system, as well as the expected trend with product development
or contract time.

    3. If any one of the previous two methods cannot be pursued,
then target reliabilities for systems in excess of 85% and preferably in
excess of 90%, subsystem reliabilities in excess of 95% and preferably
in excess of 99%, and sub-subsystem reliabilities in excess of 99.9%
may be chosen for a ball-park estimate of the target system reliability.

12.2   TARGET RELIABILITY ALLOCATION

Once the system's target reliability is established, the next step should
be to allocate it to the subsystems and sub-subsystems of the system
by one of several methods discussed in Chapter 15. Subsystem or sub-
subsystem configurations, components, and parts may then be chosen,
and the sub-subsystem and subsystem reliabilities predicted and compared with the allocated reliabilities. Several configurations, types of
components and parts, state of the art of components and parts, application stress ratings, and operation stress levels may be tried, as well

as several trade-offs among the allocated reliabilities, weights, volumes, and costs of the various sub-subsystems and subsystems, until the target reliability is obtained.

12.3 RELIABILITY PREDICTION METHODOLOGY

It is seen from the previous two sections that inevitably we are faced with the task of having to conduct reliability predictions. A reliability prediction methodology which may be carried out with varying degrees of embellishment is as follows:

1. Obtain the technical specifications of the product from company engineering specifications or contract specifications and extract all clauses pertaining to reliability and maintainability.

2. Determine all applicable company and military standards and documents.

3. Ascertain the general mission objectives, such as shown in Fig. 12.2. This example is for a spacecraft which is to land on our moon.

4. Establish the definition of complete and partial, if allowable, mission success.

5. Determine the mission events, such as those shown in Fig. 12.3.

6. Prepare a word description of the functions of all subsystems during each mission phase, tie the description to the mission events involved, and determine the subsystem inputs and outputs.

7. Prepare a functional block diagram of the system, such as shown in Figs. 12.4, 12.5, and 12.6, based on the information developed in Step 6.

8. Prepare a word description of the logic function of each subsystem, giving the flow of information through each subsystem and the functional relationships among subsystems as to sequence and mode of operation. State the functional effects on the system of the various probable modes of failure. State if a failure mode will or will not fail safe.

9. Prepare a reliability block diagram for the system, as shown in Fig. 12.7, based on the information developed in the previous step.

10. Repeat Steps 8 and 9 for *each* subsystem, as shown in Figs. 12.8 and 12.9, respectively.

11. Prepare a mission duty cycle profile for all subsystems, as shown in Fig. 12.10.

12. Prepare a mission stress profile for each subsystem, as shown in Fig. 12.11.

Flight number	Ranger number	Basic mission		Experiments	
		Primary	Secondary	Space	Lunar surface
1 & 2	13 & 14	Demonstrate a rough landing capability and postlanding survival by sending a postlanding signal.	Perform a series of postlanding experiments.	None.	Determine availability of water, oxygen, and shelter; existence of ionizing radiation and EM wave propagation characteristics, etc.
3 & 4	15 & 16	Perform postlanding experiments on lunar topological characteristics relevant to manned lunar missions.	Provide measurements related to landing capsule impact mechanics.	Perform experiments from within the spacecraft prior to capsule release that are of value to manned lunar missions.	Same as above.
5 & 6	17 & 18	Perform postlanding experiments to determine the local structural properties of the lunar surface and gross structural characteristics.	Perform experiments to provide additional data on the lunar environment or surface material composition.	None.	Perform seismic or seismic/accelerometer experiments to determine local structural properties and gross structural characteristics.

Fig. 12.2– General mission objectives for a spacecraft called the Ranger which is to land on our moon: Spacecraft Phase II.

Event no.	Event description	Event time Beginning (min)	Event time End (min)	Remarks
1	Prelaunch	L−	L	Checkout power requirements, the active systems, their level of output; their application and operation environments need to be specified for this and following events.
2	Launch	L	L+ 5 min	Acceleration, vibration, shock, acoustic, and temperature environment stress levels from firing to injection and separation of spacecraft should be specified, as well as for all events that follow.
3	Jettison nose cone	L+ min	L + 5 min	
4	Separate from Atlas	L + 5 min	L + 5.5 min	Takes place 100 miles above earth and 490 miles downrange.
5	Coast parallel to earth	L + 5.5 min	L+ min	Arcs over to parallel earth's curve and coasts for 0.5 min
6	Fire Agena D	L+ min	L+ min	Agena D and Ranger are placed in an earth orbit.
7	Coast again	L+ min	L + 25 min	Determine best moon takeoff position.
8	Fire Agena D again	L+ min (L + 25 min)	L+ min	Drives spacecraft into a lunar course at 24,000 mph.

Fig. 12.3– Mission events for a spacecraft called the Ranger which is to land on our moon: Spacecraft Phase II.

Event no.	Event description	Event time				Remarks
		Beginning		End		
9	Separate Agena D	L+	min	L+	min	It has been proposed that Agena D be kept attached to the spacecraft.
10	Remove Agena D	L+	min	L+	min	Agena D turns around and fires a retro-rocket to separate it from the Ranger. This is done to avoid reflections which might confuse the optical sensors.
11	Open solar panels	L + 48	min	L+	min	On command of the central computer and sequencer (CC&S).
12	Pivot out high-gain antenna	L+51	min	L+	min	Ranger prepares to send signals.
13	Acquire solar-lock attitude	L+	min	L + 81	min	Gas jets fire to lock solar panels to sun. CC&S is involved.
14	Begin operation of most electronic gear	L + 81	min	Impact		
15	Acquire earth-lock attitude	L+	min	L + 4	hr	High-gain antenna is aimed at earth.
16	Perform midcourse maneuver	L+	hr	L + 16	hr	Midcourse correction rocket fires from ground command, kicking spacecraft into final lunar trajectory after breaking solar lock.

Fig. 12.3–Continued.

Event no.	Event description	Event time		Remarks
		Beginning	End	
17	Resume flight attitude	L+ hr	L+ hr	Attitude control locks on sun.
18	Cruise to moon	L+ hr	L+ hr	
19	Extend gamma ray counter	L+ hr	L+ hr	Silver ball, at end of a 6-ft boom, helps identify moon materials by picking up distinctive radiation patterns.
20	Retract aluminized shroud	L+ hr	L+ hr	Exposes instrument capsule and retro-rocket.
21	Swing out omni-directional antenna	L+ hr	L+ hr	Necessary precondition for retrorocket firing.
22	Decelerate spacecraft	220,000 miles from earth	mi	Decelerates to 2,150 mph.
23	Accelerate spacecraft	L+ hr	L+ hr	Acceleration due to lunar gravity.
24	Acquire TV pictures of the moon	L+ hr	L+ hr	This mission phase is doubtful for the Block V, Phase II Ranger.
25	Perform terminal man-euver	70,000 ft	ft	
26	Activate radar altimeter	L+ hr	L+ hr	Gages distance to moon and orders firing of retrorocket.

Fig. 12.3-Continued.

Event no.	Event description	Event time		Remarks
		Beginning	End	
27	Separate payload from spacecraft	L+ hr	L+ hr	
28	Fire retrorocket	L+ hr	L+ hr	Slow sphere to momentary dead stop about 1,100 ft above the moon.
29	Jettison retrorocket	L+ hr	L+ hr	Explosive bolts break the clamp of retro-rocket and separate it from payload.
30	Land payload	L + 65 to 67 hr	L+ hr	Balsa ball strikes moon at 150 mph.
31	Fire gun in sphere	I+ hr	I+ hr	Two guns present. Bullet penetrates two fiber-glass spheres separated by 1/10 inch of Freon. Instrument package (includ-ing antenna) is locked in upright working position.
32	Transmit scientific data	I+ hr	I+ hr	Detect moon quakes, meteor crashes, and other. These data are transmitted to earth by radio transmitter via its varied antenna.
33	Control payload tempera-ture	I+ hr	I+ hr	Black and white color pattern, evapora-tion of water through a poppet valve, and payload insulation permit the function of the payload for about 30 days.

Fig. 12.3–Continued.

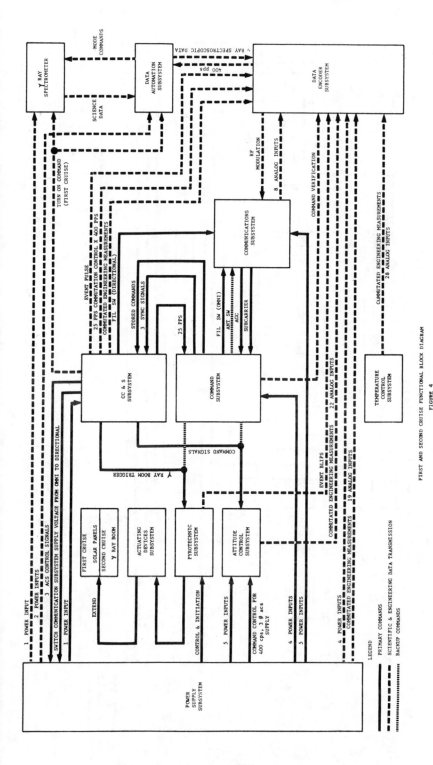

FIRST AND SECOND CRUISE FUNCTIONAL BLOCK DIAGRAM

FIGURE 4

Fig. 12.4 – First and second cruise functional block diagram.

257

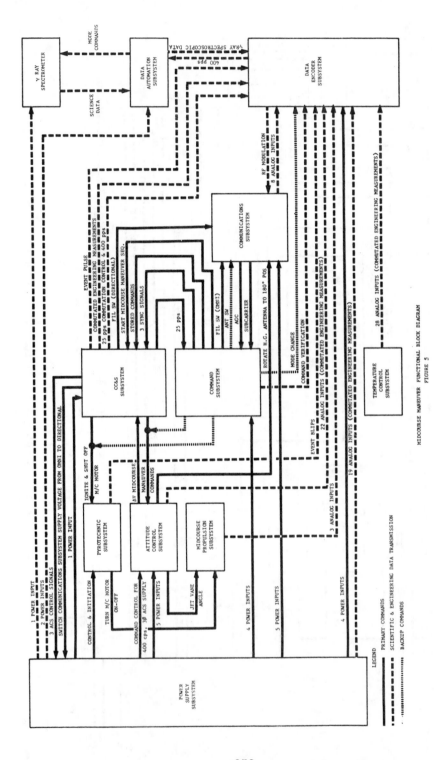

Fig. 12.5 – Midcourse maneuver functional block diagram.

258

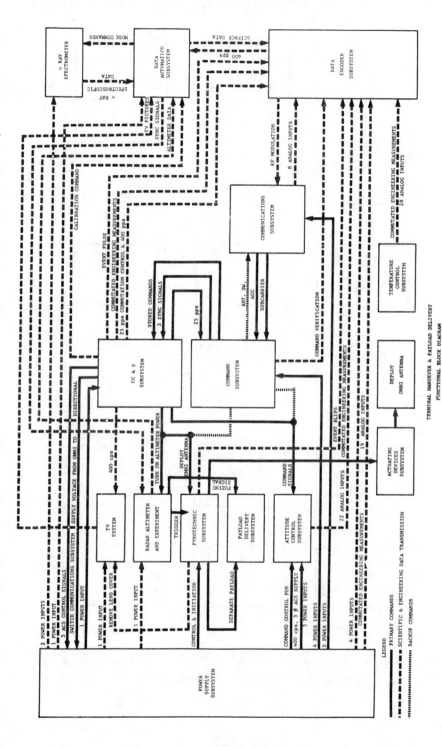

Fig. 12.6 – Terminal maneuver and payload delivery functional block diagram.

259

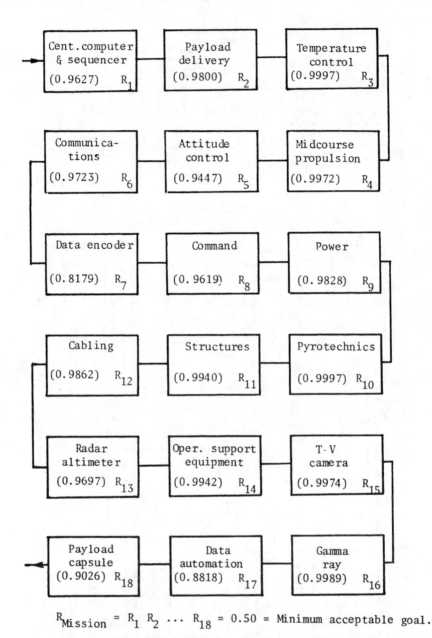

$R_{\text{Mission}} = R_1 \, R_2 \, \cdots \, R_{18} = 0.50 = \text{Minimum acceptable goal.}$

Fig. 12.7 – Spacecraft's reliability block diagram, mathematical model, and minimum acceptable goal.

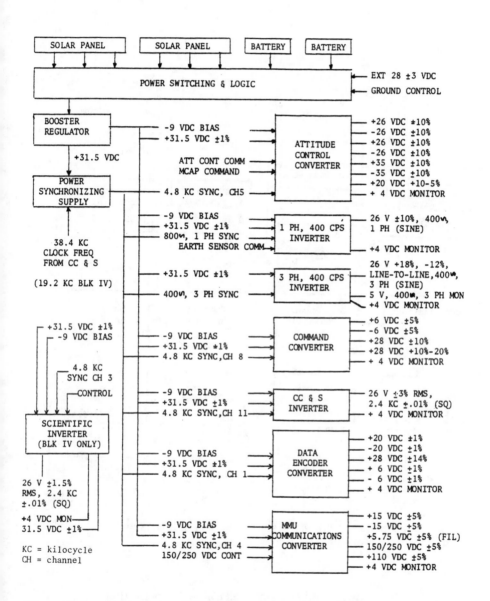

Fig. 12.8 – Schematic of the spacecraft's power subsystem.

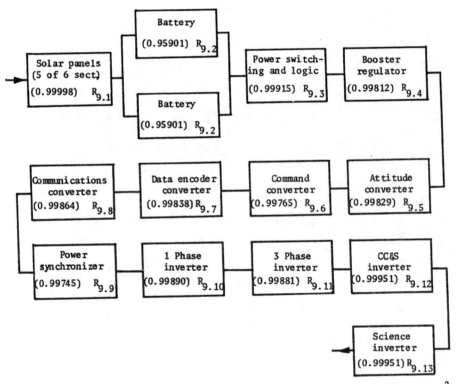

$$R_9 = R_{9.1}R_{9.3}R_{9.4}R_{9.5}R_{9.6}R_{9.7}R_{9.8}R_{9.9}R_{9.10}R_{9.11}R_{9.12}R_{9.13} \, [1 - (1 - R_{9.2})^2].$$

R_9 (Minimum acceptable goal) = 0.9828.

Fig. 12.9 – Power subsystem reliability block diagram, mathematical model, and minimum acceptable goal.

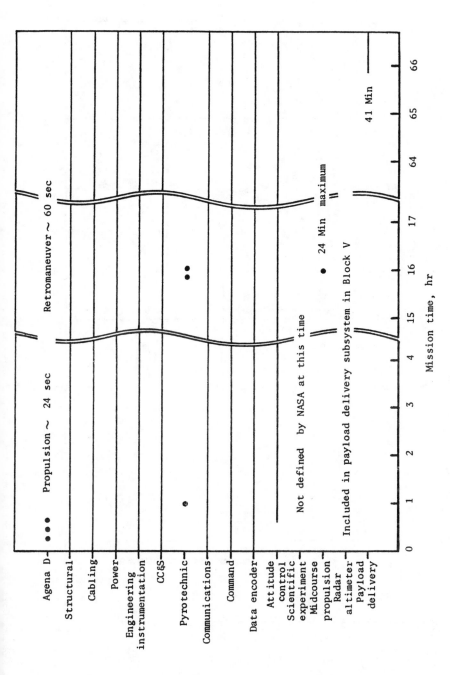

Fig. 12.10—Subsystem mission profile for the Ranger spacecraft.

263

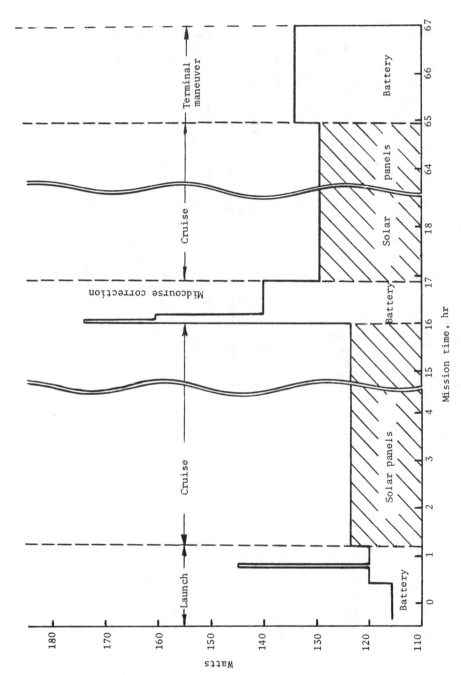

Fig. 12.11 – Power subsystem stress profile.

264

13. Repeat Steps 8, 9, 11, and 12 for *each* sub-subsystem ... down to the part level, as required.

14. Enter information called for in the System Reliability Prediction Form shown in Fig. 12.12.

Instructions for filling in this form are as follows:

Top Left Side of Form: Under *System*, enter the name of the system and the number of the system involved; such as "Spacecraft No. 13."

Under *Subsystem*, enter the subsystem involved; such as "Power."

Under *Sub-subsystem*, enter its name; such as "Power Switching and Logic."

Under *Circuit*, enter the name of the particular circuit you are working on, such as "Battery Voltage Monitor." If there is more than one circuit entered on the form, enter the names of all of them with a slant, in the order entered on the form.

Top of Form - Right Side: Under *Page,* enter first the number of pages that have been worked on to obtain the specific subsystem, and next enter the total number of pages that have been worked on to obtain the specific subsystem's failures and reliability.

Under *Analysis by,* enter legibly the name of the person, or persons, conducting the reliability prediction analysis.

Under *Dates,* enter starting and completion dates.

Under *Approved by,* have the immediate supervisor review, check, and sign each worksheet, and enter the *date* of signing.

Column 1: Enter information that identifies each part of a sub-subsystem or an assembly depending on the level of assemblage whose reliability is being predicted. For example, if a sub-subsystem's circuit reliability is being predicted, enter parts; if a subsystem's, then enter identifying information about the sub-subsystem.

Under *Symbol,* enter the identifying symbol designated on the drawing, such as R_1, for the first resistor in the circuit diagram. Parts having common entries may be grouped as a single entry.

Under *Description,* give the type of manufacture, such as wire wound (resistor), and value, such as 12 ohms.

Under *Type no., Part no., or Designation,* enter the MIL Spec. No., if known, manufacturer and manufacturer's part number, in-house company part number or other specific identification, if known, such as MIL-R-11D, RG 202F, or Dale RS 2A.

Column 2: Under Rating, list the value of watts, volts, current, cycles, speed, etc., at which the item is rated by the controlling specification or the manufacturer. This rating must be compatible with that given in MIL-HDBK-217, GIDEP, RADC, or other cources.

Column 3: In this column list the generic maximum, mean, and minimum and basic failure rate for each part, using the following sources:

Project: Spacecraft Phase II
System: Spacecraft No. 13
Subsystem: Power
Sub-subsystem: Power switching and logic
Circuit: Battery voltage monitor

1	2	3						4	5			6	7
Part or assembly description, designation	Rating	Failure rates, fr/10^6 hr											
		Source											
		1 Generic - G_{fr}			2 Basic	3 Basic							
		Max	Mean	Min	Mean*	Mean*		Applied stress	π_E	π_Q	π_{CR}	Appli-cation factor, K_a	Oper-ation factor, K_o
Symbol	Description and value	Type no. part no., or desig-nation					π_i						

*At application stress level.

Fig. 12.12 – System reliability prediction form.

8						9	10						11	12	13					
Total adjusted failure rate per part, F_r, columns 3 × 5 fr/10^6 hr						Actual hours of component operation during the mission, t_x	Failures, F, $F_r \times t_x$ = columns 8 × 9						Number of parts in series, n	Applicable redundancy factor, K_r	Total adjusted mission failures, F_A, columns 10 × 11 × 12; or failure rates, λ_A, columns 8 × 11 × 12					
Source							Source								Source					
1			2	3			1			2	3				1			2	3	
Max	Mean	Min	Mean	Mean	Mean		Max	Mean	Min	Mean	Mean	Mean			Max	Mean	Min	Mean	Mean	Mean

Sum of total adjusted mission failures $= \sum F_{A_i}$,

Sum of total adjusted mission failure rates $= \sum \lambda_{A_i}$.

Reliability $= e^{-\sum F_{A_i}} = e^{-(\sum \lambda_{A_i})t}$

Fig. 12.12 – Continued.

Source 1

Source 1 is MIL-HDBK-217, "Reliability Prediction of Electronic Equipment." Make sure to use the latest issue. In this source find the appropriate tables for the part in question. Use the ambient temperature for the subsystem being analyzed. Select the column in the part λ_b table corresponding to the stress aplied on the component as given in Column 4 and read the expected base failure rate. Check the text write-up for the particular part for additional modifying factors. These failure rates are for burned in parts. If burning in is not done, the part λ_b may be higher by a factor of from 10 to 100. However, these λ_b are not "hi-rel" parts. Such parts should exhibit very few early failures due to substandard parts. However, "hi-rel" parts are expensive. Good parts can be obtained from military-spec. parts utilizing part burn-in, resulting in at least a 10% increase in price. It is obvious that economic considerations will have a definite influence on part selection.

To obtain the three λ_b values, i.e., minimum (min), mean or expected (mean), and maximum (max), from MIL-HDBK-217, choose the stress-ratio (S), junction-temperature (T) combination that will yield the worst case or the maximum λ_b, and separately the S-T combination that will yield the expected or most likely case or the mean λ_b, and separately the S-T combination that will yield the best case or the minimum λ_b. The objective is to come up with a three-valued failure rate prediction and the corresponding three reliabilities. Then it may be concluded with approximately 95% confidence that the actual λ_b and $R(t)$ will lie within the maximum and minimum values, and most likely the mean λ_b and $R(t)$ values will be exhibited by the system.

Example 3–5 in Volume 1 illustrates the use of MIL-HDBK-217 and Example 12–2 here the three-valued approach.

Source 2

Source 2 is the "Reliability-Maintainability Analyzed Data summaries," GIDEP, Government-Industry Data Exchange Program, Reliability-Maintainability Data Bank, GIDEP Operations Center, Fleet Analysis Center, Corona, California 91720. Make sure to use the latest issue, together with their permanent issues.

Source 3

Source 3 could be the in-house failure rate and reliability data bank used in conjunction with the suppliers' and subcontractor's failure rate and reliability data bank.

Remarks on Reliability Prediction

The major utility of system reliability predictions from part and component failure rates is the estimation of system's failure rate, MTBF and reliability for feasibility studies for comparison with the goal specified, and for fairly accurate comparisons of similar systems or alternate designs of a given system. Predictions which make a system design appear to be marginally feasible should be reexamined with the greatest of care. In a marginal situation, if the predictions are of reasonable accuracy, reconsideration of the system's design or of the requirements for the system's performance may result in a prediction which is more favorable toward the feasibility of the system. It is also expected that the reliability prediction will be used as a design tool to point out areas of design weakness and of possible redesign which will improve the overall reliability of the system.

A note of caution should be made. It should be insured that failure rates from different sources are not mixed without the necessary adjustments when making reliability predictions, because the environments for which the failure rates in each source apply may be substantially different. If it is necessary to mix failure rates from different sources, one should make sure that they are reduced to a common application and operation environment stress level.

Column 4: Under *Applied stress* enter the ratio of the actual to the rated power (watts) dissipated, current drawn or voltage applied. Also enter operating cycles and speed to which the part is subjected in the circuit. If actual values are not available, enter 0.25 for a preliminary study. Later acquire actual values.

Column 5: Find the applicable π factors from MIL-HDBK-217 and enter them here.

Column 6: If the parts are not listed in MIL-HDBK-217 and you are using other sources, then it is a desirable to generate your own application stress factor, K_a. K_a is a multiplier to the generic or base failure rate to correct for the applied stress, temperature, speed, load, etc. See Fig. 3.12, Volume 1 of this Handbook, for such factors.

Column 7: If MIL-HDBK-217 does not list your parts, generate your own operation stress factor, K_{op}, which is a multiplier of the base or generic failure rates to correct for the externally applied environment stresses such as vibration, shock, acceleration, deceleration, vacuum, radiation, humidity, dust, etc. See Table 3.2, Volume 1 of this Handbook, for values of K_{op}.

Column 8: Under *Total adjusted failure rate per part,* multiply the *Max., Mean, and Min.* failure rates in Column 3 by the π factors in Column 5 or by the K_a and K_{op} values in Columns 6 and 7 to obtain the respective values for Column 8. For detailed coverage of the actual failure rate determination see Chapter 3, Volume 1 of this Handbook, Section 3.9.

Column 9: Under *Actual hours of component operation during the mission,* determine the actual hours of part operation during the mission from the mission profile. For some parts, such as those in the CC&S, for example, the operating time will be equivalent to the total mission time or approximately 66 hr. For other parts, such as those directly involved with the midcourse correction maneuver, the time will be relatively short, or about 1.46 hr, for example. If the part is a relay or a similar device with a failure rate expressed in cycles, then the total number of required operating cycles should be used instead of hours of operation.

Column 10: Under *Failures, F,* multiply the respective values in Column 8 by Column 9 to get the mission failures for the part. If the value in Column 9 is the same for all components in the circuit, sub-subsystem, or subsystem, then skip Columns 9 and 10.

Column 11: Under *Number of parts in series,* enter the number of identical parts used from the circuit diagram, or schematic, or the list of components used in series reliabilitywise. The identical parts should have identical application and operation stresses and identical failure modes, e.g., four diodes in a full-wave bridge rectifier. Two 10 kilohm resistors of identical manufacture, but used in different parts of the same circuit and dissipating different amounts of power, or two identical capacitors with differing applied voltages, etc., must be listed separately. If redundancy is involved, enter the value of one (1) in Column 11.

Column 12: Under *Applicable redundancy factor, K_r,* determine whether multiple components are used redundantly, and enter the applicable redundancy factor.

If components, circuits, or sub-subsystems are in parallel, standby, or have conditional redundancy, then the reliability, R, should be calculated, and then K_r calculated from

$$K_r = \frac{\log_e R}{\lambda t},$$

where

R = calculated reliability of the redundant parts, circuits, or sub-subsystems,

λt = product of the failure rate and the operating time of the primary part in the redundant configuration,

K_r = redundancy factor which when multiplied by λt gives the equivalent redundant circuit or sub-subsystem failures.

If two units are in parallel, then

$$R(t) = e^{-\lambda_1 t} + e^{-\lambda_2 t} - e^{-(\lambda_1 + \lambda_2)t},$$

or

$$R(t) = e^{-\lambda_{EQ} t} = e^{-K_r \lambda_1 t},$$

which is solved for K_r, and where

λ_{EQ} = equivalent failure rate of the two units in parallel strictly for computational purposes to get the total system failures yielding the correct system reliability,

λ_1 = failure rate of Unit 1, (however, any one of the two units can be used),

and

$$R(t) = e^{-K_r \lambda t}, \text{ if } \lambda_1 = \lambda_2 = \lambda.$$

If two units are in standby with perfect sensing and switching, then

$$R(t) = e^{-\lambda t}(1 + \lambda t),$$

or

$$R(t) = e^{-\lambda_{EQ} t} = e^{-K_r \lambda t},$$

which is solved for K_r. If no redundancy is involved, enter the value of one (1) in Column 12.

Column 13: Under *Total adjusted mission failures or failure rate*, multiply the values in Column 10 by Columns 11 and 12 and enter the results in the respective columns. These values are the exponent in

$$R = e^{-F_A}.$$

The values of F_A in Column 13 should be combined very carefully. Straight addition applies only when all components are reliabilitywise in series or when the proper redundancy factor has been entered into Column 12. Then the circuit, sub-subsystem, or system failures or failure rates are obtained by straight addition, and the overall reliability, R_S, is obtained from

$$R_S(t) = e^{-\sum F_{Ai}}.$$

If Columns 9 and 10 are skipped, then Column 13 is obtained from Columns $8 \times 11 \times 12$. This yields the total adjusted failure rate, λ_A. The overall failure rate then is obtained from $\sum \lambda_{Ai}$ and the corresponding reliability from

$$R_S(t) = e^{-(\sum \lambda_{Ai})t}.$$

With multiple failure modes, the applicable reliability math model should be used to arrive at the proper methodology for combining the failure rates or better yet the appropriate reliabilities. If multiple mission goals exist, then the reliability of each mission phase constituting that goal should be calculated from the failure rates and associated operating hours of the subsystems participating in a specific mission goal. The reliability of all mission phases can then be multiplied to arrive at the specific mission reliability.

It is strongly recommended that reliability predictions be programmed for the computer at the outset to reduce the tremendous labor involved, to enable speedy prediction, to facilitate revised predictions, to perform trade-off analyses, and to optimize the system's reliability.

EXAMPLE 12–1

Figure 12.13 is the schematic diagram of a rocket-borne photoflash unit. Do the following:

1. Prepare a functional diagram.

2. Draw the reliability block diagram.

3. Determine the reliability math model for this photoflash unit.

SOLUTIONS TO EXAMPLE 12–1

1. After studying Fig. 12.13 and discussing it with the designer the functional diagram of Fig. 12.14 was arrived at.

2. Figure 12.15 gives the reliability block of the photoflash unit based on Figs. 12.13 and 12.14.

3. From Fig. 12.15 the reliability math model for the photoflash unit is

$$R_T = \{[1 - (1 - R_{C1}R_{CR1})(1 - R_{C2}R_{CR2})]\} R_{S1}R_{S2}R_{M1}R_{M2}R_IR_B.$$

EXAMPLE 12–2

Table 12.1 gives the operational failure rates for the subsystems of a system whose reliability block diagram is given in Fig. 12.16. Do the following:

1. Find the system's maximum, mean, and minimum reliabilities for a 1,000-hr mission.

2. Discuss the significance of the three reliabilities.

SOLUTIONS TO EXAMPLE 12–2

1. The reliability math model for this system is

$$R_S = R_1R_5\{1 - [(1 - R_2R_3)(1 - R_4)]\}.$$

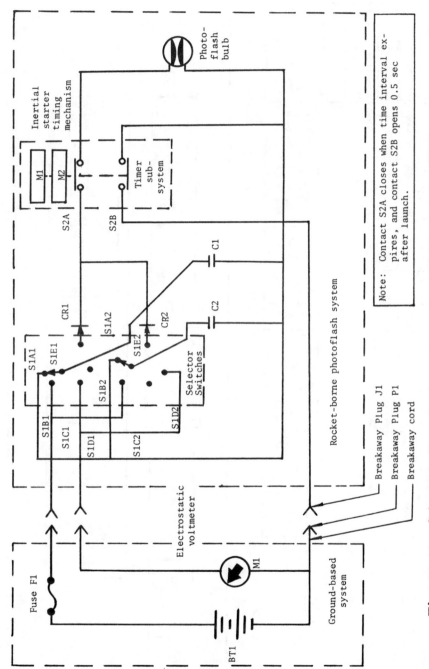

Fig. 12.13 – Schematic diagram of a rocket-borne photoflash unit of Example 12–1.

Note: Contact S2A closes when time interval expires, and contact S2B opens 0.5 sec after launch.

273

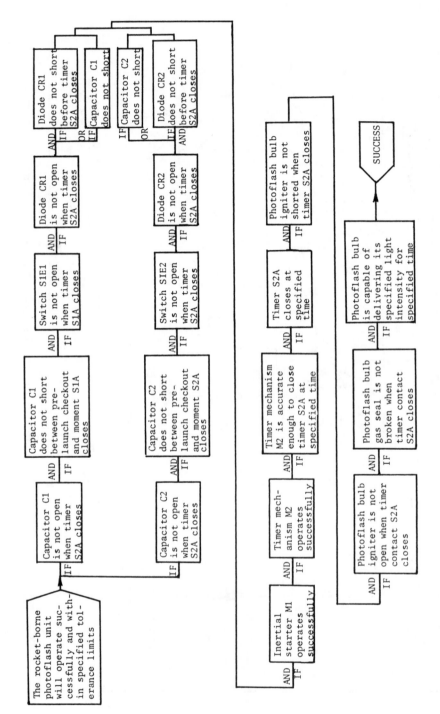

Fig. 12.14 – Reliability functional diagram of the rocket-borne photoflash unit of Example 12-1.

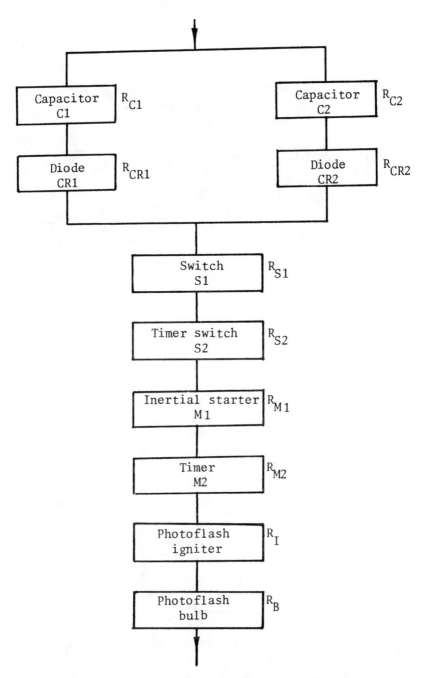

Fig. 12.15 — Reliability block diagram of the rocket-borne photoflash unit for Example 12–1.

TABLE 12.1 – Operational failure rates of the five subsystems in the system of Example 12–2.

Failure rate level	Operational failure rates, fr/10^6 hr				
	λ_1	λ_2	λ_3	λ_4	λ_5
Minimum	4	26	20	40	1
Mean	6	40	180	52	3
Maximum	13	80	350	168	9

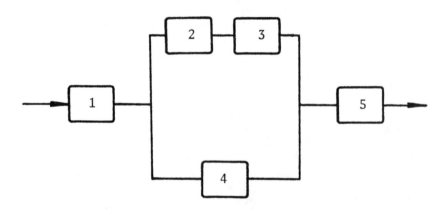

Fig. 12.16 – Reliability block diagram of the system in Example 12–2.

Substitution of the respective failure rates from Table 12.1 into the individual subsystem reliability equations yields the following subsystem reliabilities:

For the minimum λ	For the mean λ	For the maximum λ
$R_1 = e^{-4\times10^{-3}}$,	$R_1 = e^{-6\times10^{-3}}$,	$R_1 = e^{-13\times10^{-3}}$,
$= 0.996.$	$= 0.994.$	$= 0.987.$
$R_2 = e^{-26\times10^{-3}}$,	$R_2 = e^{-40\times10^{-3}}$,	$R_2 = e^{-80\times10^{-3}}$,
$= 0.974.$	$= 0.960.$	$= 0.920.$
$R_3 = e^{-20\times10^{-3}}$,	$R_3 = e^{-180\times10^{-3}}$,	$R_3 = e^{-350\times10^{-3}}$,
$= 0.980.$	$= 0.835.$	$= 0.705.$
$R_4 = e^{-40\times10^{-3}}$,	$R_4 = e^{-52\times10^{-3}}$,	$R_4 = e^{-168\times10^{-3}}$,
$= 0.960.$	$= 0.948.$	$= 0.850.$
$R_5 = e^{-1\times10^{-3}}$,	$R_5 = e^{-3\times10^{-3}}$,	$R_5 = e^{-9\times10^{-3}}$,
$= 0.999.$	$= 0.997.$	$= 0.991.$

Substitution of these subsystem reliabilities into the system's reliability math model yields the following system reliabilities for a 1,000-hr mission:

$$R_{S(max)} = (0.996)(0.999)[(0.974)(0.980) + 0.960 - (0.974)(0.980)(0.960)] = 0.995,$$

$$R_{S(mean)} = (0.994)(0.997)[(0.960)(0.835) + 0.948 - (0.960)(0.835)(0.948)] = 0.980,$$

and

$$R_{S(min)} = (0.987)(0.991)[(0.920)(0.705) + 0.850 - (0.920)(0.705)(0.850)] = 0.926.$$

Recall that the $R_{S(max)}$ value is obtained using the minimum operational subsystem λ_i, the $R_{S(mean)}$ using the mean λ_i, and the $R_{S(min)}$ using the maximum operational subsystem λ_i.

2. These reliabilities depict a meaningful picture as to a realistic estimate of the system's reliability range. The most likely estimate being $R_{S(mean)}$, somewhere between $R_{S(max)}$ and $R_{S(min)}$.

If $R_{S(min)}$ calculates out to be greater than the target system reliability, then there would be a relatively high probability of attaining the target reliability. If $R_{S(max)}$ calculates out to be smaller than the

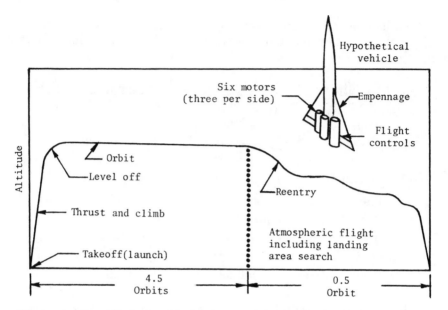

Fig. 12.17 – Flight profile for hypothetical vehicle of Example
12–3.

target system reliability, a very careful analysis of the system's design
is in order to upgrade its reliability.

EXAMPLE 12–3

A reliability analysis which consists of a probability analysis, failure cause and effect analysis, and safety analysis is performed on an
accessory power unit (APU) system. This is one of eight major systems of a hypothetical vehicle. The parameters of the vehicle, operational requirements, environmental requirements, system reliability
apportionment, APU functional description, APU probability analysis, APU failure cause and effect analysis, and APU safety analysis are
as follows:

1 – Parameters of hypothetical vehicle

The hypothetical vehicle is an unmanned, six-engine disposable
booster cluster, single ramjet flight engine, recoverable vehicle, capable
of launch, orbit, and reentry. Reentry includes normal atmospheric
flight and normal airplane-type landings. Automatic flight control is
used on all phases of flight. See Fig. 12.17. The following ground rules
are assumed:

1. The analysis is to cover the vehicle performance on only a single
   flight.

2. The vehicle is satisfactorily and completely checked out prior to

takeoff.

3. The reliability goal for one complete flight will be 90% probability of successfully completing a total flight time of 13.75 hr, which includes launch, orbit, reentry, and landing. This goal is specified by the customer and includes the reentry and landing phases for reasons of hardware cost and reusability and the desirability of recovering certain encapsulated data records.

4. The evaluation shall proceed on the premise that the vehicle itself will fail if any system or component fails, excepting, of course, local redundancy.

2 – Operational requirements

Automatic flight control is to be used to control the vehicle at all times during vehicle flight. During climb, the vehicle is stable. Upon cessation of the boost forces, due to thrust finalization and jettisoning of all boost engines, the level-off mode is entered into, followed by the orbital mode wherein the vehicle's inertial force is balanced by the gravity force. This orbital mode is an unstable one for directional control, so position control will be obtained by operating small reaction gas nozzles to maintain vehicle position, particularly with respect to the horizon. Reentry will again introduce a stabilizing medium, after suitable lowering of vehicle velocity by the use of drag brakes and by automatically dipping in and out of the atmosphere until the atmospheric flight control gear can be actuated. The vehicle is then flown to a selected landing area by use of the central ramjet engine. An instrument-controlled landing will be effected by use of the flight control gear and automatic actuation of the skid landing gear. There are assumed to be 25 minor functional systems operating during flight for this vehicle, as shown in Table 12.2. A simplified example of the operational profile for this vehicle is given in Fig. 12.18. Modes that will be used for a flight are launch, climb and level-off, orbit, reentry, atmospheric flight, and landing.

3 – Environmental requirements

Environmental situations and influences should be superimposed on one graph of time versus environmental magnitude(s), the latter being constructed to a comparative scale, where the maximum environmental condition expected is, say, equal to 100. Note in the profile of Fig. 12.19 that each environmental condition shown will reach the maximum line at least once. Thus, after construction of this profile, at any point in time the combination of environmental conditions is indicated.

TABLE 12.2 – Hypothetical vehicle functional systems of Example 12–3.

Major systems	Reliability, %	Functional system	System reliability goal	Estimated system reliability	Symbol of system
Environmental control	99.6%	Environmental control–heat	0.99980	xxxxx	R_{AC}
		Environmental control–cooling	0.99700	xxxxx	
		Environmental control–pressure	0.99900	xxxxx	
		Environmental control–humidity	0.99999	xxxxx	
Electrical distribution and power	98.7%	Accessory power unit	0.99500	See Reliability Analysis under the Solution	R_E
		Battery	0.99800	xxxxx	
		AC distribution	0.99950	xxxxx	
		DC distribution	0.99950	xxxxx	
		Hydraulic distribution	0.99494	xxxxx	
Guidance	97.7%	Stellar-inertial guidance	0.98300	xxxxx	R_G
		Landing radar	0.99400	xxxxx	
Flight control	99.2%	Flight control complex	0.99860	xxxxx	R_{FC}
		Flight control timer and computers	0.99350	xxxxx	

TABLE 12.2 – (Continued).

Major systems	Reliability, %	Functional system	System reliability goal	Estimated system reliability	Symbol of system
Hydraulics	99.6%	Flight control gear actuation	0.99800	xxxxx	R_H
		Boost motor jettison actuation	0.99976	xxxxx	
		Ram jet doors actuation	0.99990	xxxxx	
		Landing skid actuation	0.99800	xxxxx	
Landing	98.5%	Landing skid mechanism	0.99210	xxxxx	R_{LP}
		Brake mechanism	0.99310	xxxxx	
Propulsion	97.0%	Rocket boost motors	0.99955	xxxxx	R_P
		Ram jet engine and fuel system	0.98000	xxxxx	
		Retro-rocket system	0.99100	xxxxx	
Structure	99.7%	Pressure sealing system	0.99800	xxxxx	R_S
		Empennage and body structure	0.99996	xxxxx	
		Pyrographite panel complex	0.99950	xxxxx	
		Flight control gear	0.99960	xxxxx	
Product	90.0%				

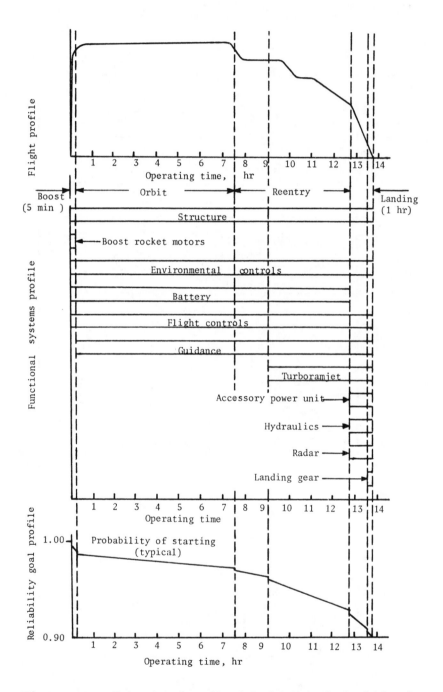

Fig. 12.18 – Operational profile of the hypothetical vehicle of Example 12–3.

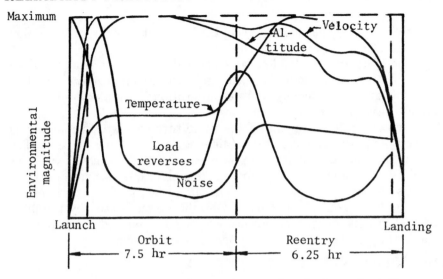

Fig. 12.19 – Environmental requirements for the hypothetical vehicle of Example 12-3.

4 – System reliability apportionment

The major systems appearing in Table 12.2 are considered to be in a series arrangement; i.e., the failure of any one system will cause failure of the vehicle. A *probability of success* of 90% was established as a total goal for the vehicle; therefore, it was necessary to apportion this goal within the eight major systems. Since each major system is equally responsible for overall vehicle reliability, it is evident that the reliability apportioned to each system must be contingent upon system performance function, complexity, state of the art, environment, and other factors. For example, the guidance system was apportioned a 97.7% reliability. This is one of the most important systems for mission accomplishment and yet it was apportioned one of the lowest reliability figures; this was necessary due to the complexity and state of the art. Complexity and state of the art advancement detract heavily from reliability. Consequently, complex developmental hardware must be accorded a low probability of success in apportionment exercises. Equivalent considerations would be used to apportion reliability to the remaining systems. Reliability apportionments should always be checked by reference to retained reliability data from previous company programs, industrial publications, government reports, and other available sources.

5 – Functional description

The system chosen for a detail example is the *accessory power unit (APU) system,* which is a part of the electrical distribution and power system. The purpose of the APU system is to generate ac, dc, and hydraulic power for the vehicle during the last hour of flight. A functional schematic is shown in Fig. 12.20.

One hour prior to landing an electrical signal fires a squib, which opens a pressure tank to force fuel into the decomposition chamber. The hot gas generated by fuel decomposition is capable of driving a double-pass turbine.

Two generators are run off the same turbine shaft, one supplying 400-cps ac power and the other 2,400-cps ac current, which is rectified to dc. In addition, hydraulic power is generated through a variable displacement pump, which is driven from the same turbine shaft.

The ac and dc power is used in the radar, landing mechanisms, and the flight control systems. Hydraulic power is used by the flight controls and landing mechanisms. The APU-generated power replaces the battery power used for all modes prior to landing.

SOLUTIONS TO EXAMPLE 12–3

RELIABILITY ANALYSIS

For successful accomplishment of the vehicle's mission, ac, dc, and hydraulic power are required during the landing phase. Therefore, the predicted reliability of the APU system is the probability of successfully providing all of these types of power for 1 hr after being subjected to 12.75 hr of the flight environment prior to operation. The APU system has a reliability goal of 0.99500 for 1 hr of operation, including the probability of starting.

A reliability block diagram of the APU system is given in Fig. 12.21. This figure lists all of the major components, showing series and parallel relationships, MTBF values, failure rates, and reliability values for instantaneously acting items.

1. The reliability of the parallel squib arrangement, R_S, depends upon the individual squib failing once in 10,000 trials; then $R_I = 0.9999$. The unreliability, Q_I, equals 0.0001 for each squib. Q total $= Q_I Q_I = (0.0001)(0.0001) = 0.00000001$. Then R total or $R_S = 1.0 - Q$ total $= 0.9_8$.

2. If it is recognized that the APU system from point A to B, C, D, and E of Fig. 12.21 represents a "common source" situation, it is convenient to treat it as four subsystems. System A to B is the turbine subsystem, B to C is the ac electrical generation subsystem, C to D is the dc electrical subsystem, and D to E is the hydraulic power subsystem. This situation conforms to the reliabilitywise series configuration. All units, except the squibs, are taken to have a constant failure rate.

3. The predicted reliability of the turbine subsystem will be:

$$\Sigma \lambda \text{ total A to B} \quad = \quad \lambda \text{ initiate } + \lambda \text{ battery } + \cdots + \lambda \text{ turbine shaft}$$

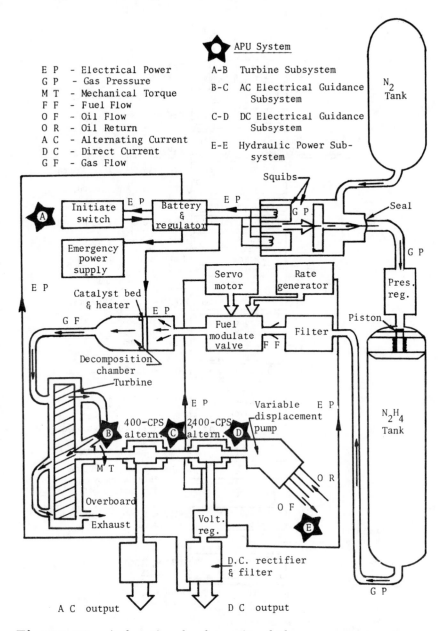

Fig. 12.20 — A functional schematic of the accessory power unit system for the hypothetical vehicle of Example 12–3.

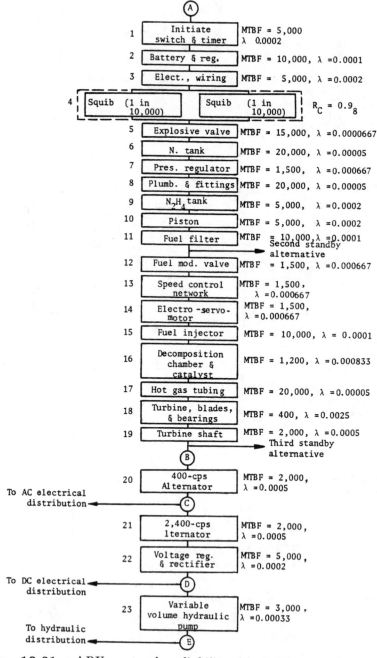

Fig. 12.21 – APU system's reliability block diagram for Example 12–3.

$$= 0.0078177 \text{ fr/hr, omitting squibs.}$$

$R_T(\text{Turbine for time} = 1 \text{ hr}) = e^{-(0.0078177) \cdot 1} = 0.992213,$

omitting squibs.

$R_{AB} = R_S \times R_T = (0.9_8)(0.992213) = 0.992213, \text{ including squibs,}$

where $R_S = 0.9_8$, the parallel squib reliability. The predicted reliability of the ac electrical generation subsystem will be

$\Sigma\lambda \text{ total B to C} = \lambda \text{ alternator} = 0.0005,$

$R_{BC} = e^{-(0.0005)1} = 0.9995.$

The predicted reliability of the dc electrical generation subsystem will be

$\Sigma\lambda \text{ C to D} = \lambda \text{ alternator} + \lambda \text{ voltage regulator and rectifier,}$
$= 0.0007,$

$R_{CD} = e^{-(0.0007)1} = 0.9993.$

The predicted reliability of the hydraulic power subsystem will be

$\Sigma\lambda \text{ total D to E} = \lambda \text{ hydraulic pump} = 0.00033,$

$R_{DE} = e^{-(0.00033)1} = 0.99967.$

Consequently, since this APU system operates as a series with regard to vehicle mission success (dc, ac, and hydraulic power are all essential), the APU system's predicted reliability will be

$R_{APU} = R_{AB} \times R_{BC} \times R_{CD} \times R_{DE},$
$= (0.992213)(0.9995)(0.9993)(0.99967) = 0.99070.$

It should be noted in the above calculations that, while all components are accounted for, no component was considered more than once. It should also be noted that the series type of operation allows the multiplication of subsystem reliability figures. The reliability goal for the

APU system is 0.9950, whereas the results of the probability analysis indicate that a reliability of 0.99070 will be obtained. The designer must determine the significance of these figures and what actions he or she should take, if any, to improve the APU's reliability.

It must be remembered that the 0.99070 figure is far from an exact indicator of reliability, especially during very early design phases. One could argue with complete veracity that a little more optimism in regard to the MTBF assignments in Fig. 12.21 might have resulted in an estimated APU reliability of 0.9950 instead of 0.99070. Assuming, however, that the MTBF assignments are reasonable, one is obliged to recognize that the 0.99070 figure is close to actuality and very useful as a tool in arriving at a decision in regard to design action.

The difference between 0.9950 and 0.99070 is 0.0043, a rather unimpressive figure. The ratio between the unreliabilities is a more useful tool. The unreliability for $R = 0.9950$ is 0.0050 and for $R = 0.9907$ is 0.0093. This results in a ratio of 1.0 to 1.86. Such a low ratio would not impell a designer to take drastic and expensive redesign actions. However, within the acknowledged realm of accuracy of data, the designer would be wise to watch the development of the APU closely and to implement methods of improving system reliability. The need for drastic action becomes progressively more acute as the relative unreliability ratio approaches or exceeds 1 to 7. If, for example, the APU reliability figure had been 0.9070 instead of 0.9907, the unreliability ratio would have been 1 to 10. In such a case, the significance of the ratio would justify the designer in recommending the implementation of every possible reliability improvement.

The steps that may be taken to improve the inherent design reliability of an item or to adjust the significance of the reliability with respect to system requirements are the following:

1. Incorporate redundancy in the design.

2. Improve the design of the components.

3. Perform trade-off studies to ascertain if the reliability of other systems can be increased.

These possibilities are examined next. Refinements in design and actions to promote manufacturing success, improve testing techniques, and delineate proper handling procedures are important, too, but are discussed later as methods to prevent degradation of inherent reliability.

SYSTEM IMPROVEMENT

Three general methods exist for improving system reliability, as noted previously. All of these methods must be examined in detail in any effective effort to improve the reliability of a system. Inasmuch as each method often presents alternate possibilities, as is shown next, it is evident that the prime objective of a reliability improvement study is to so exhaust the available possibilities that the maximum improvement can be made within the natural limitations of cost, weight, space, complexity, scheduling, range, and state of the art.

REDUNDANCY

In regard to the first method, redundancy may be undertaken with components or whole systems, depending on the situation. For example, the pressure regulator, fuel modulating valve, speed control network, electro-servo motor, decomposition chamber and catalyst, and turbine blades and bearings are all relatively low reliability items. Most of these components could be directly paralleled physically, if desired, with appreciable increases in system reliability. In many cases, an increase in redundancy at the detail part level, where cost or weight increases are minor, will gain a greater reliability increase than component redundancy at the system level. The turbine blades and bearings in our APU example, however, which have the lowest reliability in the system, could not be paralleled without paralleling the entire turbine. Three of the posible methods of paralleling the turbine are presented next for illustrative purposes. These are all instances of standby redundancy rather than parallel redundancy.

It should be noted though that such redundancy can create serious problems. Aside from weight, space, and complexity considerations, such factors as the effect of turbine redundancy on the efficiency of other systems must be evaluated before a decision is reached. For example, since turbines ordinarily require an interval of time for power buildup, some provision such as a constantly heated catalytic screen should be maintained in the standby turbine to assist initial operation.

Three possible turbine standby examples considered here are the following:

1. Provide two complete turbine systems, one in standby, including individual fuel systems and power generating equipment. One turbine system would be alternate to the other and would commence operation only upon a signal from a decision device, based on evidence of failure in the turbine or any of its power generating equipment. This type of standby redundancy would provide a system reliability of 0.999957, but would be very heavy, bulky, and complicated.

2. Provide two full-size turbines, one in standby, utilizing some elements of a common fuel system. The fuel system would be common

from the initiate switch and timer to and include the fuel filter, so fuel system elements would be duplicated from and include the fuel modulating valve onward. Each turbine would be equipped with complete power generating equipment, but one turbine would be alternate to the other and would commence operation only upon a signal from a decision device, based on evidence of failure in the turbine or any of its power generating equipment. This arrangement would provide a system reliability of 0.99815 and would be less bulky and complicated than the previous case, although it would necessarily include a fuel switching device. This arrangement, in addition to possible component improvements farther upstream, would probably yield a highly reliable system. A diagram of the required standby arrangement is shown in Fig. 12.22(a).

3. Provide two full-size turbines, one in standby, driving the same shaft through a selective coupling device. The fuel system would be common from the initiate switch through and including the hot gas tubing. The gas tubing, however, would include a hot gas switching device which would allow shutdown of the primary turbine and operation of the standby unit. The hot gas switching device would be actuated by its decision device, based on evidence of failure in the primary turbine. Inasmuch as these turbines are on one shaft and are to drive a common set of power generating equipment, a selective coupling device is required to allow operational isolation of the nonoperating turbine. This design concept would provide a system reliability of 0.98871. This alternative suffers from some disadvantages, such as vehicle vulnerability due to the common shaft, shaft bearings, and power generating equipment. A diagram of the required standby arrangement is shown in Fig. 12.22(b).

In view of the above, a decision would probably be made in favor of the second alternative, due to weight, size, expense, installation, and complexity factors, as well as reliability considerations, However, this decision would not eliminate the desirability of component design improvements.

IMPROVING THE DESIGN

As noted previously, the APU system includes a number of relatively low reliability components. A hydrazine (N_2H_4) fueled APU is not ordinarily capable of full-power checking without subsequent removal for servicing and refueling and, consequently, may never be fired in service except on a mission. Periodic checking of the turbine by running up to some nominal speed by means of compressed air is an acceptable procedure for lubrication exercise and electrical and hydraulic checking, but it does not prove out the fuel system or produce full output. Furthermore, some components in a compact APU

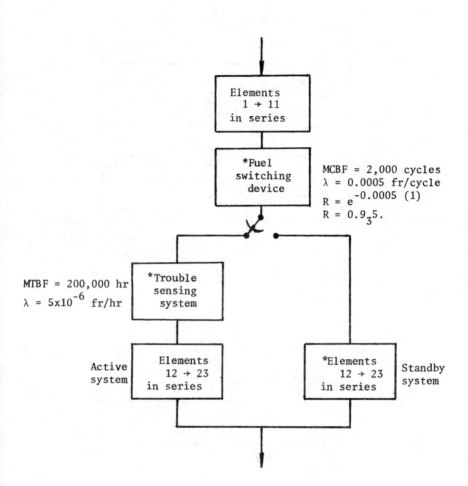

* Added to system shown in Fig. 12.21.

Fig. 12.22(a) – Reliability block diagram of the second alternative configuration for Example 12–3.

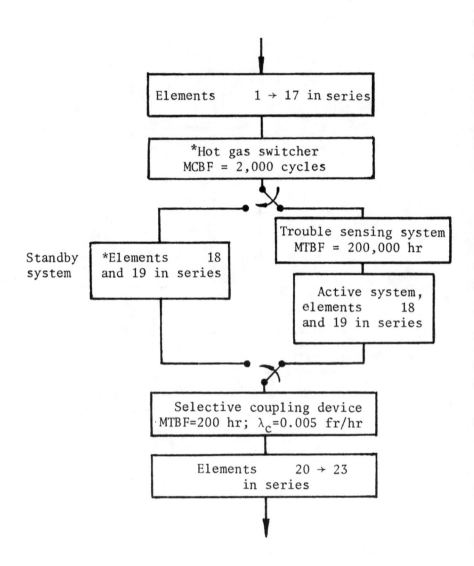

* Added to system shown in Fig. 12.21.

Fig. 12.22(b) – Reliability block diagram of the third alternative
configuration for Example 12–3.

may not be replacable without virtual disassembly of the unit. Consequently, since APU checking and maintenance would be difficult and expensive, except at the vendor's facility, it is essential that the established reliability of the assembly and its components be unimpeachable by the time the system is subjected to field service.

In a component improvement program, aspects such as the following are considered:

1. *Use Reliable Components and Parts*

   (1.1) Determine the best type by past experience and evaluative tests.

   (1.2) Determine the best vendor by past experience, screening tests and tests to failure, and vendor facility survey.

   (1.3) Control vendors with adequate procurement specification and perform qualification, lot acceptance, and requalification tests.

2. *Design for Reliability*

   (2.1) Derate components, parts.

   (2.2) Minimize component, part variation effects.

   (2.3) Use proven circuits.

   (2.4) Design for redundancy.

   (2.5) Make reliability predictions.

   (2.6) Design for environment packaging by:

   (*a*) Shock protection.
   (*b*) Humidity protection.
   (*c*) Adequate heat flow.
   (*d*) Vibration absorption.

   (2.7) Balance the packaging for:

   (*a*) Reliability.
   (*b*) Maintainability.
   (*c*) Producibility.
   (*d*) Operability.
   (*e*) Size and weight.

3. *Simplify the Design*

   Minimize the number of parts within a component and reduce the number of components within a system.

4. *Clarify Functions*

   Examine system makeup and complexities to determine if the number of components and interrelationships indicated are needed to perform the required function. Examine system relationship to the vehicle in mind to determine if the function is actually necessary.

5. *Combine Functions*

   Determine if any system(s) can perform more than one function. Single parts or components can often be utilized for dual purposes, thus eliminating the need for other parts.

6. *Reduce Functions*

   Reduce the number of functions that must be performed to allow a vehicle to complete its mission.

7. *Reduce Checks and Balances*

   The number of automatic tolerance controls on various actions resolving any function should be minimized. The design should be ingenious enough so that it does not depend on this type of control for adequate performance.

8. *Decrease Function Time*

   A reduction of system and component function time will result in a decrease in random failure possibility and a subsequent increase in reliability.

9. *Improve Resistance to Environment*

   For example, in the case of high-temperature and high-radiation effects, the most likely methods of providing increased resistance would be to provide shielding, a controlled environment, or increased design safety factors. If one cannot design to adequately resist environmental factors, due to weight limits, for example, reduce the time exposure to these conditions as much as possible.

10. *Provide for Close Control of Scheduled Replacement*

    Frequent replacement of short-lived components will achieve increased reliability. If components cannot be replaced manually before the scheduled replacement time occurs, decision devices, such as a failure sensing switch, can be provided to allow the function to be carried on by a standby component. Replacement should be scheduled to occur near the end of the chance failure phase (useful life) but prior to the wear-out phase.

RELIABILITY APPORTIONMENT TRADES

In cases in which a system presses the state of the art so closely that no realistic reliability improvement can be made in the design, it may be possible to look to other systems of the vehicle in order to help meet the overall vehicle reliability goal. In such cases, there are two recourses that may be examined and utilized individually, or combined, as seems most profitable. These are:

1. If other systems or subsystems of the vehicle are predicted to be more reliable than their apportioned goals require, it may be possible to reapportion their reliability goals so that the APU can be accepted without further improvement. This adjustment of system goals must be made so that the end item reliability requirement is still met or exceeded.

2. The second method of system reliability apportionment trading is to pursue the improvement of systems which press the state of the art less closely than the APU. When and if other such systems are improved, their cumulative reliability improvement can be used to balance the margin of unreliability that is known to exist on the APU system. An example of a major system trade to improve reliability might be to utilize the more reliable solid propellant boost rocket in place of the liquid fueled rocket motor. An example of a simple system trade would be to utilize the RL-X-96 propellant in place of RL-X-89 to improve boost rocket capability for withstanding low temperatures at the cost of a 2% maximum loss in thrust.

FAILURE CAUSE AND EFFECT ANALYSIS

A system failure cause and effect analysis considers the failures that may be expected to occur to any item within the vehicle. These failures may be in the form of a basic component failure, unstable interactions of related components which support a particular system function, uncoordinated interactions of related systems which support the vehicle in performance, or adverse installation relationships of unrelated items. The assumption is made that the checkable system under consideration will be maintained and will work as required at the start of its function. An example of a failure cause and effect analysis on the hypothetical vehicle under consideration is demonstrated on the APU system in Table 12.3. Also see Chapter 17.

TABLE 12.3–Cause and effect analysis of hypothetical APU system failures.

Failure	Effect	Means of detection	Means of reducing failure rate
Squib fails to fire.	Mission failure.	Electrical circuit can be checked but the condition of explosive squibs cannot be checked. These, however, are in parallel so should be quite reliable.	Check electric circuitry prior to launching. The initiation switch should be rugged.
Explosive valve.	Mission failure.	Valve action cannot be checked.	These valves are fairly reliable and can be designed to be even better.
N_2 pressure to N_2H_4 tank.	Mission failure.	Leakage of N_2 gas can be checked prior to launch.	Preventive maintenance. Conduct vibraton tests on APU system.
N_2H_4 tank piston.	Mission failure.	Pressure leakage, or blow-by cannot be checked prior to launch.	The piston should be made more rugged and reliable.
Fuel flow regulation.	Mission failure.	May be able to check this by independent plug-on connections prior to launch.	Component testing can determine a low failure rate for this.

TABLE 12.3—Continued.

Failure	Effect	Means of detection	Means of reducing failure rate
Speed control network.	Mission failure.	Out-of-tolerance erratic operation of the speed controlling device can result in a chain reaction failure which will affect the servomotor, the fuel injector, and eventually the turbine speed, etc.	Preventive maintenance. The alternator network and motor could be operated as a package on the bench.
Electro-servo-motor.	Mission failure.	Failure of the motor could consist of poor response, overspeeding or ceasing to run. These all result in chain reaction failures.	Preventive maintenance.
Fuel injector.	Mission failure.	Fuel injector may be clogged. No way to detect prior to launch except by running system.	Two fuel injectors could be placed in parallel.
Decomposition chamber.	Mission failure.	No means of detection except by the operation of the system prior to launch.	The catalyst screen could be made coarser to elim-inate any clogging.

TABLE 12.3–Continued.

Failure	Effect	Means of detection	Means of reducing failure rate
Tubing failure.	Mission failure.	Tubing leakage may start a fire or cause explosion in equipment compartment.	Close clamp spacing and careful installation and inspection.
Turbine failure.	Mission failure.	Blade shedding will be indicated by turbine stopping or by explosion.	Design for maximum blade and wheel strength.
Voltage regulator.	Mission failure.	Electronic equipment may not function properly or may burn out.	Install components of high proven reliability or use redundant design techniques.
Summary for APU system.	All items will result in mission failure. This is a critical situation and indicates that extreme care is required to achieve design reliability.	Most means of detection are rather inconclusive. This indicates a requirement for better design from the "ease of checkout" standpoint.	The overall goal may be met by: (1) Using two APUs operating in standby. (2) Performing trade studies to see if goal can be raised on other systems. (3) Improving the reliability of components in the APU.

SAFETY ANALYSIS

Safety analysis considers the results of the failures analyzed in the failure cause and effect analysis, with reference to other systems and any hazards to the vehicle, its maintenance crews, friendly territory, flight crew, etc. In the example of the APU, all failures could have a safety effect, since failure of the APU would cause mission failure. A safety analysis would be conducted in the following manner:

1. Utilize the prior failure analysis to locate all possible failures.

2. Review each failure and all possible consequences. Consider the physical location of the failure within the vehicle with reference to hazards.

3. Consider any chain reaction failure possibilities and secondary causes leading to major failures. Note that most accidents are of these types, rather than by single failure.

4. List the degree of explosion or fire hazard existing throughout the vehicle. What explosion proof requirements are imposed?

5. Identify items which could result in safety of flight and safety of crews.

6. Recommend to the designer methods of eliminating or reducing the consequences of failures.

RELIABILITY TESTING

The APU system for the hypothetical vehicle is also used here for the example of a mission event reliability test because of the short operating time and the relative ease of simulating environmental conditions. It should be noted, however, that funds for pure reliability testing are sometimes hard to obtain. It is often necessary to use the available data from other specialized tests to calculate the reliability. Nevertheless, the following example is presented in the interest of thoroughness.

In this example, it is planned that qualification testing of reliability will be performed on five APU systems assembled from new components to the existing design configuration, and which have been previously run only for the acceptance tests. Each APU system is mounted to simulate the mission installation and environment. The method of demonstration will be a reliability events test procedure based on repeated trials of the environmental and performance requirements. The performance criterion is defined as 1 hr of operation, with standard operating procedures being used, with normal start and normal shutdown. The environmental conditions imposed will be all operational environments and as close to flight plan as practical. Environments are to be combined where possible and the sequence of events will be in the order of environmental severity. Each APU system will undergo two events with each environment before the process is repeated. Each APU system will be cycled in a numbered sequence so that a uniform amount of testing time will be imposed on all units. After a cycle, the APU system will be serviced and acceptance tested before being submitted for further reliability cycling. If a design change becomes necessary at any point, the entire reliability test series must be run again from the start. The design operational reliability goal for the APU system is 99.5% as developed in the reliability analysis, and it is considered that this goal should be demonstrated with 90% confidence.

In establishing a reliability qualification test demonstration procedure, it is necessary to consider the cost and time for testing as well as the reliability goal and the attached confidence. In this case, assume that there are five test cells which could accommodate the five APU systems. There also are enough personnel and extra test equipment to run the five cells simultaneously. Investigation has indicated that a cost of $650.00 per test hour run is to be expected, including fuel at $3.00 per pound. The cost is estimated on the basis of an event being 1 hr. The binomial test approach for *one failure* and a reliability goal of 99.5% with 90% confidence requires 825 events or cycles of 1 hr duration.

This would result in a cost of $536,250 and 825 cycles of one hour each, or roughly 20.6 weeks at 8 hr per day, 5 days per week. Since time of testing is assumed to be critical, it is decided to use a 16-hr, 7-day week, which increases the cost of testing to $790.00 per event. This raises the possible total cost to $651,750, but reduces the time of test to approximately 7.3 weeks. Requiring the reliability test for the APU system to cycle 825 times with no more than one failure provides not only a representative sample, but also reduces the probability (risk) of accepting the APU system when the reliability is less than 99.5%. All of the APU systems are to be cycled the total number of times required to meet the test for a reliability design objective level of 99.5%, assuming each cycle is a valid sample, according to the specified event ground rules. Event ground rules for unsatisfactory and satisfactory events are based on the following:

1. Unsatisfactory events

   (a) Events which record any deviation from specified tolerances on power output.

   (b) Events in which an unscheduled shutdown of the test occurs, if the APU system is at fault.

2. Satisfactory events

(*a*) Event duration must equal or exceed 1 hr.

(*b*) Power output must be within specification tolerances.

All failures experienced in the reliability test, whether the test is passed or not, are to be thoroughly investigated and corrected if possible. The 825 events with no more than one failure and the ground rules for unsatisfactory and satisfactory events would be requirements of the specifications and monitored for accomplishment.

CONCLUSION

The usefulness of the reliability tools illustrated by this example is obvious. It should be emphasized that, while reliability figures are estimates, the numbers derived are first approximations to reliability figures. Such numbers, however, are of great use in early design phases before hardware test data are available.

The numbers derived in this example are indicators of relative reliability within the vehicle or system, particularly in view of state of the art considerations. In addition, they provide an initial indication as to whether a system can be expected to meet its reliability goal, reveal areas of low relative reliability, assist in discovering critical parts, promote an understanding of the results and consequences of failure, and provide a basis for refined reliability estimates when test data become available.

EXAMPLE 12–4

Determine the following accessory power unit system reliabilities of Example 12–3 and compare the results with those given in Example 12–3:

1. When two complete turbine systems are used, one in standby.

2. When two full-size turbines utilizing some elements of a common fuel system are used, one in standby.

3. When two full-size turbines driving the same shaft through a selective coupling device are used, one in standby.

SOLUTIONS TO EXAMPLE 12–4

1. The reliability block diagram for the system with the entire fuel system in standby is shown on Fig. 12.23(a). The reliability of a

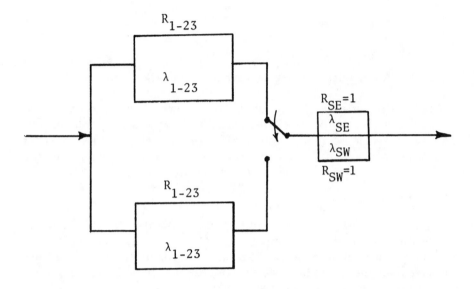

Fig. 12.23(a)–Reliability block diagram for Example 12–4, Case 1.

standby system with perfect sensing and switching, or $R_{SE} = R_{SW} = 1$, is

$$R(t) = e^{\lambda_{1-23}t}(1 + t\lambda_{1-23}).$$

The reliability of the squib subsystem is

$$R(t) = 0.99999999.$$

The equivalent failure rate is

$$\lambda_{EQ} = \frac{-\ln R(t)}{t},$$

or for $t = 1$ hr

$$\lambda_{EQ} = \frac{-\ln[0.99999999]}{1},$$

or

$$\lambda_{EQ} = 10^{-8} \text{ fr/hr.}$$

The failure rate of the system is

$$\lambda_{1-23} = \sum_{i=1}^{23} \lambda_i,$$

or

$$\lambda_{1-23} = 0.0093477 \text{ fr/hr.}$$

Thus, the system's reliability is

$$R(t) = e^{-0.0093477t}[1 + 0.0093477t],$$

and for a 1 hr mission the reliability is

$$R(t = 1 \text{ hr}) = 0.999957,$$

which agrees with the value given in Example 12–3.

2. The reliability block diagram of the power supply system when a common fuel system, as shown in Fig. 12.22(a), is used is shown in Fig. 12.23(b). The reliability is given by

$$
R(t) = R_{1-11}(t)\Big\{ R_{12-23}(t) + \int_{t_1=0}^{t} f_{12-23}(t_1) \\
\cdot R_{SW}(1 \text{ cyc})R_{SE}(t_1)R_{12-23}(t - t_1)\, dt_1 \Big\},
$$

where

$$
\begin{aligned}
R_{1-11}(t) &= e^{-\lambda_{1-11}t}, \\
R_{12-23}(t) &= e^{-\lambda_{12-23}t}, \\
f_{12-23}(t) &= -\frac{dR_{12-23}(t)}{dt} = \lambda_{12-23}e^{-\lambda_{12-23}t}, \\
R_{SW}(1) &= e^{-\lambda_{SW}},
\end{aligned}
$$

and

$$R_{SE}(t) = e^{-\lambda_{SE}t}.$$

Then

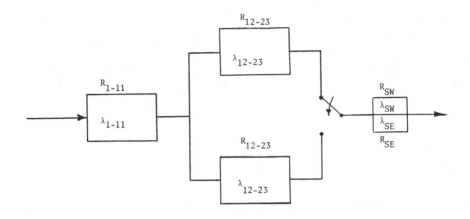

Fig. 12.23(b)–Reliability block diagram for Example 12–4, Case 2.

$$R(t) \;=\; e^{-\lambda_{1-11}t}\Big\{e^{-\lambda_{12-23}t} + \int_0^t \lambda_{12-23}e^{-\lambda_{12-23}t_1}$$
$$\cdot\, e^{-\lambda_{SW}}e^{-\lambda_{SE}t_1}e^{-\lambda_{12-23}(t-t_1)}dt_1\Big\},$$

or

$$R(t) \;=\; e^{-\lambda_{1-11}t}\Big\{e^{-\lambda_{12-23}t}$$
$$+ \frac{\lambda_{12-23}e^{-\lambda_{SW}}e^{-\lambda_{12-23}t}}{\lambda_{SE}}[1 - e^{-\lambda_{SE}t}]\Big\}.$$

The failure rates are

$$\lambda_{1-11} = \sum_{i=1}^{11}\lambda_i = 0.0018337 \text{ fr/hr},$$

$$\lambda_{12-23} = \sum_{i=12}^{23}\lambda_i = 0.007514 \text{ fr/hr},$$

$$\lambda_{SW} = 0.0005 \text{ fr/cycle},$$

$$\lambda_{SE} = 0.000005 \text{ fr/hr}.$$

Thus,

$$R(t = 1 \text{ hr}) = e^{-0.0018337(1)}\Big\{ e^{-0.007514(1)}$$

$$+ \frac{0.007514e^{-0.000005}e^{-0.007514(1)}}{0.000005}$$

$$\cdot [1 - e^{-0.000005(1)}]\Big\},$$

or

$$R(t = 1 \text{ hr}) = 0.99815,$$

which agrees with the value given in Example 12–3.

3. When two full-sized turbines are used, one in standby, as shown in Fig. 12.22(b), the reliability block diagram is as shown in Fig. 12.23(c). The reliability is given by

$$R(t) = e^{-\lambda_{1-17}t}\Big\{ e^{-\lambda_{18-19}t}$$

$$+ \frac{\lambda_{18-19}e^{-\lambda_{SW}}e^{-\lambda_{18-19}t}}{\lambda_{SE}}[1 - e^{-\lambda_{SE}t}]\Big\}$$

$$\cdot e^{-\lambda_C t}e^{-\lambda_{20-23}t},$$

where

$$\lambda_{1-17} = 0.0048177 \text{ fr/hr},$$
$$\lambda_{18-19} = 0.003 \text{ fr/hr},$$
$$\lambda_{SW} = 0.00005 \text{ fr/hr},$$
$$\lambda_{SE} = 0.000005 \text{ fr/hr},$$
$$\lambda_C = 0.005 \text{ fr/hr},$$

and

$$\lambda_{20-23} = 0.00153 \text{ fr/hr}.$$

For a 1 hr mission

$$R(t = 1 \text{ hr}) = 0.98871,$$

which agrees with the value given in Example 12–3.

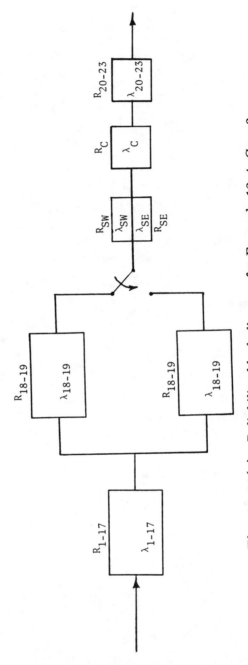

Fig. 12.23(c) —Reliability block diagram for Example 12–4, Case 3.

EXAMPLE 12-5

Three electrical generating systems for a twin-engine aircraft are shown in Fig. 12.24 in a block diagram form. The major components of the system are the engines, the generators, and the frequency changers. The generators are direct, engine driven, variable frequency machines. Their output is fed into static frequency changers which convert it into constant-frequency power.

The frequency changers in Configuration 1 are rated at 60 kVA, and in Configurations 2 and 3 at 30 kVA. In Configuration 3 the frequency changer in the middle can be switched automatically to either of two generators.

Make a comparative study of the three configurations on the basis of their reliability to supply 60 kVA of normal power and 30 kVA of emergency power.

SOLUTIONS TO EXAMPLE 12-5

Configuration 1 for 60-kVA Power Supply

Configuration 1 is a case of two paths reliabilitywise in parallel. Only one path can be allowed to fail if normal operation at 60 kVA or emergency operation at 30 kVA is to be maintained. Therefore, the system's reliability is given by

$$R_S = 1 - (1 - R_{E_1} \cdot R_{G_1} \cdot R_{F_1})(1 - R_{E_2} \cdot R_{G_2} \cdot R_{F_2}). \quad (12.1)$$

If

$$R_{E_1} = R_{E_2} = R_E,$$
$$R_{G_1} = R_{G_2} = R_G,$$

and

$$R_{F_1} = R_{F_2} = R_F,$$

then Eq.(12.1) becomes

$$R_S = 1 - (1 - R_E R_G R_F)^2,$$

where

R_E = reliability of the engine,

R_G = reliability of the generator,

and

R_F = reliability of the frequency changer.

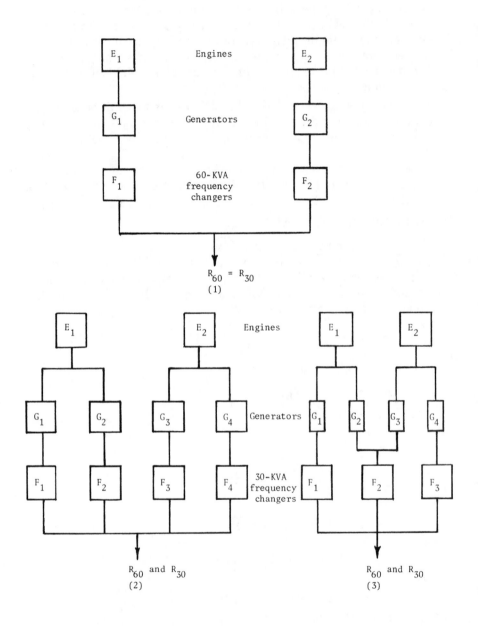

Fig. 12.24 – Three configurations of a twin-engine aircraft electricity generating system for Example 12–5.

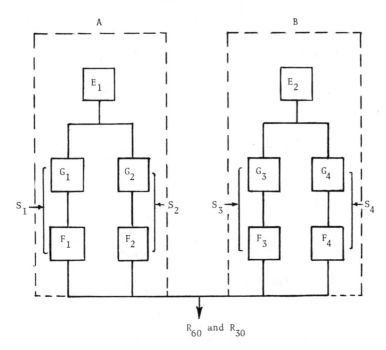

Fig. 12.25–Configuration 2 for Example 12-5.

The system's reliability of supplying 60 kVA is identical with its reliability for emergency operation; hence

$$R_{60} = R_{30} = 1 - (1 - R_E R_G R_F)^2.$$

Configuration 2 for 30-kVA Power Supply

Configuration 2, shown in Fig. 12.25, has two generators on each engine and is capable of supplying 30-kVA emergency power if at least one of the two generator-frequency sets driven by the surviving engine operates; therefore,

$$R_{30} = 1 - Q_A Q_B,$$

where

$$Q_A = \text{unreliability of the left system,}$$

and

$$Q_B = \text{unreliability of the right system.}$$

Then

$$Q_A = 1 - R_{E_1}[1 - (1 - R_{G_1}R_{F_1})(1 - R_{G_2}R_{F_2})],$$

and

$$Q_B = 1 - R_{E_2}[1 - (1 - R_{G_3}R_{F_3})(1 - R_{G_4}R_{F_4})].$$

If

and

$$R_{G_1} = R_{G_2} = R_G,$$

$$R_{F_1} = R_{F_2} = R_F,$$

then

$$R_{30} = 1 - \{1 - R_E[1 - (1 - R_G R_F)^2]\}^2.$$

Configuration 2 for 60-kVA Power Supply, Solution 1

The system's reliability of supplying 60 kVA can be calculated by the application of Bayes' Theorem. The generator-frequency-changer set on the left has a reliability of

$$R_{S_1} = R_{G_1}R_{F_1},$$

and unreliability of

$$Q_{S_1} = 1 - R_{S_1}.$$

The system's reliability is

$$R_{S_{60}} = (R_S|S_{1G})R_{S_1} + (R_S|S_{1B})(1 - R_{S_1}), \qquad (12.2)$$

where

$$R_S|S_{1G} = (R_S|E_{1G}|S_{1G})R_{E_1} + (R_S|E_{1B}|S_{1G})(1 - R_{E_1}), \qquad (12.3)$$

and

$$\begin{aligned}
R_S|E_{1G}|S_{1G} &= (R_S|S_{2G}|E_{1G}|S_{1G})R_{S_2} \\
&\quad + (R_S|S_{2B}|E_{1G}|S_{1G})(1 - R_{S_2}).
\end{aligned} \qquad (12.4)$$

But

$$R_S|S_{2G}|E_{1G}|S_{1G} = 1$$

and

$$R_S|S_{2B}|E_{1G}|S_{1G} \;=\; R_{E_2}[1 - (1 - R_{S_3})(1 - R_{S_4})],$$

or

$$=\; R_{E_2}(R_{S_3} + R_{S_4} - R_{S_3}R_{S_4}).$$

Substituting these values into Eq. (12.4), we get

$$R_S|E_{1G}|S_{1G} = R_{S_2} + R_{E_2}(R_{S_3} + R_{S_4} - R_{S_3}R_{S_4})(1 - R_{S_2}),$$

or

$$
\begin{aligned}
R_S|E_{1G}|S_{1G} \;=\;& R_{S_2} + R_{E_2}(R_{S_3} + R_{S_4} - R_{S_3}R_{S_4}) \\
& - R_{E_2}R_{S_2}(R_{S_3} + R_{S_4} - R_{S_3}R_{S_4}),
\end{aligned}
\qquad (12.5)
$$

and

$$R_S|E_{1B}|S_{1G} \;=\; R_{E_2}R_{S_3}R_{S_4}. \qquad (12.6)$$

Substituting Eqs. (12.5) and (12.6) into Eq. (12.3) yields

$$
\begin{aligned}
R_S|S_{1G} \;=\;& R_{E_1}R_{S_2} + R_{E_1}R_{E_2}(R_{S_3} + R_{S_4} - R_{S_3}R_{S_4}) \\
& - R_{E_1}R_{E_2}R_{S_2}(R_{S_3} + R_{S_4} - R_{S_3}R_{S_4}) \\
& + R_{E_2}R_{S_3}R_{S_4} - R_{E_1}R_{E_2}R_{S_3}R_{S_4}.
\end{aligned}
\qquad (12.7)
$$

The probability of survival of the system if S_1 is bad is given by

$$R_S|S_{1B} = (R_S|E_{1G}|S_{1B})R_{E_1} + (R_S|E_{1B}|S_{1B})(1 - R_{E_1}), \qquad (12.8)$$

where

$$
\begin{aligned}
R_S|E_{1G}|S_{1B} \;=\;& R_{S_2}R_{E_2}[1 - (1 - R_{S_3})(1 - R_{S_4})] \\
& + (1 - R_{S_2})(R_{E_2}R_{S_3}R_{S_4}),
\end{aligned}
$$

or

$$
\begin{aligned}
R_S|E_{1G}|S_{1B} \;=\;& R_{S_2}R_{E_2}(R_{S_3} + R_{S_4} - R_{S_3}R_{S_4}) + R_{E_2}R_{S_3}R_{S_4} \\
& - R_{S_2}R_{E_2}R_{S_3}R_{S_4},
\end{aligned}
\qquad (12.9)
$$

and

$$R_S|E_{1B}|S_{1B} = R_{E_2} R_{S_3} R_{S_4}, \tag{12.10}$$

because reliabilitywise E_2, S_3 and S_4 are in series to supply 60-kVA frequency.

Substitution of Eqs. (12.9) and (12.10) into Eq. (12.8) yields

$$R_S|S_{1B} = R_{E_1} R_{S_2} R_{E_2}(R_{S_3} + R_{S_4} - R_{S_3} R_{S_4}) + R_{E_1} R_{E_2} R_{S_3} R_{S_4}$$
$$- R_{E_1} R_{S_2} R_{E_2} R_{S_3} R_{S_4} + R_{E_2} R_{S_3} R_{S_4} - R_{E_1} R_{E_2} R_{S_3} R_{S_4}. \tag{12.11}$$

Substituting $R_S|S_{1G}$ and $R_S|S_{1B}$ from Eqs. (12.7) and (12.11), respectively, into Eq. (12.2) yields the system's reliability, or

$$\begin{aligned}
R_{S_{60}} = {}& R_{S_1} R_{E_1} R_{S_2} + R_{S_1} R_{E_1} R_{E_2}(R_{S_3} + R_{S_4} - R_{S_3} R_{S_4}) \\
& - R_{S_1} R_{E_1} R_{E_2} R_{S_2}(R_{S_3} + R_{S_4} - R_{S_3} R_{S_4}) + R_{S_1} R_{E_2} R_{S_3} R_{S_4} \\
& - R_{S_1} R_{E_1} R_{E_2} R_{S_3} R_{S_4} + R_{E_1} R_{S_2} R_{E_2}(R_{S_3} + R_{S_4} - R_{S_3} R_{S_4}) \\
& + R_{E_1} R_{E_2} R_{S_3} R_{S_4} - R_{E_1} R_{S_2} R_{E_2} R_{S_3} R_{S_4} + R_{E_2} R_{S_3} R_{S_4} \\
& - R_{E_1} R_{E_2} R_{S_3} R_{S_4} - R_{S_1} R_{E_1} R_{E_2} R_{S_2}(R_{S_3} + R_{S_4} - R_{S_3} R_{S_4}) \\
& - R_{S_1} R_{E_1} R_{E_2} R_{S_3} R_{S_4} + R_{S_1} R_{E_1} R_{S_2} R_{E_2} R_{S_3} R_{S_4} \\
& - R_{S_1} R_{E_2} R_{S_3} R_{S_4} + R_{S_1} R_{E_1} R_{E_2} R_{S_3} R_{S_4}. \tag{12.12}
\end{aligned}$$

If

$$R_{E_1} = R_{E_2} = R_E,$$

and

$$R_{S_1} = R_{S_2} = R_{S_3} = R_{S_4} = R_S,$$

Eq. (12.12) becomes

$$\begin{aligned}
R_{S_{60}} = {}& R_S^2 R_E + R_S R_E^2(2R_S - R_S^2) \\
& - R_S^2 R_E^2(2R_S - R_S^2) + R_E R_S^3 - R_E^2 R_S^3 \\
& + R_E^2 R_S(2R_S - R_S^2) + R_E^2 R_S^2 - R_E^2 R_S^3 + R_E R_S^2 \\
& - R_E^2 R_S^3 - R_E^2 R_S^2(2R_S - R_S^2) - R_E^2 R_S^3 \\
& + R_E^2 R_S^4 - R_E R_S^3 + R_E^2 R_S^3.
\end{aligned}$$

Simplification yields

$$R_{S_{60}} = (2R_E + 4R_E^2)R_S^2 - 8R_E^2 R_S^3 + 3R_E^2 R_S^4,$$

where

$$R_S = R_G R_F.$$

Therefore,

$$R_{S_{60}} = (2R_E + 4R_E^2)R_G^2 R_F^2 - 8R_E^2 R_G^3 R_F^3 + 3R_E^2 R_G^4 R_F^4. \quad (12.13)$$

Configuration 2 for 60-kVA Power Supply, Solution 2

The system's reliability can also be derived based on the binomial expansion. To supply 60-kVA of power, at least two out of four generator-frequency-changer sets must operate. Then, from the binomial expansion,

$$(R_S + Q_S)^4 = R_S^4 + 4R_S^3 Q_S + 6R_S^2 Q_S^2 + 4R_S Q_S^3 + Q_S^4 = 1, \quad (12.14)$$

where

$$R_S = R_G R_F.$$

Since at least two out of four generator-frequency-changer sets are needed, from Eq. (12.14)

$$P(2 \text{ or more sets surviving}) = R_S^4 + 4R_S^3 Q_S + 6R_S^2 Q_S^2. \quad (12.15)$$

Now consider for each term in Eq. (12.15) the ways in which it occurs. The term $6R_S^2 Q_S^2$ in Eq. (12.15) means there are six ways the system will survive with exactly two generator-frequency-changer sets operating:

1. Two ways occur when Sets 1 and 2 survive or Sets 3 and 4 survive. For each of these the respective engine must also survive; consequently, in these two cases, the probability of system survival is given by

$$2R_E R_S^2 Q_S^2.$$

2. There are four other ways two generator-frequency-changer sets survive:

(2.1) Sets 1 and 2 survive.

(2.2) Sets 2 and 3 survive.

(2.3) Sets 1 and 4 survive.

(2.4) Sets 2 and 4 survive.

For each of the above four ways, both engines must survive. So in these four cases, the probability the system survives is given by

$$4R_E^2 R_S^2 Q_S^2.$$

Thus the probability that the system survives when exactly two generator-frequency-changer sets are operating is given by

$$p_1 = (2R_E + 4R_E^4)R_S^2 Q_S^2.$$

The term $4R_S^3 Q_S$ in Eq. (12.15) gives the probability that exactly three generator-frequency-changer sets survive, and for system survival both engines need to survive. Therefore, the probability the system survives when exactly three generator-frequency-changer sets are operating is given by

$$p_2 = 4R_E^2 R_S^3 Q_S.$$

Similarly, for the cases that four generator-frequency-changer sets are operating, the probability of system survival is given by

$$p_3 = R_E^2 R_S^4.$$

Thus the system's reliability to supply 60 kVA power is given by

$$R_{S_{60}} = p_1 + p_2 + p_3,$$

or

$$R_{S_{60}} = (2R_E + 4R_E^2)R_S^2 Q_S^2 + 4R_E^2 R_S^3 Q_S + R_E^2 R_S^4. \qquad (12.16)$$

But

$$R_S = R_G R_F,$$

and

$$Q_S = 1 - R_G R_F.$$

Therefore, Eq. (12.16) becomes

$$R_{S_{60}} = (2R_E + 4R_E^2)R_G^2 R_F^2 (1 - R_G R_F)^2$$
$$+ 4R_E^2 R_G^3 R_F^3 (1 - R_G R_F) + R_E^2 R_G^4 R_F^4. \qquad (12.17)$$

Rearranging Eq. (12.17) yields

$$R_{S_{60}} = (2R_E + 4R_E^2)R_G^2 R_F^2 - 8R_E^2 R_G^3 R_F^3 + 3R_E^2 R_G^4 R_F^4, \qquad (12.18)$$

the same as Eq. (12.13).

Configuration 3 for 60-kVA Power Supply

Configuration 3 has two generators on each engine. The system's reliability can be calculated by applying Bayes' theorem. The system's unreliability to supply 60-kVA power can be written as

$$Q_{S_{60}} = (Q_S|S_{1G})R_{S_1} + (Q_S|S_{1B})(1 - R_{S_1}), \qquad (12.19)$$

with

$$R_{S_1} = R_{G_1} R_{F_1}.$$

Therefore, Eq. (12.19) becomes

$$Q_{S_{60}} = (Q_S|S_{1G})R_{G_1} R_{F_1} + (Q_S|S_{1B})(1 - R_{G_1} R_{F_1}), (12.20)$$

where

$Q_S|S_{1G}$ = probability of system failure if Set S_1, consisting of G_1 and F_1, is good,

and

$Q_S|S_{1B}$ = probability of system failure if Set S_1 is bad.

But

$$Q_S|S_{1G} = (Q_S|E_{1G}|S_{1G})R_{E_1} + (Q_S|E_{1B}|S_{1G})(1 - R_{E_1}). \qquad (12.21)$$

In Eq. (12.21)

$$Q_S|E_{1G}|S_{1G} = (Q_S|G_{2G}|E_{1G}|S_{1G})R_{G_2} + (Q_S|G_{2B}|E_{1G}|S_{1G})(1 - R_{G_2}), \qquad (12.22)$$

where

$$Q_S|G_{2G}|E_{1G}|S_{1G} = (1 - R_{F_2})(1 - R_{E_2} R_{G_4} R_{F_3}), \qquad (12.23)$$

and

$Q_S|G_{2B}|E_{1G}|S_{1G} = 1$ - P(E_2 and at least one of the generator-frequency-changer sets driven by E_2 are operating.)

Since the probability that at least one of the generator-frequency-changer sets driven by E_2 works is given by

$$R_{G_3} R_{F_2} + R_{G_4} R_{F_3} - R_{G_3} R_{F_2} R_{G_4} R_{F_3},$$

then

$$Q_S | G_{2B} | E_{1G} | S_{1G} = 1 - R_{E_2}(R_{G_3} R_{F_2} + R_{G_4} R_{F_3} \\ - R_{G_3} R_{F_2} R_{G_4} R_{F_3}). \tag{12.24}$$

Substituting Eqs. (12.23) and (12.24) into Eq. (12.22) yields

$$Q_S | E_{1G} | S_{1G} = (1 - R_{F_2})(1 - R_{E_2} R_{G_4} R_{F_3}) R_{G_2} \\ + [1 - R_{E_2}(R_{G_3} R_{F_2} + R_{G_4} R_{F_3} \\ - R_{G_3} R_{F_2} R_{G_4} R_{F_3})](1 - R_{G_2}). \tag{12.25}$$

In Eq. (12.21)

$$Q_S | E_{1B} | S_{1G} = 1 - R_{E_2} R_{G_3} R_{F_2} R_{G_4} R_{F_3}. \tag{12.26}$$

Substituting Eqs. (12.25) and (12.26) into Eq. (12.21) yields

$$Q_S | S_{1G} = R_{E_1}\{(1 - R_{F_2})(1 - R_{E_2} R_{G_4} R_{F_3}) R_{G_2} + (1 - R_{G_2}) \\ \cdot [1 - R_{E_2}(R_{G_3} R_{F_2} + R_{G_4} R_{F_3} - R_{G_3} R_{F_2} R_{G_4} R_{F_3})]\} \\ + (1 - R_{E_2} R_{G_3} R_{F_2} R_{G_4} R_{F_3})(1 - R_{E_1}) \tag{12.27}$$

The probability of system failure when S_1 is bad can be written as

$$Q_S | S_{1B} = (Q_S | E_{1G} | S_{1B}) R_{E_1} + (Q_S | E_{1B} | S_{1B})(1 - R_{E_1}), \tag{12.28}$$

where

$$Q_S | E_{1G} | S_{1B} = (Q_S | G_{2G} | E_{1G} | S_{1B}) R_{G_2} \\ + (Q_S | G_{2B} | E_{1G} | S_{1B})(1 - R_{G_2}). \tag{12.29}$$

But

$$Q_S | G_{2G} | E_{1G} | S_{1B} = 1 - R_{E_2} R_{F_2} R_{G_4} R_{F_3}, \tag{12.30}$$

and

$$Q_S | G_{2B} | E_{1G} | S_{1B} = 1 - R_{E_2} R_{G_3} R_{F_2} R_{G_4} R_{F_3}. \tag{12.31}$$

Substituting Eqs. (12.30) and (12.31) into Eq. (12.29) yields

$$Q_S | E_{1G} | S_{1B} = R_{G_2}(1 - R_{E_2} R_{F_2} R_{G_4} R_{F_3})$$

$$+ (1 - R_{G_2})(1 - R_{E_2} R_{G_3} R_{F_2} R_{G_4} R_{F_3}), \qquad (12.32)$$

and

$$Q_S|E_{1B}|S_{1B} = 1 - R_{E_2} R_{G_3} R_{F_2} R_{G_4} R_{F_3}. \qquad (12.33)$$

Therefore,

$$
\begin{aligned}
Q_S|S_{1B} = {} & R_{E_1}[R_{G_2}(1 - R_{E_2} R_{F_2} R_{G_4} R_{F_3}) \\
& + (1 - R_{G_2})(1 - R_{E_2} R_{G_3} R_{F_2} R_{G_4} R_{F_3})] \\
& + (1 - R_{E_1})(1 - R_{E_2} R_{G_3} R_{F_2} R_{G_4} R_{F_3}). \qquad (12.34)
\end{aligned}
$$

Simplifying Eq. (12.34) yields

$$
\begin{aligned}
Q_S|S_{1B} = {} & R_{E_1} R_{G_2}(1 - R_{E_1} R_{G_2} R_{G_4} R_{F_3}) \\
& + (1 - R_{E_1} R_{G_2})(1 - R_{E_2} R_{G_3} R_{F_2} R_{G_4} R_{F_3}). \qquad (12.35)
\end{aligned}
$$

The terms $Q_S|S_{1G}$ and $Q_S|S_{1B}$ are based on the requirement that at least two out of three frequency changers must operate, with their separate generators and engines, to obtain 60-kVA power.

Substituting the results of Eqs. (12.27) and (12.35) into Eq. (12.20) yields the probability of system failure, or

$$
\begin{aligned}
Q_{S_{60}} = {} & R_{G_1} R_{F_1} \Big\{ R_{E_1} \{ R_{G_2}(1 - R_{F_2})(1 - R_{E_2} R_{G_4} R_{F_3}) \\
& + (1 - R_{G_2})[1 - R_{E_2}(R_{G_3} R_{F_2} + R_{G_4} R_{F_3} \\
& - R_{G_3} R_{F_2} R_{G_4} R_{F_3})]\} \\
& + (1 - R_{E_1})(1 - R_{E_2} R_{G_3} R_{F_2} R_{G_4} R_{F_3}) \Big\} \\
& + (1 - R_{G_1} R_{F_1})[R_{E_1} R_{G_2}(1 - R_{E_2} R_{F_2} R_{G_4} R_{F_3}) \\
& + (1 - R_{E_1} R_{G_2})(1 - R_{E_2} R_{G_3} R_{F_2} R_{G_4} R_{F_3})]. \qquad (12.36)
\end{aligned}
$$

Hence the system's reliability for supplying 60-kVA power is given by

$$R_{S_{60}} = 1 - Q_{S_{60}}. \qquad (12.37)$$

Setting

$$
\begin{aligned}
R_{E_1} &= R_{E_2} = R_E, \\
R_{S_1} &= R_{S_2} = R_{S_3} = R_S, \\
R_{G_1} &= R_{G_2} = R_{G_3} = R_{G_4} = R_G,
\end{aligned}
$$

and

$$R_{F_1} = R_{F_2} = R_{F_3} = R_F,$$

in Eq. (12.36), Eq. (12.37) can be written as

$$
\begin{aligned}
R_{S_{60}} = 1 - \Big[& R_G R_F \big\{ R_E \{ R_G (1 - R_F)(1 - R_E R_G R_F) + (1 - R_G) \\
& \cdot [1 - R_E (R_G R_F + R_G R_F - R_G^2 R_F^2)] \} \\
& \cdot (1 - R_E)(1 - R_E R_G^2 R_F^2) \big\} \\
& \cdot (1 - R_G R_F)[R_E R_G (1 - R_E R_F^2 R_G) \\
& \cdot (1 - R_E R_G)(1 - R_E R_G^2 R_F^2)] \Big].
\end{aligned}
\tag{12.37'}
$$

Simplification of Eq. (12.37') yields

$$
R_{S_{60}} = R_F^2 (3 R_E^2 R_G^2 - 3 R_E^2 R_G^3 + 2 R_E R_G^2) - R_F^3 (4 R_E^2 R_G^3 - 2 R_E^2 R_G^4).
$$

$$
\tag{12.38}
$$

Configuration 3 for 30-kVA Power Supply

The system failure probability of Configuration 3 for supplying 30 kVA of power can be written as

$$
Q_{S_{30}} = (Q_S | S_{1G}) R_{S_1} + (Q_S | S_{1B})(1 - R_{S_1}),
\tag{12.39}
$$

where

$$
R_{S_1} = R_{G_1} R_{F_1}.
$$

Therefore, Eq. (12.39) becomes

$$
Q_{S_{30}} = (Q_S | S_{1G}) R_{G_1} R_{F_1} + (Q_S | S_{1B})(1 - R_{G_1} R_{F_1}),
\tag{12.40}
$$

where

$Q_S | S_{1G}$ = probability of system failure if Set S_1 is good,

and

$Q_S | S_{1B}$ = probability of system failure if Set S_1 is bad.

The terms $Q_S | S_{1G}$ and $Q_S | S_{1B}$ are based on the requirement that at least one out of three frequency changers must operate to put out 30-kVA power.

$Q_S | S_{1G}$ is calculated as follows:

$$
Q_S | S_{1G} = (Q_S | E_{1G} | S_{1G}) R_{E_1} + (Q_S | E_{1B} | S_{1G})(1 - R_{E_1});
\tag{12.41}
$$

but

$$Q_S|E_{1G}|S_{1G} = 0,$$

and

$$Q_S|E_{1B}|S_{1G} = (Q_S|E_{2G}|E_{1B}|S_{1G})R_{E_2} \\ + (Q_S|E_{2B}|E_{1B}|S_{1G})(1 - R_{E_2}). \quad (12.42)$$

In Eq. (12.42)

$$Q_S|E_{2G}|E_{1B}|S_{1G} = (1 - R_{G_3}R_{F_2})(1 - R_{G_4}R_{F_3}),$$

while

$$Q_S|E_{2B}|E_{1B}|S_{1G} = 1.$$

Substitution of these values into Eq. (12.42) yields

$$Q_S|E_{1B}|S_{1G} = (1 - R_{G_3}R_{F_2})(1 - R_{G_4}R_{F_3})R_{E_2} + (1)(1 - R_{E_2}). \quad (12.42')$$

Substitution of these values into Eq. (12.41) yields

$$Q_S|S_{1G} = (1 - R_{E_1})[(1 - R_{G_3}R_{F_2})(1 - R_{G_4}R_{F_3})R_{E_2} + (1 - R_{E_2})]. \quad (12.41')$$

Setting

$$R_{E_1} = R_{E_2} = R_E,$$
$$R_{F_2} = R_{F_3} = R_F,$$

and

$$R_{G_3} = R_{G_4} = R_G$$

in Eq. (12.41') and simplifying yields

$$Q_S|S_{1G} = (1 - R_E)[1 - R_E(2R_G R_F - R_G^2 R_F^2)]. \quad (12.41'')$$

The probability of system failure, given that S_1 is bad, is

$$Q_S|S_{1B} = (Q_S|E_{1G}|S_{1B})R_{E_1} \\ + (Q_S|E_{1B}|S_{1B})(1 - R_{E_1}), \quad (12.43)$$

where

$$Q_S|E_{1G}|S_{1B} = (Q_S|G_{2G}|E_{1G}|S_{1B})R_{G_2}$$
$$+ (Q_S|G_{2B}|E_{1G}|S_{1B})(1 - R_{G_2}),$$

$$Q_S|G_{2G}|E_{1G}|S_{1B} = (1 - R_{F_2})(1 - R_{E_2}R_{G_4}R_{F_3}),$$

$$Q_S|G_{2B}|E_{1G}|S_{1B} = 1 - \{R_{E_2}[1 - (1 - R_{G_3}R_{F_2})$$
$$\cdot (1 - R_{G_4}R_{F_3})]\},$$
$$= (1 - R_{E_2}) + R_{E_2}(1 - R_{G_3}R_{F_2})$$
$$\cdot (1 - R_{G_4}R_{F_3}), \qquad (12.44)$$

and

$$Q_S|E_{1B}|S_{1B}) = (Q_S|E_{2G}|E_{1B}|S_{1B})R_{E_2}$$
$$+ (Q_S|E_{2B}|E_{1B}|S_{1B})(1 - R_{E_2}),$$
$$= (1 - R_{G_3}R_{F_2})(1 - R_{G_4}R_{F_3})R_{E_2}$$
$$+ (1)(1 - R_{E_2}).$$

Substituting these values into Eq. (12.43) yields

$$Q_S|S_{1B} = R_{E_1}R_{G_2}(1 - R_{F_2})(1 - R_{E_2}R_{G_4}R_{F_3})$$
$$+ (1 - R_{G_2})R_{E_1}[(1 - R_{G_3}R_{F_2})(1 - R_{G_4}R_{F_3})R_{E_2}$$
$$+ (1 - R_{E_2})]$$
$$+ (1 - R_{E_1})[(1 - R_{G_3}R_{F_2})(1 - R_{G_4}R_{F_3})R_{E_2}$$
$$+ (1 - R_{E_2})],$$
$$= R_{E_1}R_{G_2}(1 - R_{F_2})(1 - R_{E_2}R_{G_4}R_{F_3})$$
$$+ [(1 - R_{E_1}) + (1 - R_{G_2})R_{E_1}][(1 - R_{G_3}R_{F_2})$$
$$\cdot (1 - R_{G_4}R_{F_3})R_{E_2}$$
$$+ (1 - R_{E_2})]. \qquad (12.43')$$

Setting
$$R_{E_1} = R_{E_2} = R_E,$$
$$R_{F_1} = R_{F_2} = R_F,$$
and
$$R_{G_1} = R_{G_2} = R_{G_3} = R_{G_4} = R_G,$$

Eq. (12.43′) simplifies to

$$Q_S|S_{1B} = R_E R_G(1 - R_F)(1 - R_E R_G R_F) + (1 - R_E R_G)$$
$$\cdot [1 - R_E(2R_G R_F - R_G^2 R_F^2)].$$

$$(12.43'')$$

Substituting the results of Eqs. (12.41″) and (12.43″) into Eq. (12.40) yields the system's unreliability for 30-kVA power, or

$$Q_{S_{30}} = R_G R_F(1 - R_E)[1 - R_E(2R_G R_F - R_G^2 R_F^2)]$$
$$+ (1 - R_G R_F)\{R_E R_G(1 - R_F)(1 - R_E R_G R_F)$$
$$+ (1 - R_E R_G)[1 - R_E(2R_G R_F - R_G^2 R_F^2)]\}, \quad (12.45)$$

and the system's reliability for 30-kVA power is given by

$$R_{S_{30}} = 1 - \Big\{ R_E R_F(1 - R_E)[1 - R_E(2R_G R_F - R_G^2 R_F^2)]$$
$$+ (1 - R_G R_F)\{R_E R_G(1 - R_F)(1 - R_E R_G R_F)$$
$$+ (1 - R_E R_G)[1 - R_E(2R_G R_F - R_G^2 R_F^2)]\}\Big\}. \quad (12.46)$$

If numerical values for failure rates of the engines, generators, and frequency changers are known, then for a particular mission the reliability of the system for the three configurations can be calculated.

It can be proved that Configuration 2 is the most reliable, but Configuration 3 may be acceptable if the reliability goal is met, as well as the safety requirements.

PROBLEMS

12-1. A lunar excursion module communication S-band power amplifier for a soft landing on the moon has the functional block diagram given in Fig. 12.26. The failure rates for each subassembly are given in Table 12.4. The characteristics of an amplitron are such that if its high-voltage coil fails or is not energized it acts as a waveguide with negligible decibel (db) loss of signal strength. Hence, when functionally connected in series, amplitrons are redundant reliabilitywise. For the problem, consider the failure rate of the amplitron low-voltage coils to be negligible. The control module is used to detect the error signal and to switch on one or the other high-voltage power supply, but not both simultaneously. In other words, only one high-voltage power suply and one amplitron can be used at a time.

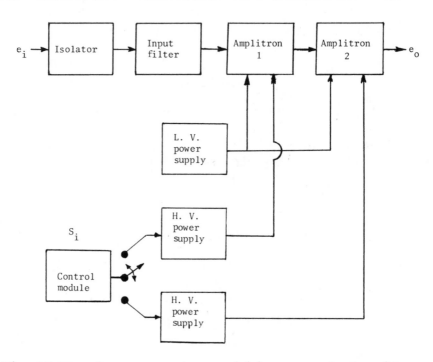

Fig. 12.26 – Lunar excursion module's communication S-band power amplifier functional diagram for Problem 12-1.

TABLE 12.4 – The failure rate of each subassembly of the lunar excursion module's communication S-band power amplifier of Problem 12-1.

Subassembly	Failure rate, fr/10^6 hr	Failure rate, fr/cycle of switching
Isolator	5	–
Input filter	5	–
Control module	10	0.1
Low-voltage power supply	20	–
High-voltage power supply	100	–
Amplitron (includes low-voltage coils)	50	–

(1) Sketch the reliability block diagram.

(2) Write the reliability mathematical model, or equation, for the S-band power amplifier.

(3) If the moon mission is planned for 1,000-hr duration, what is the probability that the second high-voltage power supply and amplitron will not be used?

(4) Will be used?

(5) What is the probability that the S-band power amplifier will survive the mission?

12-2. It is known that tubes exhibit wear-out characteristics. By extensive testing we find that in a small seven-tube communication receiver the reliabilitywise series tubes have the following wear-out life characteristics and chance failure rates:

| Tube number | Wear-out | | λ_c, fr/10^6 hr |
	Mean life, hr	Standard deviation, hr	
1	9,300	900	15
2	9,500	850	20
3	9,885	920	10
4	10,000	960	15
5	10,115	990	25
6	10,500	1,000	20
7	10,700	1,050	10

(1) If we ignore the chance failure rate of the tubes, answer the following questions:

1.1 If the receiver has already been operating 40 hr per week, 50 weeks per year for 4 1/2 years without a tube replacement, what is the probability that it will operate another 3 months without a tube replacement? Take 3 months to be 500 hr of operation.

1.2 If it has been operating for 5 1/2 years at the same schedule without a tube replacement, what is the probability that it will operate another 3 months without a tube replacement?

(2) If we include the chance failure rate, and 20 % of the tubes fail due to chance and 80 % fail due to wear-out, answer the following questions:

2.1 If the receiver has already operated for 4 1/2 years, what is the probability that it will operate another 3 months without a tube replacement?

2.2 If it has been operating for 5 1/2 years, what is the probability that it will operate another 3 months without a tube replacement?

(3) If chance failures or wear-out failures have necessitated the replacement of several tubes during the 4 1/2 year period so that the tubes in the receiver are of mixed age, resulting in an overall average failure rate approaching 145.5 fr/10^6 hr, answer the following questions:

3.1 If the receiver has already operated for 4 1/2 years, what is the probability that it will operate another 3 months?

3.2 If it has been operating for 5 1/2 years, what is the probability that it will operate another 3 months?

12-3. There are 1,000 display systems in use. Each display system has the following subsystems and respective failure rates in "bits":

Quantity	Subsystem	Failure rate, $\lambda \times 10^{10}$ fr/hr
1	Input interface and data distributor	2,500
1	Output interface and data commutator	1,300
3	Buffers and associated logic	7,600 each
5	Display consoles	44,000 each

If all three buffers are needed, but only three out of the five display consoles are needed for full data display capability, do the following:

(1) Draw the reliability block diagram.

(2) Write the reliability math model.

(3) Compute the MTBF of the system, assuming constant failure rates.

(4) What is the probability that the radar will last 1,000 hr without emergency shutdown for repairs?

(5) What is the expected number of shutdowns over a 1-year period if the system is operated an average of 3,000 hr per year?

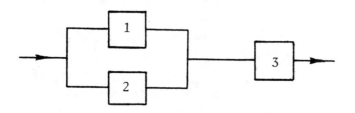

Fig. 12.27 – The subsystem of Problem 12-4.

(6) What is the expected number of subsystem failures (i.e., repairs) for the same period? Assume a decaying population.

12-4. A subsystem is composed of three units which are arranged reliabilitywise as shown in Fig. 12.27.

Unit 1 has a generic chance failure rate of 2 fr/10^6 hr, an application factor of $K_{ap} = 0.6$, a mean wear-out life of $\bar{T} = 9,910$ hr and a standard deviation of $\sigma_T = 2,080$ hr. Assume a normal time-to-wear-out-failure distribution. Take $N_c/N = 0.80$ and $N_w/N = 0.20$.

Unit 2 exhibits a Weibull time-to-early-failure distribution given by

$$f_e(T) = \frac{\beta_e}{\eta_e} \left(\frac{T - \gamma_e}{\eta_e} \right)^{(\beta_e - 1)} e^{-\left[(T - \gamma_e)/\eta_e \right]^{\beta_e}},$$

where

$$\gamma_e = 0.0 \text{ hr}, \quad \beta_e = 0.2, \quad \eta_e = 48 \text{ hr}.$$

The generic chance failure rate is 10 fr/10^6 hr with $K_{ap} = 1$. The unit exhibits a Gaussian wear-out time-to-failure distribution with $\bar{T} = 10,000$ hr and a standard deviation of $\sigma_T = 2,000$ hr. Take $N_e/N = 0.05$, $N_c/N = 0.80$, and $N_w/N = 0.15$.

Unit 3 is functioning during its useful life for the mission. It is energized 65 % of the time and is de-energized 34 % of the time, while it is switched on and off for the remainder of the mission 100 times. Its energized failure rate is 10 fr/10^6 hr, its de-energized failure rate is 0.1 fr/10^6 hr, and its switching failure rate is 5 fr/10^6 cycles. Its $K_{ap} = 1$.

What is the subsystem's reliability for a mission of 300 hr when Unit 1 starts the mission at the age of 3,000 hr, Unit 2 starts the mission at the age of 2 hr, and Unit 3 starts the mission at the age of 500 hr? The whole system is in an operating environment with a K_{op} of 10.

Assume that the early and wear-out pdf's are those exhibited in a $K_{op} = 10$ operating environment.

12-5. An electronic computer contains 1,000 ceramic capacitors of MIL-C-11015A in series in addition to a number of resistors, transistors, diodes, and other parts.

(1) What is the expected number of capacitors that will fail during a useful life period of 100,000 hr if the capacitors are used at their full-rated voltage and in an ambient temperature of 80°C. See Fig. 12.28 for the failure rates.

(2) How many will fail if at full-rated voltage but in an ambient temperature of 45°C?

(3) How many will fail at 0.6 of rated voltage and 60°C?

(4) How many will fail at 0.2 of rated voltage and 35°C?

(5) How many will fail at 0.2 of rated voltage and 125°C?

(6) If your opinion were asked by design engineers, what derating level and ambient temperature stress limits would you prescribe for good reliability engineering practice?

(7) If the capacitors are used at their full-rated voltage and in an ambient temperature of 80°C, what is the probability that no capacitors will fail during a 10-hr period?

(8) What is their reliability for a 100-hr period?

(9) Repeat Cases 7 and 8 if the capacitors are used at 0.2 of rated voltage and 35°C.

12-6. It is necessary to build a highly reliable power source for an aircraft electrical system. The maximum electrical load to be supplied is assumed to be 10 kW, the average load 7 kW, and the vital functions alone may be run with only 5 kW. Three different generating means are being considered: use of a single 10-kW generator, use of two 5-kW generators with their outputs in parallel, or three 3.5-kW generators with parallelled outputs. Three modes of system operation exist: full rated power, average power, and emergency power, as given in Table 12.5.

(1) Derive the reliability expression for each mode-generator combination and write the end result in the appropriate box in Table 12.5, assuming the reliability of the 10-kW generator is R_1, of the 5-kW generator is R_2, and of the 3.5-kW generator is R_3.

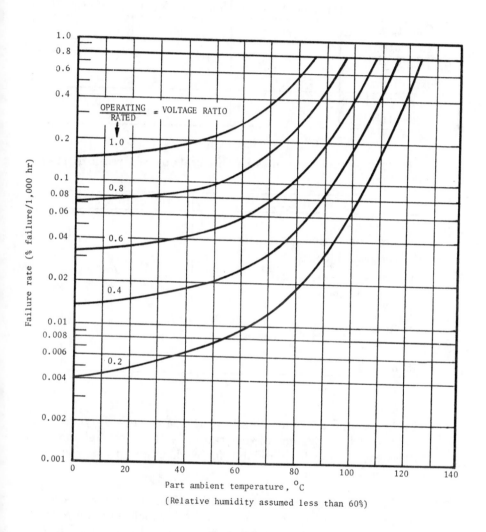

Fig. 12.28 — Failure rates for the MIL-C-11015A ceramic capacitors of Problem 12-5.

328 RELIABILITY PREDICTION AND TARGET RELIABILITY

TABLE 12.5 – Power source requirements for the aircraft electrical system of Problem 12-6.

System	Maximum load mode (10 kW)	Average load mode (7 kW)	Emergency mode (5 kW)
1. One 10-kW generator			
2. Two 5-kW generators			
3. Three 3.5-kW generators			

TABLE 12.6 – Components in the system of Problem 12-7.

Components	Number in system
1. Capacitors (fixed, mica, button)	5
2. Resistors (fixed composition)	10
3. Diodes (silicon)	15
4. Transistors (silicon)	20
5. Transformers	2
6. Motors	5
7. Blowers	3
8. Relays	5
9. Switches	5

(2) Assuming all generators have the same reliability, R_1, find out which system of generators is the most reliable and which is the least reliable for each mode of electrical system operation.

12-7. Predict the failure rate and reliability of the system composed of the nine components listed in Table 12.6, which function reliabilitywise in series, using the failure rates found in MIL-HDBK-217.

ASSUMPTIONS

1. General

(a) Under normal operating conditions.

(b) Operational environment: ground support equipment.

(c) Ambient temperature: $T_a = 50°C$.

(d) Stress ratio: $S = 25\%$.

(e) For all components use the highest quality level.

2. *Specific:*

Capacitors (MIL-C-10950)

(a) MIL-C-10950; style CB, 1,260 pF.

(b) Relative humidity less than 60%.

(c) 100% rated voltage at 125°C.

Resistors (MIL-R-11)

(a) Relative humidity less than 60%.

(b) From 56 to 470 kohms.

(c) 100% rated power at 70°C.

Diodes and Transistors (MIL-S-19500)

(a) Derating temperature interval, 150° to 25°C.

(b) Relative humidity less than 60%.

(c) Low power (less than 1 watt).

(d) 100% rated power at 25°C.

(e) Metallurgically bonded.

(f) PNP.

(g) Switch.

(h) $S_2 = 0.30$.

(i) Current rating, 3.5 amps.

Transformers (MIL-T-27)

In addition to the general assumptions listed in 1.

General:

(a) Class R insulation.

(b) Hermetically sealed construction.

(c) Types of application: audio, power and high power/pulse.

(d) Hot-spot temperature, $T_H = T_a + T_r$.
   T_a = ambient temperature = 50°C.
   T_r = temperature rise = 30°C.

Motors and Blowers

4,000-rpm brush-type, dc general-purpose motor with blower, class A insulation. Hot-spot temperature, 80°C. Motor operating period = 500 hr. The failure rate of the blowers is $\lambda_p = 5$ fr/10^6 hr.

Relays (MIL-R-19523)

In addition to the general assumptions listed in 1.

(a) Contact form: double pole, double throw.

(b) Six actuations per hour.

(c) For power relay above 10 amps.

(d) Resistive load, ratio: $S_2 = 30\%$.

(e) 100% rated at 125°C.

(f) Balanced armature relay.

Switches (MIL-S-3786)

In addition to the general assumptions listed in 1.

(a) Rotary switch, medium power wafer, one deck, and six contacts.

(b) The load current is 25% of rated current and is resistive.

(c) Six actuations per hour.

12-8. Design a hydraulically actuated elevation and azimuth control system for the airborne vehicle shown in Fig. 12.29.

Requirements

1. Contractual reliability goal of 0.975.

2. Total "use," or mission, time of 10.0 hr.

3. Under specified environments.

Assumptions

1. We have investigated all sources and these are the most reliable components available. The failure rate of each component is given in Table 12.7.

2. The failure of any single component will cause system failure.

3. We cannot eliminate any component.

4. Number of lines and fittings:
   A to B = 12,
   B to C = 6,
   and in the remainder of the system = 44.

12-9. The hydraulic system diagram in Fig. 12.30 shows a hydraulic pressure supply to three subsystems considered essential for the completion of a successful aircraft landing. Two engine-driven pumps and related components supply normal pressure to the subsystems. Either pump system is singly capable of supplying the systems should one pump system fail. In the event both engine systems fail to supply pressure, an auxiliary hydraulic

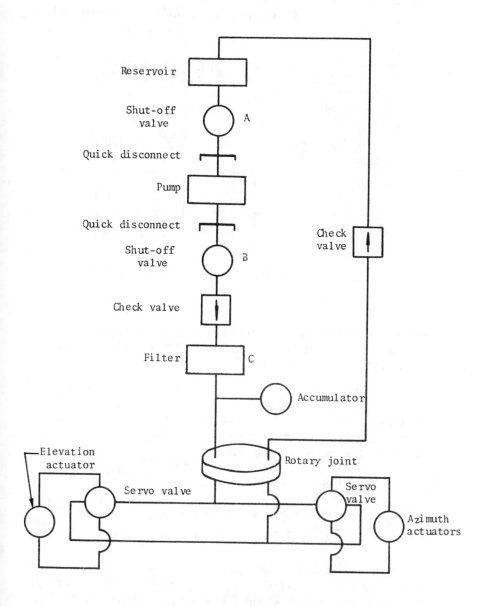

Fig. 12.29 — The hydraulically actuated elevation and azimuth control system of Problem 12-8.

TABLE 12.7 – The quantity and the failure rates of the components in the elevation and azimuth control system of Problem 12-8.

Component	Quantity	λ, fr/10^6 hr
Reservoir	1	50
Shut-off valve	2	15
Pump	1	1,875
Quick disconnects	2	120
Check valves	2	30
Filter	1	70
Accumulator	1	170
Rotary joint	1	205
Servo valves	2	360
Actuators	2	270
Each line and fitting	–	10

pressure system is on standby to supply the three essential subsystems. It is assumed that electrical power is always available to the auxiliary pump motor.

All reliabilities are based on an operating time of 10 hr. The system's reliability is for 10 hr also. It is assumed that any component in the system, except the standby elements, may fail at any time, whether in operation or not. The failure rate of each component is given in Table 12.8.

Do the following:

(1) Draw the reliabilty block diagram.

(2) Determine the reliabilty of the system.

12-10. The radio-communication system of an aircraft is required to have a reliability of 99.625% for a 10-hr flight. The communication system consists of the following subsystems: A receiver-transmitter, antenna control, antenna coupler, filter, and an antenna of a specified performance. The reliability block diagram of the basic system is shown in Fig. 12.31.

The past-experience-based and predicted failure rates are given in Table 12.9. Do the following:

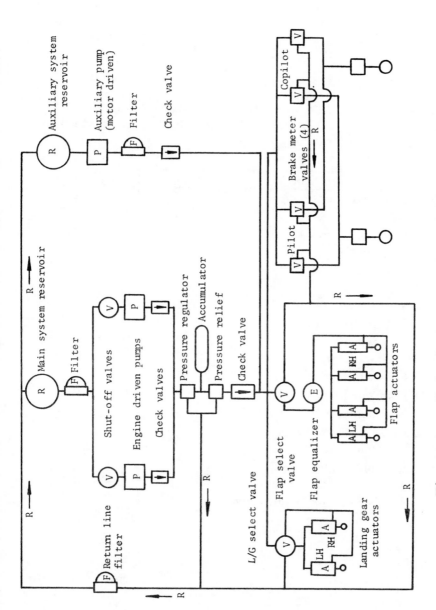

Fig. 12.30 – Hydraulic system diagram of Problem 12-9.

TABLE 12.8 – The quantity and the failure rates of the components in the hydraulic system of Problem 12-9.

Component	Quantity	λ, fr/10^6 hr
Main reservior	1	203
Filter	3	67
Pressure regulator	1	1,210
Accumulator	1	400
Pressure relief valve	1	62
Check valve	4	26
Shut-off valve	2	74
Engine driven pump	2	1,291
L/G select valve	1	62
L/G actuator	2	393
Flap select valve	1	62
Flap equalizer	1	1,210
Flap actuator	4	156
Brake meter valve	4	341
Brake actuator	2	1,136
Pressure reducer brake	2	170
Auxiliary system reservoir	1	23
Auxiliary pump (motor driven)	1	1,040

TABLE 12.9 – Historical and predicted failure rates of the basic radio-communication system of Problem 12-10.

Unit number	Unit	Historical failure rate, fr/10^6 hr		Predicted failure rate, λ_i, fr/10^6 hr
1	Radio-transmitter	Source A:	15,000	14,000
		Source B:	6,000	
		Source D:	17,000	
2	Antenna control	Source B:	100	200
		Source C:	400	
3	Antenna coupler	Source C:	5,000	500
4	Filter	Source A:	30	100
		Source D:	200	
5	Antenna	Source B:	–	200
		Source C:	100	
		Source A:	200	

$$\sum_{i=1}^{5} \lambda_i = 15,000 \text{ fr}/10^6 \text{ hr}$$

(1) Calculate the reliability of the system based on the predicted failure rates given in Table 12.9.

(2) If the target reliability is not met, analyze the subsystem that dominates the overall failure rate and redesign it using lower failure rate parts, or by derating them. Make use of the latest issue of MIL-HDBK-217 wherever possible.

(3) If the target reliability is still not met, try several configurations with redundancy.

The radio-transmitter is composed of 73 transistors, 617 resistors, 584 capacitors, 24 relays, 51 transformers and coils, and 11 switches, as shown in Fig. 12.32.

12-11. A twin-engine aircraft has a hydraulic-type brake system, as shown in Fig. 12.33. The constant failure rate of each engine is 500 fr/10^6 hr. Pumps, connections, and fittings have the constant failure rates of 100, 400, and 23 fr/10^6 hr, respectively. Pumps operate in parallel while the connections and fittings in the utility

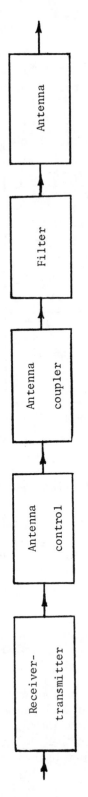

Fig. 12.31 — Reliability block diagram of the basic radio-communication system for Problem 12-10.

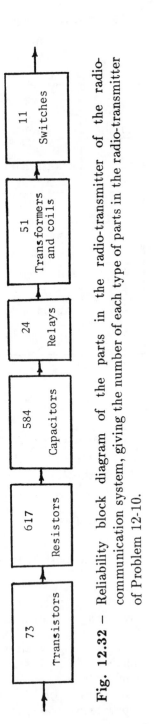

Fig. 12.32 — Reliability block diagram of the parts in the radio-transmitter of the radio-communication system, giving the number of each type of parts in the radio-transmitter of Problem 12-10.

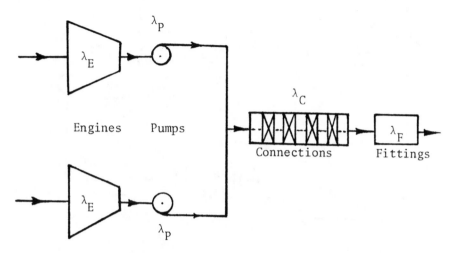

Fig. 12.33 – Hydraulic pressure supply to the twin-engine aircraft's brake system of Problem 12-11, Case 1.

hydraulic pressure system are subject to potential leakage points and operate in series. Do the following:

(1) Find the reliability of the hydraulic brakes for a flight of 12 hr.

(2) To increase the reliability of the hydraulic brakes, an auxiliary hydraulic system powered by an electric-motor driven pump is added in parallel with the other hydraulic system, as shown in Fig. 12.34. A specific check valve allows the hydraulic fluid to reach the brakes as needed. The constant failure rates of the electric motor, pump, connections, fittings and check valve are 400, 100, 210, 13, and 8 fr $/10^6$ hr, respectively. Find the reliability of the new configuration.

(3) An automatic standby configuration is also considered, as shown in Fig. 12.35, where the auxiliary hydraulic system of Case 2 is in standby configuration. This requires a sensing and switching subsystem, which in this case is considered to be functioning perfectly, or its failure rate is zero, and the standby system does not fail in the quiescent mode; i.e., $\lambda_{2Q} = 0$.

(4) Rework Case 3 utilizing the following additional failure rates:

λ_{2Q} = constant, quiescent failure rate of the standby unit,

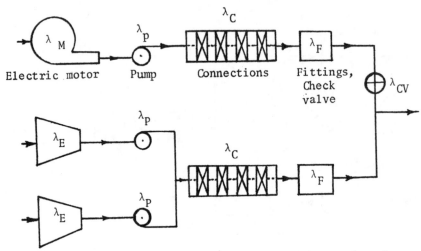

Fig. 12.34 – Hydraulic pressure supply to a twin-engine aircraft's brake system utilizing a parallel hydraulic
motor driven, hydraulic pressure supply system
of Problem 12-11, Case 2.

$$= 14 \text{ fr}/10^6 \text{ hr},$$

λ_{SE} = constant, sensing failure rate of the sensing
subsystem,

$$= 5 \text{ fr}/10^6 \text{ hr},$$

λ_{SWQ} = constant, quiescent, switching subsystem
failure rate,

$$= 0.15 \text{ fr}/10^6 \text{ hr},$$

λ_{SWE} = constant, energized switching subsystem
failure rate, considering only one cycle of
operation,

$$= 1.0 \text{ fr}/10^6 \text{ cycle},$$

and

λ_{SWO} = switch failing open failure rate,
= 0.20 fr/10^6 hr.

(5) Comparatively discuss the results obtained in all previous
cases.

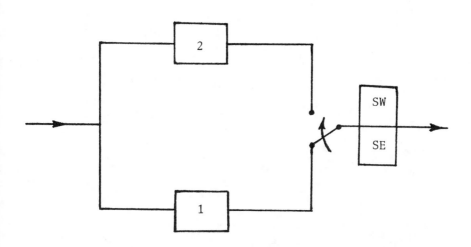

Fig. 12.35 (a) Reliability block diagram of the standby system in Case 3 of Problem 12-11.

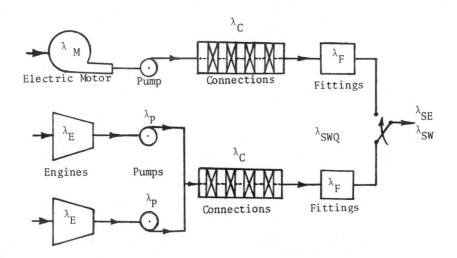

Fig. 12.35(b) Hydraulic pressure supply to a twin-engine aircraft brake system utilizing a standby auxiliary hydraulic pressure supply system of Problem 12-11, Case 3.

Chapter 13

LIMIT LAW OF THE TIME–TO–FAILURE DISTRIBUTION OF A COMPLEX SYSTEM: DRENICK'S THEOREM

13.1 DRENICK'S THEOREM

Consider a complex system with n units connected in series reliabilitywise and each unit has its own pattern of malfunction and replacement. Further assume that (1) the components are independent, (2) every unit failure causes system failure, and that (3) a failed unit is replaced immediately with a new one of the same kind. Then, under some reasonably general conditions, the distribution of the time between failures of the whole system tends to the exponential as the complexity and the time of operation increase.

The proof of this result is partially based on the results of renewal theory. As time goes on, if a particular unit fails, the sequence of failures and restorations to satisfactory function of this unit constitutes a renewal process. Let $N_i(t)$ denote the number of renewals of the ith unit within $[0, t]$, then

$$N(t) = \sum_{i=1}^{n} N_i(t)$$

is the number of renewals of the whole system within $[0, t]$. The system's overall failure pattern is therefore a superposition of the n individual renewal processes.

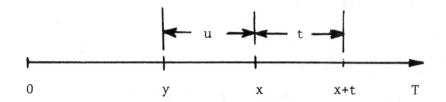

Fig. 13.1–The residual life of a component.

In a renewal process, let T be the time to failure of the unit, $R(T)$ be the reliability of this unit, and T_x be the residual life of this unit at an arbitrarily selected instant x as shown in Fig. 13.1. The event $\{T_x \geq t\}$, which is the event that this unit will survive for a period t beyond x, is the union of the following two independent events:

1. No renewal occurs before x, and the first unit fails some time beyond $x + t$. Denote this event by A_1; then

$$A_1 = \{T \geq x+t| \text{ there is no renewal before } x\}.$$

2. Some renewals occur before x, the last renewal before x occurs in the time interval $[y, y + dy]$ for all $y \in [0, x]$, and the last repaired or replaced unit does not fail before $x + t$. Denote this event by A_2; then

$$A_2 = \{T \geq x + t - y| \text{ the last renewal before x occurs at y}$$
$$\text{for all } 0 \leq y \leq x\}.$$

Therefore,

$$\{T_x \geq t\} = A_1 \cup A_2,$$

and

$$P(T_x \geq t) = P(A_1) + P(A_2).$$

It may be seen that

$$P(A_1) = R(x + t),$$

and $P(A_2)$ can be obtained as follows:

Let h(y) be the renewal density of this unit; then the probability that a renewal will occur in a small interval $[y, y + dy]$ can be approximated by $h(y) \, dy$, and

$$P(T \geq x + t| \text{ last renewal occurred in } [y, y + dy]) = R(x - y + t)h(y) \, dy.$$

Since y can take any value between 0 and x,

$$P(A_2) = \int_0^x R(x - y + t)h(y) \, dy,$$

or

$$P(A_2) = \int_0^x R(u + t)h(x - u) \, du,$$

where $u = x - y$, which is the time from the last renewal till x, as shown in Fig. 13.1.

Now let $G(t)$ be the residual survival probability of this unit in the equilibrium state, which is the probability that the unit will survive for a period t beyond some arbitrarily selected instant in the equilibrium state; then

$$G(t) = \lim_{x \to \infty} P(T_x \geq t),$$

or

$$G(t) = \lim_{x \to \infty} [R(x + t) + \int_0^x R(u + t)h(x - u) \, du].$$

Since

$$\lim_{x \to \infty} R(x + t) = 0,$$

and by the key renewal theorem [1, pp. 61–64]

$$\lim_{x \to \infty} \int_0^x R(u+t)h(x-u)\,du \;=\; \frac{1}{m} \int_0^\infty R(x+u)\,du,$$

$$=\; \frac{1}{m} \int_t^\infty R(\tau)\,d\tau, \qquad (13.1)$$

where m is the mean life of the unit; then

$$G(t) = \frac{1}{m} \int_t^\infty R(\tau)\,d\tau.$$

Let m_i and $G_{ci}(t)$ be the mean life and the residual survival probability of the ith unit, respectively, with $i = 1, 2, ..., n$, then we can define two quantities for the whole system:

1. The residual survival probability of a series system, $G_c(t)$, defined by

$$G_c(t) = \prod_{i=1}^n G_{ci}(t). \qquad (13.2)$$

2. $\bar{m}_n$ defined by

$$\frac{1}{\bar{m}_n} = \sum_{i=1}^n \frac{1}{m_i}, \qquad (13.3)$$

which can be viewed to be the system's equivalent mean life.

Drenick [2] has provided the following theorem: Consider the following conditions:

(1) $\displaystyle \lim_{n \to \infty} \sup_{1 \le i \le n} (\bar{m}_n/m_i) = M,$

where sup stands for supremum or least upper bound, and it is the number M, such that

(a) for any $X \in \{(\bar{m}_n/m_i), i = 1, 2, ..., n\}$, $X \le M$,

where $\in$ means that any X is contained in this set, and

(b) for any $\varepsilon > 0$, there exists an $X' \in \{(\bar{m}_n/m_i), i = 1, 2, ..., n\}$,

such that

$$X' > M - \varepsilon.$$

(2) $Q_i(t) \leq At^\delta, A > 0, \delta > 0$, as $t \to 0, i = 1, 2, ..., n$.

Then

$$\lim_{n \to \infty} G_c(\bar{m}_n \tau) = e^{-\tau}, \text{ for } \tau > 0,$$

where τ is, as will be seen in the next section, the equivalent number of failures of the system. Proof of this theorem is presented in the next section.

These two conditions are sufficient. Condition (1) requires that $\bar{m}_n \to 0$ as $n \to \infty$. It also requires, roughly speaking, that there not be any very "poor" components in the system, i.e., components with mean life so small that $\bar{m}_n / m_i$ increases with n even though $\bar{m}_n$ decreases. Condition (2) controls the behavior of the failure law near $t = 0$. It excludes failure rates which go to infinity as $t \to 0$, at least as rapidly as $1/t$, but includes all failure laws that are commonly used. This implies that early failures as represented by the Weibull distribution are also included.

This means that at the system level the times-to-failure pdf is exponential regardless of the nature of the times-to-failure pdf's of the components; consequently, the system's failure rate tends to constancy when the number of the components, n, contained in the system increases. From the practical viewpoint, $n \to \infty$ may be interpreted as $n \geq 200$ components. Also, at least three different kinds of components should preferably be replaced with burnt-in and broken-in ones for the stabilization of the system's failure rate to be approached sooner.

13.2 PROOF OF DRENICK'S THEOREM

Denote

$$t/\bar{m}_n = \tau,$$

where it can be seen that τ is the equivalent number of failures of the system, or it can be written as

$$t = \bar{m}_n \tau.$$

Then from Eq. (13.1)

$$\begin{aligned} 1 - G_{ci}(t) &= \frac{1}{m_i} \int_0^t R_{ci}(u) \, du, \\ &= \frac{1}{m_i} \int_0^{\bar{m}_n \tau} R_{ci}(u) \, du, \end{aligned} \tag{13.4}$$

which is a function of τ, and we denote it by

$$\phi_{in}(\tau) = 1 - G_{ci}(t),$$

or just by

$$\phi_{in} = 1 - G_{ci}(t) > 0, \tag{13.5}$$

dropping the argument τ.

Taking the logarithm of Eq. (13.2) yields

$$\log_e G_c(\bar{m}_n \tau) = \sum_{i=1}^{n} \log_e G_{ci}(\bar{m}_n \tau). \tag{13.6}$$

Proving

$$\lim_{n \to \infty} G_c(\bar{m}_n \tau) = e^{-\tau}$$

is equivalent to proving that

$$\lim_{n \to \infty} \log_e G_c(\bar{m}_n \tau) = \lim_{n \to \infty} \sum_{i=1}^{n} \log_e G_{ci}(\bar{m}_n \tau) = -\tau.$$

From Eq. (13.4)

$$\begin{aligned}
\phi_{in} &= \frac{1}{m_i} \int_0^{\bar{m}_n \tau} [1 - Q_i(u)] \, du, \\
&= \frac{\bar{m}_n \tau}{m_i} - \frac{1}{m_i} \int_0^{\bar{m}_n \tau} Q_i(u) \, du \\
&\leq \frac{\bar{m}_n \tau}{m_i},
\end{aligned}$$

where

$$Q_i(u) = 1 - R_i(u).$$

Therefore, as $n \to \infty$, by condition (1)

$$\sup_{1 \leq i \leq n} \phi_{in} \leq \sup_{1 \leq i \leq n} \frac{\bar{m}_n \tau}{m_i} \to 0.$$

Thus we can assume that, from some n onward, $\phi_{in} < 1/2$ for all $i = 1, 2, ..., n$.

From Eq. (13.4)

$$G_{ci}(\bar{m}_n \tau) = 1 - \phi_{in},$$

or

$$\log_e G_{ci}(\bar{m}_n\tau) = \log_e(1 - \phi_{in}), \tag{13.7}$$

and by Taylor's series expansion

$$\log_e(1 - \phi_{in}) = -(\phi_{in} + \frac{\phi_{in}^2}{2} + \cdots + \frac{\phi_{in}^n}{n} + \cdots).$$

Consequently,

$$
\begin{aligned}
|\log_e G_{ci}(\bar{m}_n\tau) + \phi_{in}| &= |\log_e(1 - \phi_{in}) + \phi_{in}|, \\
&= \sum_{k=2}^{\infty} \frac{\phi_{in}^k}{k}, \\
&\leq \frac{1}{2}\sum_{k=2}^{\infty} \phi_{in}^k,
\end{aligned}
$$

and using the geometric series equivalent yields

$$|\log_e G_{ci}(\bar{m}_n\tau) + \phi_{in}| = \frac{1}{2}\frac{\phi_{in}^2}{1 - \phi_{in}} < \phi_{in}^2. \tag{13.8}$$

From Eq. (13.5)

$$
\begin{aligned}
|\log_e G_c(\bar{m}_n\tau) + \sum_{i=1}^{n}\phi_{in}| &= |\sum_{i=1}^{n}[\log_e G_{ci}(\bar{m}_n\tau) + \phi_{in}]|, \\
&\leq \sum_{i=1}^{n}|\log_e G_{ci}(\bar{m}_n\tau) + \phi_{in}|,
\end{aligned}
$$

and from Eq. (13.8)

$$
\begin{aligned}
|\log_e G_c(\bar{m}_n\tau) + \sum_{i=1}^{n}\phi_{in}| &< \sum_{i=1}^{n}\phi_{in}^2, \\
&\leq (\sup_{1\leq i\leq n}\phi_{in})\sum_{i=1}^{n}\phi_{in}.
\end{aligned}
$$

Since

$$\lim_{n\to\infty}\sup_{1\leq i\leq n}\phi_{in} = 0,$$

and

$$
\begin{aligned}
\sum_{i=1}^{n}\phi_{in} &\leq \sum_{i=1}^{n}\frac{\bar{m}_n\tau}{m_i}, \\
&= \bar{m}_n\tau\sum_{i=1}^{n}\frac{1}{m_i} = \tau,
\end{aligned}
$$

using Eq. (13.3), then

$$\lim_{n \to \infty} |\log_e G_c(\bar{m}_n \tau) + \sum_{i=1}^{n} \phi_{in}| = 0,$$

or

$$\lim_{n \to \infty} \log_e G_c(\bar{m}_n \tau) = \lim_{n \to \infty} (- \sum_{i=1}^{n} \phi_{in}). \qquad (13.9)$$

But

$$\begin{aligned}
\sum_{i=1}^{n} \phi_{in} &= \sum_{i=1}^{n} \frac{1}{m_i} \int_0^{\bar{m}_n \tau} [1 - Q_i(u)] \, du, \\
&= \sum_{i=1}^{n} \frac{\bar{m}_n \tau}{m_i} - \sum_{i=1}^{n} \frac{1}{m_i} \int_0^{\bar{m}_n \tau} Q_i(u) \, du, \\
&= \tau - \sum_{i=1}^{n} (\frac{1}{m_i} \int_0^{\bar{m}_n \tau} Q_i(u) \, du), \qquad (13.10)
\end{aligned}$$

and from condition (2)

$$\begin{aligned}
\sum_{i=1}^{n} (\frac{1}{m_i} \int_0^{\bar{m}_n \tau} Q_i(u) \, du) &\le \sum_{i=1}^{n} [\frac{1}{m_i} \int_0^{\bar{m}_n \tau} A u^\delta \, du], \\
&= \sum_{i=1}^{n} \frac{A}{m_i} \frac{1}{\delta + 1} (\bar{m}_n \tau)^{\delta+1}, \\
&= \frac{A}{\delta + 1} \bar{m}_n^\delta \tau^{\delta+1} (\bar{m}_n \sum_{i=1}^{n} \frac{1}{m_i}), \\
&= \frac{A}{\delta + 1} \bar{m}_n^\delta \tau^{\delta+1} \to 0
\end{aligned}$$

as $n \to \infty$, from condition 1 as presented in Section 13.1. Therefore, from Eq. (13.10)

$$\lim_{n \to \infty} \sum_{i=1}^{n} \phi_{in} = \tau.$$

Consequently, from Eq. (13.9)

$$\lim_{n \to \infty} \log_e G_c(\bar{m}_n \tau) = -\tau,$$

or

$$\lim_{n \to \infty} G_c(\bar{m}_n \tau) = e^{-\tau},$$

which was to be proved.

REFERENCES

1. Cox, D.R., *Renewal Theory*, John Wiley & Sons, Inc., 142 pp., 1962.

2. Drenick, R.F., "The Failure Law of Complex Equipment," *J. Soc. Indust. Appl. Math.*, Vol. 8, No. 4, pp. 680–689, Dec. 1960.

Chapter 14

RELIABILITY OF COMPONENTS WITH A POLICY OF REPLACING THOSE THAT FAIL BY A PRESCRIBED OPERATING TIME

14.1 METHODOLOGY

Consider the case of identical components operating in different equipment subjected to the same application and operation stresses. Let us find how many of these components will fail, on the average, if they operate a prescribed period of time or cycles, n_1, from age zero; those that are found to have failed after n_1 cycles of operation are replaced by fresh ones, and the replaced and nonreplaced components operate n additional cycles. The reliability of N_o such components for the first n_1 cycles of operation is $R(n_1)$, and the number that will fail by n_1 cycles of operation is

$$N_{F-R}(n_1) = N_o Q(n_1) = N_o[1 - R(n_1)]. \tag{14.1}$$

These are replaced by fresh ones and they operate n cycles thereafter. The number of these replaced components that will fail after n additional cycles of operation, using Eq. (14.1), would be

$$N_{F-R}(n) = N_{F-R}(n_1)Q(n) = N_o[1 - R(n_1)][1 - R(n)]. \tag{14.2}$$

The number of those that do not fail by n_1 and function an additional n cycles is

$$N_S(n_1) = N_o R(n_1). \tag{14.3}$$

The number of these components that will fail while operating n additional cycles, using Eq. (14.3), would be

$$N_{F-NR}(n_1, n) = N_S(n_1)[1 - R(n_1, n)],$$

or

$$N_{F-NR}(n_1, n) = N_o R(n_1)[1 - \frac{R(n_1 + n)}{R(n_1)}]. \tag{14.4}$$

Consequently, the total number of such components that will fail by $(n_1 + n)$ cycles of operation, under the condition that those that fail by n_1 are replaced, is given by the sum of Eqs. (14.1), (14.2), and (14.4), or

$$\begin{aligned} N_{F-T}(n_1 + n) &= N_o[1 - R(n_1)] + N_o[1 - R(n_1)][1 - R(n)] \\ &+ N_o R(n_1)[1 - \frac{R(n_1 + n)}{R(n_1)}]. \end{aligned} \tag{14.5}$$

Rearrangement of Eq. (14.5) yields

$$N_{F-T}(n_1+n) = N_o\{[1-R(n_1)][2-R(n)]+[R(n_1)-R(n_1+n)]\}. \tag{14.5'}$$

Simplification of Eq. (14.5′) yields

$$N_{F-T}(n_1+n) = N_o[2-R(n_1)-R(n)+R(n_1)R(n)-R(n_1+n)]. \tag{14.5''}$$

Equation (14.5″) gives the average number of spares that should be provided for N_o such components with the replacement policy considered here.

Let us now find the *total combined reliability* of such components. Of those that fail by n_1 cycles of operation and are replaced, the number that will survive after an additional n cycles of operation is

$$N_{S-R}(n) = N_{F-R}(n_1)R(n) = N_o[1 - R(n_1)]R(n). \tag{14.6}$$

Of those that do not fail by n_1 cycles of operation and operate n additional cycles, the number surviving is

$$N_{S-NR}(n_1 + n) = N_o R(n_1)R(n_1, n) = N_o R(n_1)\frac{R(n_1 + n)}{R(n_1)}, \tag{14.7}$$

or

$$N_{S-NR}(n_1 + n) = N_o R(n_1 + n). \qquad (14.7')$$

Out of the N_o that started out, the number of those that survive, after replacement of those that failed by n_1, is given by the sum of Eqs. (14.6) and (14.7'), or

$$N_{S-R}(n_1 + n) = N_o\{[1 - R(n_1)]R(n) + R(n_1 + n)\}. \qquad (14.8)$$

The total combined reliability of such components, under the replacement policy considered here, is therefore

$$R_{T-R}(n_1 + n) = \frac{N_{S-R}(n_1 + n)}{N_o}. \qquad (14.9)$$

Substitution of Eq. (14.8) into Eq. (14.9) and simplification yields

$$R_{T-R}(n_1 + n) = [1 - R(n_1)]R(n) + R(n_1 + n). \qquad (14.10)$$

A study of Eq. (14.10) reveals that the total combined reliability of such components, under the replacement policy considered here, is given by the probability that either the components fail by n_1 cycles, are replaced, and function successfully for n cycles thereafter, or they do not fail by n_1 cycles and thus function successfully the full $(n_1 + n)$ cycles.

In Eqs. (14.5), (14.6), and (14.10), the quantities $R(n_1), R(n)$, and $R(n_1 + n)$ need to be calculated.

If it has been established that the times-to-failure distribution of such components is Weibullian, then

$$R(n_1) = e^{-[(n_1-\gamma)/\eta]^\beta}, \qquad (14.11)$$

$$R(n) = e^{-[(n-\gamma)/\eta]^\beta}, \qquad (14.12)$$

$$R(n_1 + n) = e^{-[(n_1+n-\gamma)/\eta]^\beta}, \qquad (14.13)$$

and

$$R(n_1, n) = \frac{R(n_1 + n)}{R(n_1)} = \frac{e^{-[(n_1+n-\gamma)/\eta]^\beta}}{e^{-[(n_1-\gamma)/\eta]^\beta}}. \qquad (14.14)$$

EXAMPLE 14-1

Identical aluminum spindles are operating at an alternating stress level of 25,000 psi. Each spindle has completed 500,000 revolutions (cycles) successfully. Their cycles-to-failure distribution is lognormally distributed with parameters $\bar{n}' = 5.827$ and $\sigma_{n'} = 0.124$.

1. If 1,000 of such spindles are operating, how many will survive after 500,000 cycles of operation and how many will fail?

$$n_1 = 500,000 \text{ cycles,}$$

$$n_1' = \log_{10} n_1 = \log_{10} 500,000 = 5.69897,$$

$$z(n_1') = \frac{n_1' - \bar{n}'}{\sigma_{n'}} = \frac{5.69897 - 5.827}{0.124} = -0.103250,$$

then

$$R(n_1) = \int_{z(n_1')}^{\infty} \phi(z)\, dz = \int_{-1.03250}^{\infty} \phi(z)\, dz, = 0.849081.$$

From Eq. (14.3), the number of spindles that will survive is

$$N_S(n_1) = N_S(500,000) = N_o R(n_1) = 1,000 \times 0.849081 \cong 849.$$

From Eq. (14.1), the number of spindles that will fail is

$$N_{F-R}(n_1) = N_o[1 - R(n_1)] = 1,000(1 - 0.849081),$$

or

$$N_{F-R}(n_1) = 1,000 \times 0.150.919 \cong 151.$$

2. If the failed spindles are replaced by new ones and all of them operate for $n = 330,000$ cycles thereafter, how many of the fresh ones will fail? From Eq. (14.2)

$$N_{F-R}(n) = N_o[1 - R(n_1)][1 - R(n)],$$

$$R(n) = \int_{z(n')}^{\infty} \phi(z)\, dz,$$

$$z(n') = \frac{\log_{10} 330,000 - \bar{n}'}{\sigma_{n'}} = \frac{5.518514 - 5.827}{0.124} = -2.48779,$$

and

$$R(330,000) = \int_{-2.48779}^{\infty} \phi(z)\, dz = 0.993573.$$

Then

$$N_{F-R}(330,000) = 1,000 \times (1 - 0.849081)(1 - 0.993573),$$

or

$$N_{F-R}(330,000) = 1,000 \times 0.1590919 \times 0.006427 = 0.97 \cong 1.$$

Therefore, one spindle out of the fresh ones will fail.

3. Of those that survived n_1 cycles how many will fail during the additional n cycles? From Eq. (14.4)

$$N_{F-NR}(n_1, n) = N_{F-NR}(500,000; 330,000),$$

$$= N_o R(n_1)[1 - \frac{R(n_1 + n)}{R(n_1)}].$$

From Case 1

$$N_o R(n_1) = N_S(n_1) = 849,$$

$$R(n_1 + n) = \int_{z(n_1+n)}^{\infty} \phi(z)\, dz,$$

$$z(n_1 + n)' = \frac{\log_{10}(n_1 + n) - \bar{n}'}{\sigma_{n'}},$$

and

$$z(500,000 + 330,000) = \frac{\log_{10} 830,000 - 5.827}{0.124} = 0.742565.$$

Therefore,

$$R(500,000 + 330,000) = \int_{0.742565}^{\infty} \phi(z)\, dz = 0.228872,$$

and

$$N_{F-R}(500,000; 330,000) = 849(1 - \frac{0.228872}{0.849081}) = 849 \times 0.730447,$$

or

$$N_{F-R}(500,000; 330,000) = 620.15.$$

Consequently, 620 of these spindles will fail.

4. What is the total number of spindles that will fail by $(n_1 + n)$ cycles given that 1,000 start at age zero and that those that fail by n_1 cycles are replaced? From Eq. (14.5)

$$N_{F-T}(n_1 + n) = N_o[2 - R(n_1) - R(n) + R(n_1)R(n) - R(n_1 + n)],$$

$$
\begin{aligned}
N_{F-T}(500,000 + 330,000) &= 1,000[2 - 0.849081 - 0.993573 \\
&\quad + (0.849081)(0.993573) - 0.228872], \\
&= 1,000 \times 0.772098,
\end{aligned}
$$

or

$$N_{F-T}(500,000 + 330,000) = 772.098.$$

Consequently, 772 such spindles will fail by $n_1 + n = 830,0000$ cycles of operation. The same answer can also be obtained from

$$
\begin{aligned}
N_{F-T}(830,000) &= N_{F-T}(500,000) + N_{F-T}(330,000) \\
&\quad + N_{F-NR}(500,000; 330,000),
\end{aligned}
$$

or

$$N_{F-T}(830,000) = 151 + 1 + 620 = 772 \text{ spindles.}$$

5. What is the total number of components that survive after $(n_1 + n)$ cycles when we follow the policy of replacing the failed ones by n_1 cycles? From Eq. (14.8)

$$N_{S-R}(n_1 + n) = N_o\{[1 - R(n_1)]R(n) + R(n_1 + n)\},$$

$$
\begin{aligned}
N_{S-R}(500,000 + 330,000) &= 1,000\{[1 - 0.849081](0.993573) \\
&\quad + (0.228872)\}, \\
&= 1,000[(0.150919)(0.993573) \\
&\quad + (0.228872)], \\
&= 1,000 \times 0.37882, \\
&= 378.82,
\end{aligned}
$$

or

$$N_{S-R}(830,000) \cong 379.$$

Therefore, a total of 379 such spindles will survive. The same answer can be obtained from

$$
\begin{aligned}
N_{S-R}(830,000) &= N_o + N_{F-R}(n_1) - N_{F-T}(830,000), \\
&= 1,000 + 151 - 772,
\end{aligned}
$$

or

$$N_{S-R}(830,000) = 379.$$

6. What is the total combined reliability of the components under this replacement policy? From Eq. (14.10)

$$R_{T-R}(n_1 + n) = [1 - R(n_1)]R(n) + R(n_1 + n),$$

$$
\begin{aligned}
R_{T-R}(500,000 + 330,000) &= (1 - 0.849081)(0.9935373) \\
&\quad + (0.228872), \\
&= (0.150919)(0.993573) \\
&\quad + (0.228872),
\end{aligned}
$$

or

$$R_{T-R}(830,000) = 0.37882.$$

Also,

$$R_{T-R}(n_1 + n) = \frac{N_{S-R}(n_1 + n)}{N_o},$$

$$R_{T-R}(500,000 + 330,000) = R_{T-R}(830,000),$$
$$= \frac{378.82}{1,000},$$

or

$$R_{T-R}(830,000) = 0.37882.$$

Therefore, the combined reliability of these components is 0.37882.

7. What is the reliability for $(n_1 + n)$ cycles without a replacement policy? Compare the answer with that of Case 6.

$$R(n_1 + n) = \int_{z(n_1+n)'}^{\infty} \phi(z)\,dz,$$
$$z(n_1 + n)' = \frac{\log_{10}(n_1 + n) - \bar{n}'}{\sigma_{n'}},$$

$$z(500,000 + 330,000)' = \frac{\log_{10} 830,000 - 5.827}{0.124},$$
$$= \frac{5.91908 - 5.827}{0.124},$$

or

$$z(830,000)' = 0.742565.$$

Then

$$R(500,000 + 330,000) = R(830,000),$$
$$= \int_{0.742565}^{\infty} \phi(z)\,dz,$$

or

$$R(830,000) = 0.228871.$$

Consequently,

$$R(830,000| \text{ without replacement}) < R(830,000| \text{ with replacement}),$$
$$0.228871 < 0.37882,$$

or $R(830,000)$ with replacement is 65% larger than without replacement!

EXAMPLE 14–2

Identical types of bearings have a Weibull time-to-failure distribution with the following parameters:

$$\beta = 2.0, \quad \eta = 2,000 \text{ hr, and } \gamma = 0 \text{ hr.}$$

1. If 100 of such bearings are operating in identical equipment at the same application and operation stress level, how many will survive if each operates $T_1 = 1,300$ hr at 675 rpm, and how many will fail?

$$R(T_1) = e^{-[(T_1-\gamma)/\eta]^\beta},$$

$$R(1,300 \text{ hr}) = e^{-(\frac{1,300}{2,000})^2},$$

or

$$R(1,300 \text{ hr}) = 0.522046.$$

The number of bearings that will survive is

$$N_S(T_1) = N_S(1,300 \text{ hr}) = N_o R(T_1) = 100 \times 0.522046,$$

or

$$N_S(1,300 \text{ hr}) \cong 52.$$

The number of such bearings that will fail is

$$
\begin{aligned}
N_{F-R}(T_1) &= N_o Q(T_1) = N_o[1 - R(T_1)], \\
&= 100 \times (1 - 0.522046),
\end{aligned}
$$

or

$$N_{F_R}(1,300 \text{ hr}) \cong 48.$$

2. If the failed bearings are replaced by new ones and all of them operate for $T = 700$ hr thereafter, how many of the fresh ones will fail? From Eq. (14.2)

$$N_{F-R}(T) = N_o[1 - R(T_1)][1 - R(T)],$$

where

$$R(T) = e^{-(\frac{T}{\eta})^\beta} = e^{-(\frac{700}{2,000})^2},$$

or

$$R(700 \text{ hr}) = 0.884706.$$

Then

$$
\begin{aligned}
N_{F-R}(T = 700 \text{ hr}) &= 100(1 - 0.522046)(1 - 0.884706), \\
&= 100(0.47954)(0.115294) = 5.51,
\end{aligned}
$$

or

$$N_{F_R}(700 \text{ hr}) \cong 6.$$

Therefore, six out of the fresh bearings will fail while operating 700 hr after replacement.

3. Of those that survived T_1 hours, how many will fail during the additional T hours of operation? From Eq. (14.4)

$$N_{F_N R}(T_1, T) = N_o R(T_1)[1 - \frac{R(T_1 + T)}{R(T_1)}],$$

and from Case 1

$$N_o R(T_1) = N_S(T_1) = 52,$$
$$R(T_1 + T) = e^{-(\frac{T_1+T}{\eta})^\beta} = e^{-(\frac{1,300+700}{2,000})^2} = 0.367879.$$

Then

$$
\begin{aligned}
N_{F-NR}(1,300 \text{ hr}; 700 \text{ hr}) &= 52[1 - \frac{0.367879}{0.522046}], \\
&= 52 \times 0.295313,
\end{aligned}
$$

$$N_{F-R}(1,300 \text{ hr}; 700 \text{ hr}) = 15.36 \cong 15.$$

Therefore, 15 of these bearings will fail.

4. What is the total number of bearings that will fail by $(T_1 + T)$ hours of operation, given that 100 start at age zero and that those that fail by T_1 hours are replaced? From Eq. (14.5)

$$N_{F-T}(T_1 + T) = N_o[2 - R(T_1) - R(T) + R(T_1)R(T) - R(T_1 + T)],$$

$$
\begin{aligned}
N_{F-T}(1,300 \text{ hr} + 700 \text{ hr}) = \ &100[2 - 0.522046 - 0.884706 \\
&+ (0.522046)(0.884706) - 0.367879],
\end{aligned}
$$

or

$$N_{F-T}(1,300 \text{ hr} + 700 \text{ hr}) = 100(0.687226) = 68.72 \cong 69.$$

Consequently, 69 such bearings will fail by $T_1 + T = 2,000$ hr.

The same answer can be obtained from

$$
\begin{aligned}
N_{F-T}(2,000 \text{ hr}) = \ &N_{F-R}(1,300) + N_{F-R}(700) \\
&+ N_{F-NR}(1,300; 700),
\end{aligned}
$$

or

$$N_{F-T}(2,000 \text{ hr}) = 48 + 6 + 15 = 69.$$

5. What is the total number of bearings that survive after $(T_1 + T)$ hours when we follow the policy of replacing the failed ones by T_1 hours? From Eq. (14.8)

$$N_{S-R}(T_1 + T) = N_o\{[1 - R(T_1)]R(T) + R(T_1 + T)\},$$

$$N_{S-R}(1,300 \text{ hr} + 700 \text{ hr}) = 100\{[1 - 0.522046](0.884706) + (0.367879)\},$$

$$N_{S-R}(2,000 \text{ hr}) = 100[(0.477954)(0.884706) + (0.367879)],$$

or

$$N_{S-R}(2,000 \text{ hr}) = 100 \times 0.790728 = 79.07 \cong 79.$$

Therefore, a total of 79 of these bearings will survive.

The same answer can be obtained from

$$
\begin{aligned}
N_{S-R}(2,000 \text{ hr}) &= N_o + N_{F-R}(T_1) - N_{F-T}(T_1 + T), \\
&= 100 + 48 - 69,
\end{aligned}
$$

or

$$N_{S-R}(2,000 \text{ hr}) = 79.$$

6. What is the total combined reliability of these components under the replacement policy used? From Eq. (14.10)

$$
\begin{aligned}
R_{T-R}(T_1 + T) &= [1 - R(T_1)]R(T) + R(T_1 + T), \\
R_{T-R}(1,300 \text{ hr} + 700 \text{ hr}) &= (1 - 0.522046)(0.884706) + (0.367879), \\
&= 0.790728,
\end{aligned}
$$

or

$$R_{T-R}(T_1 + T) = \frac{N_{S-R}(T_1 + T)}{N_o},$$

$$R_{T-R}(1,300 \text{ hr} + 700 \text{ hr}) = R_{T-R}(2,000) = \frac{79.07}{100} = 0.7907.$$

Therefore, the combined reliability is 0.7907.

7. What is the reliability for $(T_1 + T)$ hours with a no-replacement policy? Compare the answer with that of Case 6.

$$R(T_1 + T) = e^{-[(T_1+T)/\eta]^\beta},$$
$$R(1,300 \text{ hr} + 700 \text{ hr}) = e^{-(\frac{1,300+700}{2,000})^2},$$

or

$$R(2,000 \text{ hr}) = e^{-1} = 0.367879.$$

Consequently,

$$R(2,000 \text{ hr} | \text{ without replacement}) < R(2,000 \text{ hr} | \text{ with replacement})$$
$$0.367879 < 0.790728,$$

or $R(2,000 \text{ hr})$ with replacement is more than double that without replacement!

PROBLEMS

14-1. Derive the equation for calculating the *total combined reliability* if the age of the units which have been put into operation at the beginning of the mission is $T_0 \neq 0$.

14-2. Find the *total combined reliability* for the following replacement policy: At the prescribed replacement time T_1, those that are found to have failed by T_1 are replaced by fresh ones, and p percent of those that have not failed by T_1 are also replaced by fresh ones.

14-3. Find the optimal value of the p given in Problem 14-2 in terms of cost, assuming that

C_1 = replacement cost per unit,

C_2 = unit cost,

C_3 = salvage value of an unfailed unit at replacement time T_1,

and

C_4 = failure cost at the end of the mission.

14-4. Give an alternative derivation for Eq. (14.10).

14-5. Identical units have a Weibull time-to-failure distribution with the following parameters: $\beta = 1.75$, $\eta = 3,000$ hr, and $\gamma = 0$ hr. The replacement policy is as follows: Those that are found to have failed at 1,500 hours are replaced with units which have a Weibull time-to-failure distribution with the following parameters: $\beta = 2.50$, $\eta = 1,500$ hr, and $\gamma = 0$ hr. Do the following:

(1) If 1,000 such units are operating in identical equipment at the same application and operation stress level, how many will fail if each one operates $T_1 = 1,500$ hr?

(2) If the failed ones are replaced according to the given policy, how many will fail when operating for $t = 500$ hr thereafter?

(3) Of those that survive $T_1 = 1,500$ hr, how many will fail during the additional $t = 500$ hr of operation?

(4) What is the total combined reliability of these components under the replacement policy of Case 3?

Chapter 15

RELIABILITY ALLOCATION: APPORTIONMENT

15.1 INTRODUCTION

Reliability apportionment encompasses the problem of assigning the correct numerical reliability to each sublevel of an item in such a manner that the overall item reliability is equal to its reliability goal. It can also be thought of as the assignment or allocation to each sublevel of the allowable levels of unreliability. Reliability goal could also be set for every part, such as screw, wiring, connector, etc.; however, this is usually not practical. In designing for reliability, goals must be assigned to every major component for which the designer is responsible. Once these sublevel reliabilities are established, the designer can select the materials, configurations, and types so that the overall reliability requirement can be achieved.

Reliability goals are usually established by either the customer or proposed by the manufacturer for customer approval. The assignment of reliability goals by the customer is usually based on studies of strategic necessities, logistic requirements, mission accomplishment, operational readiness, maintainability, safety, budgetary limitations, and other factors of intrinsic importance. Apportionment is required when the manufacturer is obliged to decide how reliable various subsystems must be in order to design an end product that will meet the specified reliability requirement. It should be noted that, as reliability approaches perfection, each advance becomes very much more expensive. Morever, the manufacturer should be able to advise the customer as to what a given reliability requirement will cost in terms of time, performance, and money.

15.2   WHY RELIABILITY ALLOCATION?

Key reasons for reliability allocations are the following:

1. The well-meaning but ineffectual philosophy often applied to reliability-"we will do the best we can"-should be replaced by a contractual obligation in the form of quantitative system reliability requirements that forces contractors to consider reliability equally with other system parameters, such as performance, weight, cost, maintainability, etc. Given the system's reliability goal, the designer needs to know what reliability each sublevel should be designed to. This is achieved via reliability allocation.

2. Since an allocation program forces contractors to meet specified goals, improved design, procurement, manufacturing, and testing procedures result. This situation would not only ensure a reliable system, but should improve the state of the art of all facets of the associated program.

3. Reliability allocation focuses attention on the relationship between component, equipment, subsystem, and system reliability, thus leading to a more complete understanding of the basic reliability problems inherent in the design.

4. Requirements determined through an allocation procedure would be more realistic, consistent, and economically attainable than those obtained through subjective or haphazard methods, or those resulting from crash programs initiated after unfortunate field experiences.

5. Reliability allocation could lead to optimum system reliability since the program would provide for handling such factors as essentiality, complexity, state of the art, effect of operation environment, cost, maintenance, weight, and space.

6. The overall cost of such a program would be more than counterbalanced by a saving in the amount of time and money usually expended in meeting specified reliability goals by the first design. Without allocation, a system fails repeatedly and gets fixed over and over again, costing much in time and money. Through reliability allocation, subtantial reductions of operating, maintenance, and management costs would be realized by eliminating most of these problems in the beginning of the design stage.

15.3 HOW AND WHEN CAN RELIABILITY ALLOCATION BE BEST USED?

Apportionment is required when the manufacturer is obliged to decide how reliable various subsystems must be in order to design an end product that will meet the specified reliability requirement. In a series type of reliability relationship, in which the failure of any component results in mission failure, every component must be much more reliable than the goal for the mission, and every part must be more reliable than its component. In cases where the manufacturer has established its reliability goal, the process usually involves the determination of the part, component, subsystem, and system goals in this sequence.

Since apportionment is normally required early in the program when little or no hardware information is available, the apportionment must be updated periodically. Some points in the project at which reapportionment should take place are:

1. At the conclusion of the subsystem development testing phase.

2. At the conclusion of qualification testing for the majority of the functional components of the system.

3. At the initiation of any major design revision to the system.

For the initial apportionment, it is necessary to adopt a design capability apportionment procedure. A design capability apportionment is a complete set of design plans expressed in quantitative terms by which an engineering operation may be directed and by which its goals can be reached by the most profitable and timely course. In providing the design capability apportionment function, the following individual level assignments are performed:

1. *Formulation of design capability apportionment intention.* Design capability apportionment must be treated as a tool which allows an engineering program manager to concentrate on future design actions without resorting to nebulous generalizatons. This type of apportionment also serves as a major coordination tool by providing project leaders their work tasks in mutually understandable quantitative terms. Because of its sensitivity to personal feelings, apportionments must be developed through realistic predictions by design and reliability engineers at the lower levels of the organization, but they must be supported by the program manager through his or her policy decisions. The form of the apportionment must be aligned to the product being designed, the organization structure, and the method of testing and data reporting.

2. *Scheduling of design capability apportionment.* In the process of constructing a comprehensive apportionment, basic product factors, traits, and figures of merit must be first identified. These particular product capabilities, usually the critical or limiting ones specified by the customer, serve as starting points in the building of the apportionment structure. The period of time the apportionment is to cover must be short enough to permit fairly accurate prediction, but long enough to raise significant problems of policy, strategy, and procedure. The length of the apportionment period also depends on the availability of factual information and the stability of the customer's requirements. The apportionment period must correspond with preestablished program milestones so that actual results may be more realistically compared with apportioned values. Short-term apportionments permit closer alignment of the design effort to changing resources and conditions. Conversely, the apportionment values may be adjusted (assuming the nature of the product and customer requirements allow this) to permit more realistic alignment with design results. This is called a moving apportionment where the apportionment is planned for, say, a year; but at the end of each month the apportioned values are revised for the next 12 months. Moving apportionments are useful in extremely long development programs where precision in forecasting is not possible.

3. *Integration of design capability apportionment effort.* Because the apportionment is, among other things, a device to control the design outcome through people, emphasis must be placed on the human relations aspects of the operation. Apportionment parameters must be carefuly and diligently classified to avoid duplication of effort or arbitrary allocation of apportionments among different design groups.

4. *Conduct of design capability apportionment.* The preparation of the original design capability apportionment, or the making of revisions to it, is an area which requires cooperation between the design engineers and the reliability engineers.

5. *Survey of design capability apportionment results.* Engineering work is performed more effectively and economically when the functional activity constituting this work is based on quantitative research and methods. Setting of design goals requires adequate historical performance, endurance, cost, and schedule data if these goals are to be knowledgeably established and clearly expressed. Quantitative apportionment must also be established to permit rapid and authentic comparison of actual results with

predetermined design characteristics and capability values.

6. *Analysis of design capability apportionment misunderstandings.* Once completed, design apportionments must be reviewed to ensure subsequent effort will result in trouble-free and effectively executed design prediction and apportionment operation.

7. *Realignment of design apportionment effort.* Design apportionment cannot realistically be made too far in advance, and it is unwise to attempt the perfection of the design apportionment process all at once at the outset of the program. After minimum apportionment requirements are established initially, these requirements must be continually aligned to match program developments.

Apportionment has its limitations. As it is frequently applied, its mathematics are based on the asssumptions of statistical independence and on the form of the mathematical model which describes the relationship between the components and the system. It is not necessary in apportionment mathematics for independence to exist, but the mathematics become increasingly difficult if independence does not hold. It can be seen that these arguments apply only when we are referring to reliability numbers estimated from separate component or subsystem tests. When we are referring to subsystem tests as part of the system tests, then correct mathematical and statistical interrelationships are automatically present and the objection does not hold. As an example, it is perfectly valid (though it may not be feasible) to require an igniter to demonstrate 99.9% reliability on rocket engine tests, where 99.9% is the apportionment for the igniter of the complete engine's required reliability. On the other hand, it is not valid (in this example) to state that we have demonstrated our apportioned reliability by igniter tests which are conducted independently of the engine. The reason is that we do not know the extent of the interaction between the igniter and the propellant. In general, electronic equipment would have weak mutual interdependency.

15.4 RELIABILITY ALLOCATION: APPORTIONMENT METHODS

15.4.1 BASIC METHOD FOR SERIES SYSTEMS

The allocation of system reliability to its subsystems involves solving the basic inequality [1, pp. 184-196]

$$f(R_1^*, R_2^*, \ldots, R_N^*) = R_S^*, \tag{15.1}$$

where

$R_i^* =$ reliability allocated to the ith subsystem, $i = 1, 2, \ldots, N,$

$R_S^* =$ system reliability requirement or goal,

and

$f(\cdot) =$ functional relationship between subsystem and system
reliability.

For a simple reliabilitywise series system in which the R_i represent the probability of survival for t hours, Eq. (15.1) becomes

$$R_1^*(t) R_2^*(t) \ldots R_N^*(t) = R_S^*(t). \tag{15.2}$$

The basic method is based on subsystem failure rates which are assumed to be constant. The other necessary assumptions are that subsystem failures are independent and that failure of any subsystem will result in system failure; i.e., the system is composed of subsystems which are reliabilitywise in series. If

$\lambda_i^* =$ allocated failure rate of the ith subsystem in the system,

and

$\lambda_S^* =$ failure rate goal of the system,

then, from Eq. (15.2),

$$e^{-\lambda_1^* t} e^{-\lambda_2^* t} \ldots e^{-\lambda_N^* t} = e^{-t \sum_{i=1}^{N} \lambda_i^*} = e^{-\lambda_S^* t} = R_S^*(t). \tag{15.3}$$

Now two approaches may be used:

1. The reliability goal, $R_S^*(t)$, may be converted to a failure rate goal, λ_S^*, and then this λ_S^* can be allocated to the N subsystems.

However, how much of λ_S^* should be allocated to each subsystem is unknown. There exists an infinite number of combinations of λ_i^* that will satisfy Eq. (15.3). A good combination would be to allocate λ_S^* in proportion to known failure rates of similar subsystems, which is quite realistic.

2. $R_S^*(t)$ may be allocated directly to each subsystem using the allocation weights to be developed for the previous approach.

The first approach is made up of the following steps:

1. If the system's reliability goal is $R_S^*(t)$ and if the system's failure rate is constant, then

$$R_S^*(t) = e^{-\lambda_S^* t}.$$

Consequently, the system's failure rate goal is found from

$$\lambda_S^* = -\frac{\log_e R_S^*(t)}{t}.$$

2. For a system with reliabilitywise series subsystems

$$\lambda_S^* = \sum_{i=1}^{N} \lambda_i^*.$$

The λ_i^* can now be determined by allocating λ_S^* in proportion to known similar subsystem λ_i based on previous experience or predictions, using the following weights:

$$W_i = \frac{\lambda_i}{\sum\limits_{i=1}^{N} \lambda_i}. \tag{15.4}$$

Consequently, the allocated subsystem failure rate goals may be obtained from

$$\lambda_i^* = W_i \lambda_S^*. \tag{15.5}$$

It may be seen that

$$\sum_{i=1}^{N} \lambda_i^* = \sum_{i=1}^{N} W_i \lambda_S^* = \sum_{i=1}^{N} \frac{\lambda_i}{\sum\limits_{i=1}^{N} \lambda_i} \lambda_S^* = \frac{\lambda_S^*}{\sum\limits_{i=1}^{N} \lambda_i} \sum_{i=1}^{N} \lambda_i = \lambda_S^*.$$

Therefore, Eq. (15.4) does provide the correct allocation weights.

3. After the failure rate allocation is completed, check to make sure

$$\sum_{i=1}^{N} \lambda_i^* = \lambda_S^*.$$

The second approach allocates the system's reliability goal directly to its subsystems as reliabilities, rather than as failure rates, which is the first approach. To accomplish this, Eq. (15.3) may be rewritten as

$$R_S^*(t) = e^{-W_1 \lambda_S^* t} \; e^{-W_2 \lambda_S^* t} \ldots e^{-W_N \lambda_S^* t}, \qquad (15.6)$$

using Eq. (15.5) for λ_i^*, and

$$R_i^*(t) = e^{-W_i \lambda_S^* t} = (e^{-\lambda_S^* t})^{W_i} = [R_S^*(t)]^{W_i}. \qquad (15.7)$$

This approach is made up of the following steps:

1. Find the allocation weights using Eq. (15.4).

2. Find the reliabilities allocated to each subsystem directly using Eq. (15.7), or

$$R_i^*(t) = [R_S^*(t)]^{W_i}. \qquad (15.8)$$

3. Check to make sure the reliability allocation is correct from

$$R_S^*(t) = \prod_{i=1}^{N} R_i^*(t).$$

EXAMPLE 15-1

A system's reliability goal is specified as

$$R_S^*(t) = 0.95,$$

for a mission of $t = 5$ hr. The system consists of four, reliabilitywise-in-series subsystems. Their predicted failure rates are

$$\lambda_1 = 0.002 \text{ fr/hr}, \quad \lambda_2 = 0.003 \text{ fr/hr},$$
$$\lambda_3 = 0.004 \text{ fr/hr}, \quad \lambda_4 = 0.007 \text{ fr/hr}.$$

Allocate $R_S^*(t)$ to its subsystems in terms of their failure rates.

SOLUTION TO EXAMPLE 15-1

Given $R_S^*(t = 5 \text{ hr}) = 0.95$, the system's failure rate goal is

$$\lambda_S^* = \frac{-\log_e 0.95}{5} = 0.010 \text{ fr/hr}.$$

The allocated subsystem failure rates are

$$\lambda_1^* = \frac{\lambda_1}{\sum\limits_{i=1}^{4} \lambda_i}(\lambda_S^*) = \frac{0.002}{0.002 + 0.003 + 0.004 + 0.007}(0.010),$$

or

$$\lambda_1^* = \frac{0.002}{0.016}(0.010) = (0.1250)(0.010) = 0.001250 \text{ fr/hr},$$

$$\lambda_2^* = \frac{0.003}{0.016}(0.010) = (0.1875)(0.010) = 0.001875 \text{ fr/hr},$$

$$\lambda_3^* = \frac{0.004}{0.016}(0.010) = (0.2500)(0.010) = 0.002500 \text{ fr/hr},$$

and

$$\lambda_4^* = \frac{0.007}{0.016}(0.010) = (0.4375)(0.010) = 0.004375 \text{ fr/hr}.$$

As a check

$$\sum_{i=1}^{4} \lambda_i^* = 0.001250 + 0.001875 + 0.002500 + 0.004375,$$

or

$$\sum_{i=1}^{4} \lambda_i^* = 0.010000 \text{ fr/hr}.$$

Consequently, the failure rate allocation has been accomplished successfully.

EXAMPLE 15-2

The system of Example 15-1 is to be designed. Find the reliability goal to which each of its subsystems should be designed.

SOLUTION TO EXAMPLE 15-2

The allocation weights, from Example 15-1, are

$$W_1 = 0.1250, W_2 = 0.1875, W_3 = 0.2500, W_4 = 0.4375.$$

Using Eq. (15.8), the allocated subsystem reliability goals that should be designed into them are

$$R_1^*(t = 5 \text{ hr}) = (0.95)^{0.1250} = 0.99361,$$

$$R_2^*(t = 5 \text{ hr}) = (0.95)^{0.1875} = 0.99043,$$

$$R_3^*(t = 5 \text{ hr}) = (0.95)^{0.2500} = 0.98773,$$

and

$$R_4^*(t = 5 \text{ hr}) = (0.95)^{0.4375} = 0.97781.$$

As a check

$$\begin{aligned}
R_S^*(t = 5 \text{ hr}) &= \prod_{i=1}^{4} R_i^*(t = 5 \text{ hr}), \\
&= (0.99361)(0.99043)(0.98773)(0.97781),
\end{aligned}$$

or

$$R_S^*(t = 5 \text{ hr}) = 0.950.$$

EXAMPLE 15-3

A system consists of six communication sites linked in tandem by five radio subsystems. The user-to-user circuits are to provide both voice and digital data capability on a continuous 24-hr per day basis with equipment preventive maintenance normally scheduled every five days. With these basic assumptions, the following system reliability objectives were used in the specification:

1. The minimum acceptable system reliability shall be 0.90 based on the preventive maintenance interval of 120 hr.

2. The radio path reliability shall be designed to be no less than 0.999 for each individual radio link.

3. Each site shall be comprised of at least four units of major transmission equipment.

Determine the following:

1. The site reliability.

2. The equipment reliability.

3. The equipment MTBF.

SOLUTIONS TO EXAMPLE 15-3

1. Total system reliability is based on radio path reliability times site reliability. Using the individual path reliability of 0.999 and the minimum system reliabilty of 0.90, the minimum individual site reliability can be calculated from

$$R_T = R_p^n \times R_s^m,$$

where

R_T = system reliability = 0.90,

R_p = path reliability = 0.999,

R_s = site reliability,

n = number of paths = 5,

and

m = number of sites = 6.

Substituting the given values yields the site reliability, or

$$0.90 = (0.999)^5 (R_s)^6,$$
$$R_s = (\frac{0.90}{0.999^5})^{1/6},$$

or

$$R_s \cong 0.985.$$

This means that each site must function successfully for 985 out of 1,000 five-day missions in order to meet the total system reliability requirement of 0.90.

2. Site reliability is in turn based on the product of individual site equipment reliability. Using the site reliability of 0.985, an equally apportioned equipment reliability can be calculated from

$$R_s = (R_e)^p,$$

where

R_e = equipment reliability,

and

p = number of units per site = 4.

Substituting the given values yields the equipment reliability, or

$$0.985 = (R_e)^4,$$
$$R_e = (0.985)^{1/4},$$

or

$$R_e \cong 0.996.$$

This means that each item of equipment must function successfully for 996 out of 1,000 five-day missions in order to meet the site reliability requirement of 0.985.

3. Based on $R_e \cong 0.996$ and a 120-hr preventive maintenance interval, the equipment's MTBF can be calculated as follows:

$$R_e = e^{-\frac{t}{MTBF}},$$

$$0.996 = e^{-\frac{120}{MTBF}},$$

$$MTBF = -\frac{120}{\log_e 0.996},$$

or

$$MTBF \cong 30,000 \text{ hr.}$$

The above equipment's reliability and the corresponding MTBF must be achieved through effective apportionment techniques and built-in reliability features.

15.5   AGREE ALLOCATION METHOD

15.5.1   DESCRIPTION OF METHOD

The AGREE allocation method is based on unit or subsystem complexity rather than unit failure rates. The importance or essentiality of the unit defines quantitatively the relationship between unit and system failure and is considered explicitly in the AGREE allocation formula.

The allocation formula is used to determine a minimum acceptable mean life for each unit to satisfy a minimum acceptable system reliability. The assumption is made that units within the system have independent failure rates and operate in series with respect to their effect on mission success.

Unit complexity is defined in terms of the number of modules and their associated circuitry, where a module is an electron tube, a transistor, or a magnetic amplifier. Diodes represent a half-module. AGREE states that for digital computers, where the module count is high, the module count should be reduced to allow for the fact that failure rates for digital parts are generally far lower than for radio-radar types.

The importance factor for a unit is defined in terms of the probability of system failure if that particular unit fails. If the importance factor of a unit equals 1, the unit must operate satisfactorily for the system to operate satisfactorily; if the factor equals 0, then failure of the unit has no effect on system operation.

The specific basis of the allocation is to require that each module make an equal contribution to system success. An equivalent requirement is that each module have the same mean life or failure rate.

15.5.2 MATHEMATICAL MODEL FOR THE METHOD

With the approximation that

$$e^{-x} = 1 - x,$$

where x is small and less than 1, the allocated failure rate of the jth unit is shown in the AGREE report to be [2, pp. 52-57]

$$\hat{\lambda}_j = \frac{n_j[-\log_e R^*(T)]}{N E_j t_j}, \qquad (15.9)$$

where

n_j = number of modules in the jth unit,

N = total number of modules in the system,

E_j = importance factor of the jth unit,

and

t_j = number of hours the jth unit will be required to operate in T system hours ($0 < t_j \leq T$).

The allocated reliability for the jth unit for t_j unit operating hours, $\hat{R}(t_j)$, is given by

$$\hat{R}(t_j) = 1 - \frac{1 - [R^*(T)]^{n_j/N}}{E_j}. \qquad (15.10)$$

The AGREE report cautions against the use of the allocation formula for units of very low importance, which, if included, will distort the allocation.

15.5.3 APPLICATION TO THE SERIES SYSTEM

The system to be used as an example is the early warning radar set of Fig. 15.1. The primary mission of the system is the detection of aircraft. A system reliability requirement of 0.90 for 12 hr of continuous operation is desired. The subsystems are the following:

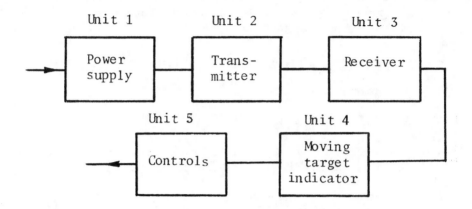

Fig. 15.1 – An early warning radar set in a reliabilitywise
    series arrangement.

1. Power supply.

2. Transmitter.

3. Receiver.

4. Moving target indicator.

5. Controls.

The first step is the determination of subsystem complexity. This
is accomplished by counting the number of modules in each subsystem.
The second step is the determination of the operating hours of each
subsystem. The third step is the determination of the essentiality,
E_j, of each unit. From the operating requirements on the radar set,
it is determined that all units except the moving target indicator are
required. If the moving target indicator fails an estimated 25% of the
target may be lost. Since unit importance is defined as the probability
that the system will fail to accomplish its mission if the unit fails while
all others are satisfactory, then

$$E_1 = E_2 = E_3 = E_4 = 1.0, \text{ and } E_5 = 0.25.$$

Table 15.1 summarizes the necessary inputs to the formulas for allocated acceptable failure rate or for allocated reliability. The allocated

TABLE 15.1 – Allocation inputs and allocated reliabilities.

System: AEW radar set. Primary mission: Aircraft detection. Reliability requirement, $R^*(12 \text{ hr}) = 0.90$.				
Subsystem	Number of modules, n_j	Operating time, hr, t_j	Essentiality, E_j	Allocated reliability, $R_i(t_j)$
1. Power supply	35	12	1.0	0.993
2. Transmitter	91	12	1.0	0.982
3. Receiver	88	12	1.0	0.983
4. Control	231	12	1.0	0.955
5. Moving target indicator	88	6	0.25	0.932
	$N = 533$			

acceptable unit failure rates, from Eq. (15.9), are as follows:

$$\hat{\lambda}_1 = \frac{35(-\log_e 0.90)}{(533)(1.0)(12)} = 575 \times 10^{-6} \text{ fr/hr},$$

$$\hat{\lambda}_2 = \frac{91(-\log_e 0.90)}{(533)(1.0)(12)} = 1,495 \times 10^{-6} \text{ fr/hr},$$

$$\hat{\lambda}_3 = \frac{88(-\log_e 0.90)}{(533)(1.0)(12)} = 1,445 \times 10^{-6} \text{ fr/hr},$$

$$\hat{\lambda}_4 = \frac{231(-\log_e 0.90)}{(533)(1.0)(12)} = 3,790 \times 10^{-6} \text{ fr/hr},$$

and

$$\hat{\lambda}_5 = \frac{88(-\log_e 0.90)}{(533)(0.25)(6)} = 11,560 \times 10^{-6} \text{ fr/hr}.$$

The allocated unit reliabilities, from Eq. (15.10), are as follows:

$$\hat{R}_1(12) = 1 - \frac{1 - (0.90)^{35/533}}{1.0} = 0.993,$$

$$\hat{R}_2(12) = 1 - \frac{1 - (0.90)^{91/533}}{1.0} = 0.982,$$

$$\hat{R}_3(12) = 1 - \frac{1 - (0.90)^{88/533}}{1.0} = 0.983,$$

$$\hat{R}_4(12) = 1 - \frac{1 - (0.90)^{231/533}}{1.0} = 0.955,$$

$$\hat{R}_5(6) \ = 1 - \frac{1 - (0.90)^{88/533}}{0.25} = 0.932,$$

$$\hat{R}_5(12) = \hat{R}_5(6) + [1 - \hat{R}_5(6)](1 - E_5) = 0.983.$$

As a check,

$$R_{set}(12 \ hr) = \hat{R}_1 \hat{R}_2 \hat{R}_3 \hat{R}_4 [\hat{R}_5(12)] = 0.8998 \cong 0.90.$$

The allocation inputs together with the apportioned reliability for each subsystem are presented in Table 15.1.

15.5.4 APPLICATION TO A PARALLEL SYSTEM

The AGREE report briefly dicusses allocation when redundancy exists, but the formulas given are very poor approximations at best. Any interested reader should read [2]. Where redundancy exists, it may be advisable to employ other methods of reliability allocation which are more accurate.

15.6 KARMIOL METHOD USING PRODUCT OF EFFECTS FACTORS

15.6.1 DESCRIPTION OF THE METHOD

This Karmiol method [3] utilizes the following factors of influence:

1. Complexity.

2. State of the art.

3. Operational profile.

4. Criticality.

   Consider a typical mission for a system having a number of subsystems functioning both individually and together to fulfill a set of mission objectives. The guidelines for determining the factors influencing the achievement of mission objectives are the following:

1. *Complexity.* This is the relative number of discrete mission functions which must be performed by each subsysten. This factor recognizes the limitations imposed by complexity on reliability attainment and will, therefore, vary inversely with the apportionment.

2. *Criticality.* This is the relative effect of each subsystem upon the achievement of the primary mission objectives. The criticality of the subsystem has a direct relationship with the apportionment.

3. *State of the art.* This is the state of present engineering progress in relevant fields related to the design and development phase of each subsystem.

4. *Operational profile.* This is the level of environmental exposure and it also takes into account the operational time requirements.

This method yields an apportioned reliability and an apportioned failure rate. The first step of the method is the rating of the influence factors. For complexity, state of the art, and operational profile the most severe requirement is rated 10 and the least severe is rated 1. For example, a subsystem that is the most complex would have a complexity rating of 10, and the least complex subsystem would have a complexity rating of 1. All other subsystems would have complexity ratings between 1 and 10. Conversely, the most critical subsystem would have a criticality of 1, and the least critical would have a rating of 10. Table 15.2 gives a compilation of the units and the weighting factors.

In addition to the influence factors mentioned above, the safety of the crew may be considered. One approach is to consider the safety of the crew (if there is any) and mission criticality together as one factor. Another is to measure the complexity of the subsystem by the interaction dependency. The *interaction dependency* shows the total number of systems which are dependent upon the successful operation of the subsystem in question (including itself). The product of the influence factors gives the *combined effects factor, n_i,* for each subsystem.

15.6.2 MATHEMATICAL MODEL FOR THE METHOD

Consider the system shown in Fig. 15.2. If the desired system reliability is R, and we let the combined effects factor and number of functions

TABLE 15.2– Determination of weighting factors for reliability apportionment by the Karmiol product of effects method.

		System reliability requirement = 0.95						
No.	Subsystem	Complexity or interaction dependency	State of the art	Operation profile	Criticality	No. of functions, F_j	Combined effects factor, n_j	Apportioned reliability
1	Propulsion	2	9	10	1	2	180	0.9961
2	Ordnance	3	7	1	1	1	21	0.9805
3	Guidance	8	8	10	2	6	1,280	0.9945
4	Flight controls	5	8	9	4	12	1,440	0.9895
5	Structures	7	6	9	8	8	3,024	0.9931
6	Electrical power	1	10	10	1	1	100	0.9952
						30	6,045	

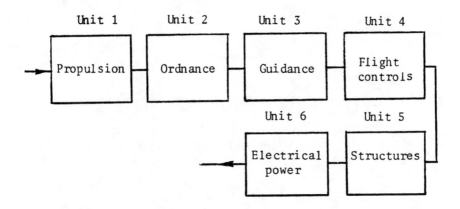

Fig. 15.2 – A system's reliability block diagram.

apply equally to the apportionment, then for n subsystems we have

$$R_G = \prod_{j=1}^{n} R_G^{\left[\left(\frac{\sum_{i=1}^{n} n_i}{2Nn_j}\right) + \left(F_j/2 \sum_{j=1}^{n} F_j\right)\right]},$$

$$N = \sum_{j=1}^{n} \frac{\sum_{i=1}^{n} n_i}{n_j},$$

where

R_G = reliability goal,

n_j = combined effects factor for each subsystem,

F_j = number of functions for each subsystem, which is determined by the number of operations it performs during the mission.

By this methodology, the apportioned reliabilities are obtained as follows:

Subsystem 1:

$$R_1 = R_G^{\left[\left(\sum_{i=1}^{n} n_i/2Nn_1\right) + \left(F_1/2 \sum_{j=1}^{n} F_j\right)\right]},$$

Subsystem 2:

$$R_2 = R_G^{\left[\left(\sum\limits_{i=1}^{n} n_i/2Nn_2\right)+\left(F_2/2\sum\limits_{j=1}^{n} F_j\right)\right]} \text{, etc.}$$

15.6.3 APPLICATION TO A SERIES SYSTEM

Consider the system of Fig. 15.2. The system has six subsystems reliabilitywise in series. Table 15.2 gives the units and the weighting factors. Let the system's reliability requirement be $R_G = 0.95$. Then

$$\sum_{i=1}^{6} n_i = 6,045, \qquad \sum_{i=1}^{6} F_i = 30,$$

$$N = \frac{6,045}{180} + \frac{6,045}{21} + \frac{6,045}{1,280} + \frac{6,045}{1,440}$$

$$N = 392.81.$$

The apportioned reliability for Subsystem 1 is

$$
\begin{aligned}
R_1 &= R_G^{\left[\left(\sum\limits_{i=1}^{n} n_i/2Nn_1\right)+\left(F_i/2\sum\limits_{j}^{n} F_j\right)\right]} \\
&= (0.95)^{\left[\frac{6,045}{2\times392.81\times180}+\frac{2}{2\times30}\right]} \\
&= (0.95)^{0.04275+0.33333}
\end{aligned}
$$

or

$$R_1 = 0.9961.$$

The apportioned reliabilities for Subsystems 2, 3, 4, 5, and 6 are calculated similarly. All values are listed in Table 15.2.

15.6.4 APPLICATION TO A PARALLEL SYSTEM

Consider the system illustrated in Fig. 15.3. Take each parallel subsytem to be a single unit with the following weighting factors:

$$F_p = \frac{1}{k}\frac{\sum\limits_{j=1}^{k} F_j}{k} \quad \text{and} \quad n_p = \frac{\sum\limits_{j=1}^{k} n_j}{k},$$

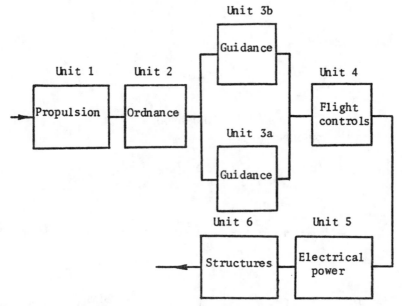

Fig. 15.3 – Reliability block diagram of a system with parallel redundancy.

where n_j and F_j are the weighting factors of each unit in parallel, and k is the number of units in this parallel subsystem.

For each unit calculate

$$x_i = \frac{\sum\limits_{i=1}^{n} n_i}{2N n_i} + \frac{F_i}{2 \sum\limits_{i=1}^{n} F_i},$$

and for each unit in series calculate $R_i = R_G^{x_i}$.

For each unit in the parallel subsystem calculate

$$y_j = \frac{1}{2n_j \sum\limits_{j=1}^{k} 1/n_j} + \frac{F_j}{2 \sum\limits_{j=1}^{k} F_j}, \text{ and } R_{ij} = 1 - \left(1 - R_G^{x_i}\right)^{y_j}.$$

By applying these rules, and using the data given in Table 15.2 and Fig. 15.3, the characteristics of the parallel subsystem become

$$F_p = \frac{\sum\limits_{j=1}^{k} F_j}{k^2} = \frac{6+6}{4} = 3 \text{ and } n_p = \frac{\sum\limits_{j=1}^{k} n_j}{k} = \frac{1280 + 1280}{2} = 1280,$$

$$N = \sum\limits_{j=1}^{n} \frac{\sum\limits_{i=1}^{n} n_i}{n_j} = 392.81,$$

$$x_1 = \sum_{i=1}^{6} n_i/2Nn_1 + F_1/2\sum_{i=1}^{6} F_i = 0.0798,$$

and
$$R_1 = R_G^{x_1} = 0.95^{0.0798} = 0.995960.$$
For Unit 3a the exponent y_a is

$$y_a = 1/2n_a \sum_{j=1}^{2} \frac{1}{n_j} + F_a/2\sum_{j=1}^{2} F_j = 0.5,$$

$$x_3 = \frac{\sum_{i=1}^{6} n_i}{2Nn_3} + \frac{F_3}{2\sum_{i=1}^{6} F_i} = 0.061567.$$

Thus

$$R_{3a} = 1 - \left(1 - R_G^{x_3}\right)^{y_a} = 1 - (1 - 0.95^{0.061567})^{0.5} = 0.943849.$$

Similarly,
$$R_2 = 0.980450, \quad R_{3b} = 0.943849, \quad R_4 = 0.988395,$$
$$R_5 = 0.992300, \quad \text{and} \quad R_6 = 0.995115.$$
Finally the reliability of the system is

$$R_s = R_1R_2[1 - (1 - R_{3a})(1 - R_{3b})]R_4R_5R_6 = 0.950000 = R_G.$$

Hence the system with this configuration meets its reliability goal.

15.7 KARMIOL METHOD UTILIZING SUM OF WEIGHTING FACTORS

This Karmiol method utilizes weighting factors for complexity, state of the art, operational profile, and criticality. These factors are added to obtain the total weighting factor for each subsystem functioning reliabilitywise in series, to apportion the unreliability first and subsequently the reliability and the failure rate.

The application of this method is illustrated for a specific system in Tables 15.3 and 15.4, where the system's reliability and failure rate are apportioned to its subsystems. The reapportionment to each component of a subsystem using this method is illustrated in Tables 15.5 and 15.6. The determination of the weighting factors is discussed next.

TABLE 15.3 – Determination of the weighting factors for each subsystem utilizing the Karmiol sum of weighting factors method.

Subsystem	Complexity	State of the art	Operational profile	Criticality	Total weighting factor
A-1 Command link	7	6	2	1	16
A-2 Tracking	6	5	2	8	21
A-3 Vehicle operation	10	7	2	6	25
A-4 Power supply	1	2	2	1	6
B-Payload operation	1	5	2	7	15
C-Monitoring	8	3	2	7	20
					$\Sigma = 103$

TABLE 15.4 – Apportionment of the unreliability, reliability, and failure rate to each subsystem for the case of Table 15.3.

Subsystem	Weighting factor ratio	Ratio x mission unreliability*	Apportioned unreliability, $Q_A(T)$	Apportioned reliability, $R_A(T)$	Apportioned failure, rate**, fr/10^6 hr
A-1 Command link	16/103	0.155x 0.10	0.0155	0.9845	156
A-2 Tracking	21/103	0.204 x 0.10	0.0204	0.9796	206
A-3 Vehicle operation	25/103	0.243 x 0.10	0.0243	0.9757	246
A-4 Power supply	6/103	0.058x 0.10	0.0058	0.9942	60
B-Payload Operation	15/103	0.146 x 0.10	0.0146	0.9854	147
C-Monitoring	20/103	0.194 x 0.10	0.0194	0.9806	202

  * Assume a subsystem reliability requirement of 0.90, and an unreliability requirement of 0.10.

  ** Assume a mission time of 100 hr and
$\lambda_A = -\log_e R_A(T)/T$, where T = 100 hr.

TABLE 15.5 – Determination of the weighting factors for each component within Subsystem A-1 for the case of Table 15.4.

Component	Complexity	State of the art	Operational profile	Criticality	Total weighting factor
Receiver	8	6	1	2	17
Translator	4	6	1	1	12
Programmer	2	2	1	3	8
Baroswitch	8	3	1	1	13
					50

TABLE 15.6 – Apportionment of the unreliability, reliability, and failure rate to the components within Subsytem A-1 for the case of Table 15.5.

Component	Weighting factor ratio	Ratio × mission unreliability*	Apportioned unreliability	Apportioned reliability	Apportioned failure rate, fr/10^6 hr
Receiver	17/50	0.34× 0.0155	0.00527	0.99473	54
Translator	12/50	0.24× 0.0155	0.00372	0.99628	38
Programmer	8/50	0.16× 0.0155	0.00248	0.99752	26
Baroswitch	13/50	0.26× 0.0155	0.00403	0.99597	42**

* Apportioned unreliability for function A-1 is 0.0155.
** The baroswitch is a cyclic device with a mission of 100 cycles.

15.8 DETERMINATION OF THE WEIGHTING FACTORS FOR UNRELIABILITY AND SUBSEQUENTLY FOR RELIABILITY APPORTIONMENT

15.8.1 COMPLEXITY FACTOR

The complexity is greater if the subsystem has more parts and components and a higher assembled intricacy of these parts and components. The subsystem with the greatest complexity is scored 10 and with the least 1. Thus the more complex subsystem is apportioned a higher unreliability and consequently a lower reliability.

15.8.2 STATE OF THE ART FACTOR

The state of the art factor considers the present state of engineering progress. The subsystem with the most advanced state of the art is scored 10 and with the least 1. This procedure apportions the lowest reliability to the subsystem having the highest state of the art, the implication being that its reliability cannot be improved significantly any more if it had to be increased.

15.8.3 OPERATIONAL PROFILE FACTOR

The operational profile involves the duty cycle or percent of mission time it operates. The subsystem with the more severe duty cycle and operating time requirements is scored 10 and with the least 1.

15.8.4 CRITICALITY FACTOR

Criticality considers the effect of subsystem failure on mission success. The most critical subsystem is scored 1 and the least 10, because failure aborts the mission. Therefore, the most critical subsystem is apportioned a lower unreliability and as a consequence a higher reliability, as it should be.

15.9 THE BRACHA METHOD OF RELIABILITY ALLOCATION

15.9.1 DESCRIPTION OF THE METHOD

The Bracha method [4] of reliability allocation uses four factors:

1. *State of the art.* The stage of present engineering progress of each sublevel is evaluated and compared.

2. *Sublevel complexity.* Complexity is the estimated number of parts making up the sublevel and is also judged by the assembled intricacy of these parts.

3. *Environmental conditions.* Environmental conditions to which each sublevel will be exposed are rated, depending on relative severity.

4. *Operating time.* Relative operating time during the complete functional period is determined for each sublevel.

The method is based on an evaluation of a unique index for each sublevel. This index represents the sum of the effect of each of the four factors given above.

15.9.2   MATHEMATICAL MODEL FOR THE METHOD

Given a reliability goal, $R_s(t)$, for an item with n sublevels in series, and assuming that the exponential time-to-failure distribution holds, the reliability goal apportioned to sublevel i is given by

$$R_i(t) = [R_s(t)]^{W_i},$$

where

$$W_i = \frac{I_i}{\sum\limits_{j=1}^{n} I_j},$$

and for any sublevel

$$I = A(C + E + T),$$

where

$A = $ state of the art index,

$C = $ complexity index,

$E = $ environment index,

and

$T = $ operating time index.

The following definitions are required to compute these indexes:

f = unit stress.

Estimates of the unit stress due to environmental conditions are made on a 0 to 100 scale. Complete breakdown can be expected at $f = 100$. The stress level at which no failure is expected to occur is set at $f = 0$.

$$K_i = Z_i K_{b_i} / \sum_{j=1}^{n} Z_j K_{b_j},$$

$$K_{b_i} = 10 n_{bi} / n_{bc},$$

n_{bi} = number of parts in sublevel i,

n_{bc} = number of parts in the most complex sublevel,

$$K_{pi} = 10 n_{pi} / n_{pc},$$

n_{pi} = number of redundant parts in sublevel i,

n_{pc} = number of redundant parts in the most complex sublevel,

T_m = total mission time of the item,

T_o = number of years since a given base date,

T_u = operating time of the sublevels,

T_w = number of years during which work has been done on the sublevel since the given base date,

a = effect of the state of the art on the reliability of the sublevel in the year in which the calculation is performed.

$$\bar{a} = a + T_o \, \Delta a,$$

and

λ_i = failure rate for the sublevel i.

The steps required to assign a goal to a sublevel by this method are:

1. Verify and satisfy assumptions.

2. Review availability of information required to compute indexes.

3. Make necessary estimates and calculations to establish the basic parameters $\bar{a}, K_b, K_p, \lambda, T_o, T_m, T_u$, and f.

4. Calculate the four sets of indexes for each sublevel.

5. Calculate the sublevel index.

6. Calculate the sublevel goal assessment.

7. Countercheck to assure the multiplicity of sublevel goals equals the goal for the item.

15.9.2.1-INDEX OF THE STATE OF THE ART

The state of the art index is given by

$$A = K^v,$$

where

$$v = a^{-T_w},$$

$A = 0$ when $K = \bar{m}$ and $\bar{a} \neq 0$,

$A = 1$ when $K \neq 0$ and $\bar{a} = 0$,

A is undefined only when $\bar{a} = K = 0$.

For example, from a reseach study, the relative effects of the state of the art on reliability for an existing system for a base date of January 1, 1987 were estimated to be as follows:

1. Mechanical systems: $a = 0.9842(\Delta a = 0.0017$ per year).

2. Electronic systems: $a = 0.90(\Delta a = -0.0130$ per year).

3. Electromechanical systems: $a = 0.9905(\Delta a = 0.0017$ per year).

4. Hydromechanical systems: $a = 0.80(\Delta a = 0.0001$ per year).

15.9.2.2-INDEX OF COMPLEXITY

The complexity index is given by

$$C = 1 - e^{-K_b + 0.6K_p}.$$

Both K_b and K_p should be estimated at the initial stages of development.

15.9.2.3-INDEX OF ENVIRONMENT

The environment index is given by

$$E = 1 - \frac{1}{f}.$$

The selection of f is difficult particularly in the proposal stage when no data are available. The stress considered is the externally applied effective stress. For example, if one were concerned with structures that were unaffected by radiation, the stress due to radiation would be disregarded. Radiation effects on electronics, however, could not be disregarded. The steps necessary to accomplish the stress level assignment are:

1. List all applicable external stresses.

2. Considering each stress separately, define the stress level at which complete failure is expected. Assign it a value of 100.

3. Considering each stress separately, define the stress level at which no failure due to stress may be expected. Assign it a value of 0.

4. Considering each stress separately and using Steps 2 and 3 above, assign a value of between 0 and 100 to the stress requirements of the design.

5. Considering the levels set in Step 4, set a best judgement value for the overall stress requirement (considering all stresses simultaneously) on the item.

15.9.2.4-INDEX OF OPERATING TIME

The operating time index is given by

$$T = \frac{T_m}{T_u}.$$

15.9.2.5-GENERAL PROCEDURE

Regardless of the type of the system (series, parallel, standby, etc.), a reliability prediction is formulated according to the following procedure:

1. A functional system schematic for each system in each mode of operation is prepared. The schematic will show all the components, the function of each, and the arrangement required to

accomplish the specified function. A reliability block diagram is then prepared from the functional schematic. This indicates the functional reliability arrangement of all components in series, parallel, standby, etc.

2. Failure rate data and failure pdf curves for all items listed on the block diagram are obtained. These data may be in the form of mean time between failures or mean cycles between failures, etc.

3. Components should be arranged into series, and parallel combinations. When items are arranged in series a failure of any item will result in system failure. When items are in parallel, the failure of any one item results in the absorption of some overload by the remaining items, and the system will continue to function without serious degradation.

4. System reliability should be determined by combining the component reliabilities. If the predicted system reliability is less than the apportioned reliability objective, it will be necessary to review and improve the design when the reliability deficiency cannot be traded off with the reliability excess of another component or subsystem.

15.9.3   APPLICATION TO A SERIES SYSTEM

Consider a system whose reliability goal has been specified by the customer to be 0.80 for mission accomplishment. The system contains the following subsystems:

1. Propulsion.

2. Ordnance.

3. Guidance.

4. Flight controls.

5. Structures.

6. Electrical power.

These subsystems are considered to be in series as shown in Fig. 15.2. The assumptions required for the apportionment have been satisfied, and information has been reviewed and tabulated as shown in Table 15.7. These values have been selected as representative of values found in practice. K_p is equal to zero for convenience only. Base time is January 1, 1991, and calculations were made for January 1,

TABLE 15.7 – Data for system evaluation and allocation.

Subsystem	T_m	T_u	T_w	a	$T_o \Delta a$	K_b	K_p	Z	f
Propulsion	0.6	1	3	0.9842	0.0068	2.736	0	0.02	95
Ordnance	0.3	1	4	0.90	0.052	5.586	0	4.20	40
Guidance	0.8	1	1	0.90	0.052	10.000	0	4.28	40
Flight controls	1.0	1	1	0.80	0.0004	1.372	0	10.00	80
Structures	1.0	1	1	0.9842	0.0068	0.403	0	0.00	70
Electric power	1.0	1	2	0.9905	0.0068	1.735	0	0.99	90

Subsystem	T	$\exp(-K_b + 0.6K_p)$	c	$1/f$	E
Propulsion	0.6	0.0648	0.9352	0.010526	0.989474
Ordnance	0.3	0.0038	0.9962	0.02500	0.9750
Guidance	0.8	0.00005	0.99995	0.02500	0.9750
Flight controls	1.0	0.2521	0.7479	0.01250	0.98750
Structures	1.0	0.6680	0.3320	0.014286	0.98571
Electric power	1.0	0.1764	0.8236	0.01111	0.98889

Subsystem	v	ZK_b	K	$\log_e K$	$\log_e K^v$	A
Propulsion	0.9732	0.054	0.00066	6.81954-10	-3.09522	0.000803
Ordnance	0.8214	23.46	0.28696	9.45773-10	-0.44542	0.3586
Guidance	0.9520	42.8	0.52353	9.71892-10	-0.26759	0.5401
Flight controls	0.8004	13.72	0.16782	9.22479-10	-0.62048	0.2396
Structures	0.9910	0.0	0.0			0.0
Electric power	0.9946	1.7176	0.02101	8.32243-10	-1.66851	0.02145

$$\sum_{j=1}^{n} z_j K_{bj} = 81.752$$

Subsystem	$C + E + T$	I	W_i	$R_i(t)$
Propulsion	2.52467	0.002028	0.00067	0.99850
Ordnance	2.2712	0.814452	0.26871	0.94129
Guidance	2.77495	1.49875	0.49448	0.89553
Flight controls	2.7354	0.6554	0.21624	0.95290
Structures	2.3177	0.0	0.0	1.00000
Electric power	2.81249	0.060328	0.01991	0.99557

$$\sum_{i=1}^{n} I_i = 3.03096$$

Checking the results: $\prod_{i=1}^{n} R_i(t) = 0.7985$. This compares well with the goal of 0.80.

1995, which means that T_o is equal to 4.0. T_w can be no greater than four years, and $\bar{a}$ is computed from base year $a + 4 \, \Delta a$. The step by step procedure for allocating the reliability goal of 0.80 to each of the six subsystems is given in Table 15.7. The last column gives these allocated reliabilities. Checking the results yields:

$$R_s(t) = \prod_{i=1}^{6} R_i(t),$$

$$= (0.99850)(0.94129)(0.89553)(0.95290)(1.00000)(0.99557),$$

or

$$R_s(t) = 0.7985.$$

This compares well with the system's reliability goal of 0.80.

15.9.4   APPLICATION TO A PARALLEL SYSTEM

Considering the system of Section 15.6.4, the guidance subsystem has been duplicated as illustrated in Fig. 15.3. The calculations of Table 15.7 also apply, but the reliability of the system is now given by

$$R_s = R_1 R_2 R_4 R_5 R_6 [1 - (1 - R_{3a})(1 - R_{3b})],$$

where the reliabilities of the two parallel subsystems are R_{3a} and R_{3b}. The subsystem reliabilities are determined by the procedure shown in Table 15.7, and the index I remains at

$$\sum_{i=1}^{n} I_i = 3.03096.$$

The combined index of the parallel guidance system is equal to 0.89553 as before; then

$$0.89553 = 1 - (1 - R_{3a})(1 - R_{3b}).$$

For the simple case where they are equal, R_{3a} and R_{3b} are determined by direct solution, or

$$R_{3a} = R_{3b} = 0.676782.$$

Therefore, the total combined reliability of the system is equal to 0.7985, which compares well with a target reliability of 0.80.

TABLE 15.8 – Vehicle subsystem observed or estimated failure rates for Problem 15-1.

Subsystem number	Subsystem	Observed or estimated failure rate, $fr/10^6$ miles
1	Engine	0.065
2	Chassis, suspension, and drive line	0.045
3	Transmission	0.055
4	Electrical	0.035

15.9.5 APPLICATION TO AN INACTIVE REDUNDANT (STANDBY) SYSTEM

The same procedure applies as shown in Section 15.6.4 for two identical units in a redundant configuration. An alternative is to consider the degree of redundancy from

$$K = \frac{\sum_{i=1}^{n} I_{r_i}}{\sum_{i=1}^{n} I_i},$$

where r represents the number of redundant units and I has the same meaning as in Section 15.6.4.

15.9.6 MORE COMPLEX SYSTEM RELIABILITY ALLOCATION

For more complex system reliability allocation techniques, the reader is referred to [5].

PROBLEMS

15-1. A vehicle has a reliability goal of 0.99994 for a 500-mile mission. The failure rates of its subsystems, as observed from similar vehicles or estimated, are as given in Table 15.8.

(1) Allocate this vehicle's failure rate goal to its subsystems using the basic method, knowing that these subsystems are reliabilitywise in series.

(2) Allocate this vehicle's reliability goal to its subsystems.

15-2. Using the complexity, state of the art, operational profile, and criticality weighting factors given in Table 15.9, do the following:

(1) Reallocate the reliability goal of the vehicle in Problem 15-1 to its subsystems.

(2) Compare the allocated reliabilities with those obtained in Problem 15-1.

(3) Reallocate the failure rate goal of the vehicle in Problem 15-1 to its subsystems.

(4) Compare the allocated failure rates with those obtained in Problem 15-1.

15-3. Based on the information given in Table 15.10, allocate the system's reliability goal of 0.95, using the AGREE allocation method. The system is composed of five units reliabilitywise in series.

15-4. Reapportion the reliability goal of 0.99994 for the system given in Problem 15-2 if the following additional information is provided: The number of functions of the engine is 9; the number of functions of the chassis, suspension, and drive line is 2; the number of functions of the transmission is 7; and the number of electrical functions is 18.

15-5. Given the indexes for a four-subsystem system in Table 15-11, allocate the reliability goal of 0.975 to its subsystems using the Karmiol product of weighting factors method.

15-6. Rework Problem 15-5 using the Karmiol sum of weighting factors method, and compare these results with those obtained in Problem 15-5.

REFERENCES

1. ARINC Research Corp., *Reliability Engineering,* Prentice Hall, Englewood Cliffs, N.J., 593 pp., 1964.

2. Advisory Group of Reliability of Electronic Equipment (AGREE), *Reliability of Military Electronic Equipment,* Office of the Assistant Secretary of Defense Research and Engineering, Washington, D.C., 377 pp., June 4, 1957.

TABLE 15.9 – Weighting factors for complexity, state of the art, operational profile, and criticality of Problem 15-2, for allocating the vehicle reliability and failure rate goal to its subsystems.

Subsystem No.	Subsystem	Complexity weighting factor	State of the art weighting factor	Operational profile weighting factor	Criticality weighting factor
1	Engine	10	8	4	1
2	Chassis, suspension, and drive line	5	4	3	3
3	Transmission	6	7	3	2
4	Electrical	4	6	4	1

TABLE 15.10 – Allocation inputs for Problem 15-3.

Unit, j	Number of modules, n_j	Operating time, hr, t_j	Essentiality, E_j
1	23	20	1.0
2	37	20	1.0
3	131	10	0.5
4	52	20	1.0
5	9	20	1.0

TABLE 15.11 – Indexes for a four-subsystem system for Problem 15-5.

Subsystem	Complexity weighting factor	State of the art weighting factor	Operational profile weighting factor	Criticality weighting factor
1	6	5	2	3
2	10	9	3	7
3	5	2	3	9
4	2	4	2	2

3. Karmiol, E.D., *Reliability Apportionment,* Preliminary Report EIAM-5, Task II, General Electric, Schenectady, N.Y., pp. 10-22, Apr. 8, 1965.

4. Bracha, V.J., "The Methods of Reliability Engineering," *Machine Design,* pp. 70-76, July 30, 1964.

5. Ballaban, H.S., and Jeffers, H.R., *The Allocation of System Reliability: Development of Procedures for Reliability Allocation and Testing,* Vols. I, II, Technical Documentary Report No. ASD-TDR-62-20, ARINC Research Corporation, 2551 Riva Road, Annapolis, Maryland 21401, 156 pp., June 1962.

Chapter 16

RELIABILITY GROWTH

16.1 INTRODUCTION

Reliability growth [1-5] is either a projection of the reliability of a system or an equipment to some future development time from information available now from predictions or prior experience on identical or similar systems or equipment; or the monitoring of the reliability, of the MTBF, or of the failure rate of the system or equipment to establish the trend in the increase of the reliability, the increase of the MTBF, or the decrease in the failure rate with engineering, research, development, test, analyze, and fix (TAAF) and further test time until it passes its acceptance tests or is delivered and operated by the user. Such reliability growth curves are shown in Figs. 16.1, 16.2, and 16.3. Two major types of reliability growth trends are shown in Fig. 16.4. One type of an MTBF growth curve and one type of a failure rate improvement curve are shown in Fig. 16.5.

Reliability growth studies are necessary to insure that, from information available at the beginning of a project, the reliability, MTBF, or failure rate goals are capable of being met by acceptance or delivery and use time. A growth model is used and projected to the project completion date. If this projected R, m, or λ is equal to or beyond the specified goal, then the project manager would be confident that she or he will be able to meet the project's R, m, or λ requirements. Otherwise, the manager will have to reassess the reliability prediction techniques or refine them in the hope of exceeding the goal; or the manager will have to improve the design, or use more derating, or use more redundancy or more reliable components; or allocate a greater proportion of the contract's resources to design engineering, reliability and maintainability engineering, research and development, manufacturing, purchasing, quality control, inspection, and testing; and perhaps

401

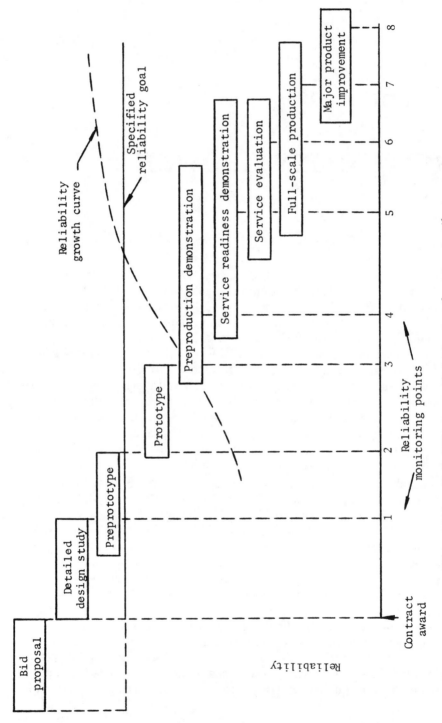

Fig. 16.1 – Reliability growth in terms of contract milestones.

402

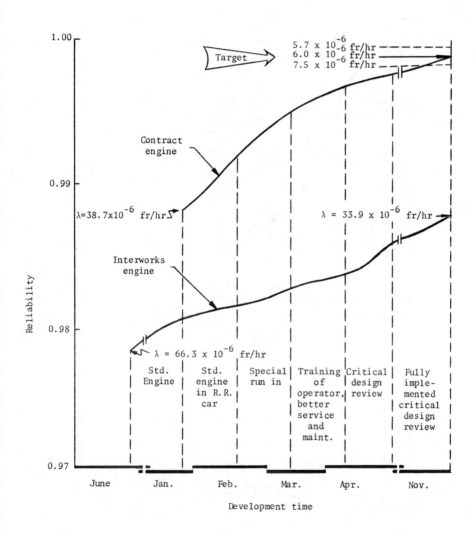

Fig. 16.2 – Diesel engine reliability growth curves.

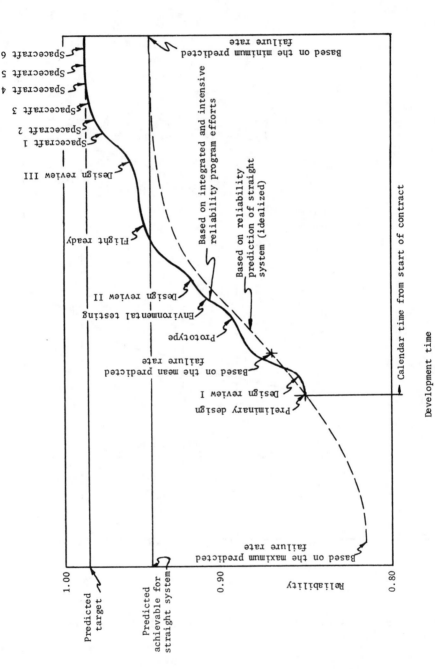

Fig. 16.3 – Ranger spacecraft reliability growth curve.

404

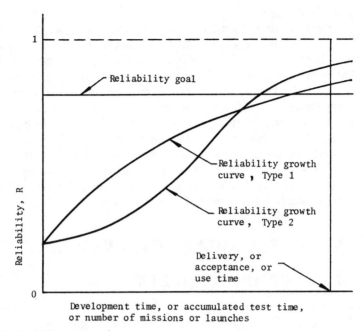

Fig. 16.4 – Reliability growth curves.

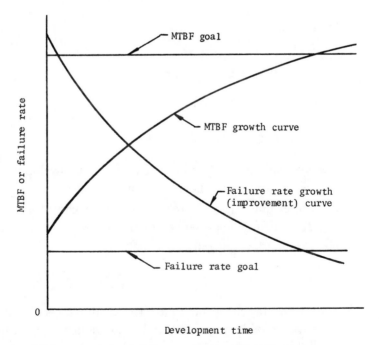

Fig. 16.5 – MTBF growth and failure rate improvement curves.

provide for more test units.

Determining R, m, or λ during various milestones of the project, and various development and test-analyze-fix-and-test-again times, and plotting these on the predicted and applicable growth curves establishes the relative trends between the predicted and the actual R, m, or λ. If the trend of the actual curve is favorable, then a good job in predicting and pursuing the growth has been done. If not, then the project manager can decide what steps should be taken, prior to going into full-scale production, to insure that the R, m, or λ goals will be met at the desired time. Otherwise, either the system or equipment will be rejected by the customer, or much time, money, personnel, and material resources will have to be expended after production to correct the problems and retrofit the equipment.

16.2 RELIABILITY GROWTH MATH MODELS

16.2.1 GOMPERTZ MODEL

If the reliability growth trend, from past experience, is similar to the Type 1 reliability growth curve shown in Fig. 16.4, then a good model to use is the Gompertz growth model given by

$$R = ab^{c^{T_a}},$$ (16.1)

where

$0 < a \leq 1$, if R is in decimals,

$0\% < a \leq 100\%$, if R is in percent,

$0 < b < 1$, $0 < c < 1$, and $T_a > 0$,

$R =$ system's or equipment's reliability at development time T_a, or at launch number T_a, or mission number T_a,

$a =$ upper limit that the reliability approaches asymptotically as $T_a \to \infty$ (as a special case, if $R \to 1$ as $T_a \to \infty$, then $a = 1$),

$ab =$ original level of reliability at $T_a = 0$,

and

$c =$ establishes the growth pattern (small values of c model rapid early reliability growth, and large values of c model slow reliability growth).

TABLE 16.1 – Design and development time versus demonstrated reliability data for a device.

Group number	Growth time, T_a, months	Reliability, $R, \%$	$\log_{10} R^*$
	0	56.0	1.748
1	1	64.0	<u>1.806</u>
			$S_1 = 3.554$
	2	70.5	1.848
2	3	76.0	<u>1.881</u>
			$S_2 = 3.729$
	4	80.0	1.903
3	5	83.0	<u>1.919</u>
			$S_3 = 3.822$

*Any logarithm base may be used as long as the same base is used consistently throughout all of the calculations involved.

To estimate the values of the parameters $a, b,$ and c, do the following:

1. Arrange the currently available data in terms of T_a and R as in Table 16.1. The T_a values should be chosen at *equal intervals* and increasing in value by 1, such as one month, one hour, etc. For other intervals or increments, see Appendix 16A.

2. Calculate $\log R$.

3. Divide the column of values for $\log R$ into *three equal size* groups, each containing n items. There should always be *three* groups. Each group should always have the *same number*, n, of items, measurements, or values.

4. Add the values of $\log R$ in each group, obtaining the sums identified as S_1, S_2, and S_3, starting with the lowest values of $\log R$.

5. Calculate c from

$$c = \left(\frac{S_2 - S_3}{S_1 - S_2}\right)^{\frac{1}{n}}. \tag{16.2}$$

6. Calculate a from

$$a = \text{antilog } \frac{1}{n}\left(S_1 - \frac{S_1 - S_2}{1 - c^n}\right). \tag{16.3}$$

7. Calculate b from

$$b = \text{antilog } \frac{(S_1 - S_2)(1 - c)}{(1 - c^n)^2}. \tag{16.4}$$

8. Write the Gompertz reliability growth equation.

9. Substitute the value of T_a, the time at which the reliability goal is to be achieved, to see if indeed the reliability is to be attained if not even exceeded by T_a.

See Appendix 16A for the derivation of Eqs.(16.2), (16.3), and (16.4) for any time increment of the growth period.

EXAMPLE 16–1

It is required that a device have a reliability of 90% at the end of a 10-month design and development period.

1. What will the reliability be at the end of this 10-month period if the data given in the second and third columns of Table 16.1 are obtained for the first 5 months of the design and development period?

2. What will the maximum achievable reliability be if the reliability program plan pursued during the first 5 months is continued?

3. How do the predicted reliability values compare with the actual?

SOLUTIONS TO EXAMPLE 16–1

1. Having completed Steps 1 through 4 by preparing Table 16.1 and calculating the last column to find S_1, S_2, and S_3, proceed as follows:

a. Find c from Eq. (16.2), or

$$c = \left(\frac{3.729 - 3.822}{3.554 - 3.729}\right)^{\frac{1}{2}} = 0.729.$$

b. Find a from Eq. (16.3), or

$$a = \text{antilog}_{10}\left[\frac{1}{2}\left(3.554 - \frac{3.554 - 3.729}{1 - 0.729^2}\right)\right],$$
$$a = \text{antilog}_{10}(1.964),$$

or

$$a = 92.0\%.$$

This is the upper limit for the reliability as $T_a \rightarrow \infty$.

c. Find b from Eq. (16.4), or

$$b = \text{antilog}_{10}\left[\frac{3.554 - 3.729)(1 - 0.729)}{(1 - 0.729^2)^2}\right],$$

$$b = \text{antilog}_{10}(-0.216),$$

or

$$b = 0.608.$$

d. The Gompertz reliability growth curve now is

$$R = 92.0(0.608)^{0.729^{T_a}}. \tag{16.5}$$

e. The achievable reliability at the end of the 10-month period of design and development is

$$R = 92.0(0.608)^{0.729^{10}},$$

or

$$R = 90.2\%.$$

The required reliability is 90.0%. Consequently, from the previous result, this requirement will barely be met. Every effort should therefore be expended to implement the reliability program plan fully, and perhaps augment it slightly to assure that the reliability goal will be met.

2. The maximum achievable reliability from Step b, or from the value of a, is 92.0%.

3. The predicted reliability values, as calculated from the Gompertz equation, Eq. (16.5), found in Step d are compared with the actual data in Table 16.2. It may be seen in Table 16.2 that the Gompertz curve appears to provide a good fit for the data used, since the equation reproduces the available data with less than 1% error. Equation (16.5) is shown plotted in Fig. 16.6 and identifies the type of reliability growth curve this equation represents.

16.2.2 LLOYD-LIPOW MODEL

Lloyd and Lipow [6] considered a situation in which a test program is conducted in N stages, each stage consisting of a certain number of trials of the item under test. All tests in a given stage of testing involve similar items. The results of each stage of testing are used to

TABLE 16.2 – Comparison of the predicted reliabilities with the actual data.

Growth time, T_a, months	Reliability calculated from the Gompertz equation, %	Reliability calculated from the available data, %
0	56.0	56.0
1	64.0	64.0
2	70.6	70.5
3	75.9	76.0
4	80.0	80.0
5	83.0	83.0
6	85.4	
7	87.1	
8	88.4	
9	89.4	
10	90.1	

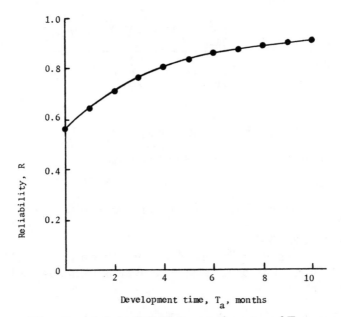

Fig. 16.6 – The plot of the reliability growth curve of Example 16–1 and of Eq.(16.5).

improve the item for further testing in the next stage. They imposed on the data a reliability growth function of the form

$$R_k = R_\infty - \frac{\alpha}{k},\qquad(16.6)$$

where

R_k = reliability during the kth stage of testing,

R_∞ = ultimate reliability, a parameter,

and

α = a parameter.

The least squares estimates of R_∞ and α are

$$\hat{R}_\infty = \frac{(\sum\limits_{k=1}^{N} 1/k^2)(\sum\limits_{k=1}^{N} S_k/N_k) - (\sum\limits_{k=1}^{N} 1/k)(\sum\limits_{k=1}^{N} S_k/kN_k)}{N(\sum\limits_{k=1}^{N} 1/k^2) - (\sum\limits_{k=1}^{N} 1/k)^2},(16.7)$$

and

$$\hat{\alpha} = \frac{(\sum\limits_{k=1}^{N} 1/k)(\sum\limits_{k=1}^{N} S_k/N_k) - N(\sum\limits_{k=1}^{N} S_k/kN_k)}{N(\sum\limits_{k=1}^{N} 1/k^2) - (\sum\limits_{k=1}^{N} 1/k)^2}, \tag{16.8}$$

where

N = total number of test stages,

N_k = number of test items in kth stage,

and

S_k = number of successful tests in kth stage.

EXAMPLE 16-2

After a 20-stage reliability development test program, 20 groups of success-failure data were obtained and are given in Table 16.3. Do the following:

1. Fit these data to the Lloyd-Lipow model.

2. Plot the reliabilities predicted by the Lloyd-Lipow model and the observed reliabilities, and comparatively discuss the results.

SOLUTIONS TO EXAMPLE 16-2

1. In Eqs. (16.7) and (16.8)

$$\sum_{k=1}^{N} \frac{1}{k} = \sum_{k=1}^{20} \frac{1}{k} = 3.59775,$$

$$\sum_{k=1}^{N} \frac{1}{k^2} = \sum_{k=1}^{20} \frac{1}{k^2} = 1.59616,$$

$$\sum_{k=1}^{N} \frac{S_k}{N_k} = \sum_{k=1}^{20} \frac{S_k}{N_k} = 16.165,$$

and

$$\sum_{k=1}^{N} \frac{S_k}{kN_k} = \sum_{k=1}^{20} \frac{S_k}{kN_k} = 2.36757.$$

Substituting into Eqs. (16.7) and (16.8) yields

$$\hat{R}_{\infty} = \frac{(1.59616)(16.165) - (3.59775)(2.36757)}{(20)(1.59616) - (3.59775)^2},$$

or

TABLE 16.3 – The test results, the reliabilities of each stage calculated from raw data, and the predicted reliability from the Lloyd-Lipow model for Example 16–2.

Number of test stage, k	Number of tests in kth stage, N_k	Number of successful tests in kth stage, S_k	Reliability calculated from raw data, $\hat{\hat{R}}_k = S_k/N_k$	Predicted reliability, $R_k = R_\infty - \alpha/k$
1	10	5	0.500	0.447
2	8	5	0.625	0.637
3	9	6	0.667	0.700
4	9	7	0.778	0.732
5	10	6	0.600	0.751
6	10	7	0.700	0.764
7	10	8	0.800	0.773
8	10	7	0.700	0.780
9	10	6	0.600	0.785
10	11	7	0.636	0.789
11	10	9	0.900	0.793
12	11	10	0.909	0.795
13	12	9	0.750	0.798
14	10	8	0.800	0.800
15	10	7	0.700	0.802
16	10	8	0.800	0.803
17	10	9	0.900	0.805
18	10	9	0.900	0.805
19	10	10	1.000	0.807
20	10	9	0.900	0.808

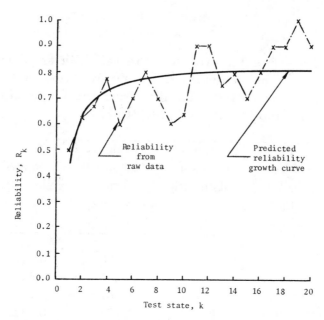

Fig. 16.7 – Comparison of the predicted reliability and the reliability from raw data for Example 16–2.

$$\hat{R}_{\infty} = 0.82657,$$

and

$$\hat{\alpha} = \frac{(3.59775)(16.165) - (20)(2.36757)}{(20)(1.59616) - (3.59775)^2},$$

or

$$\hat{\alpha} = 0.3798.$$

Therefore, the Lloyd-Lipow reliability growth model is

$$R_k = 0.82657 - \frac{0.3798}{k}, \qquad (16.9)$$

where k is the number of the test stage.

2. The reliabilities from the raw data and the reliabilities predicted from Eq. (16.9) are given in Columns 4 and 5, respectively, of Table 16.3. The plots are given in Fig. 16.7 comparatively. The model provides a good tracking of the raw data and appears to be an acceptable one for this equipment.

16.3   METHODS TO ESTIMATE RELIABILITY GROWTH FROM ATTRIBUTE DATA

For many kinds of equipment, especially missiles and space systems, only success and failure attribute data are obtained to analyze the trend of reliability growth. Methods are given in this section to handle such data.

METHOD 1

Conservatively, the cumulative reliability can be used to estimate the trend of reliability growth. The cumulative reliability is given by

$$\hat{\bar{R}}(N) = \frac{N-r}{N},$$

(16.10)

where

N = current number of trials,

and

r = number of failures.

It must be emphasized that the instantaneous reliability of the developed equipment is increasing as the test-analyze-fix-and-test process continues, and the instantaneous reliability is higher than the cumulative reliability. Therefore, the reliability growth curve based on the cumulative reliability can be thought of as the lower bound of the true reliability growth curve.

EXAMPLE 16–3

In an aerospace program the launch vehicle was successful in 15 out of the 22 launches. The specific success and failure launch data are given in Table 16.4. Do the following based on the cumulative reliability:

1. Find the Gompertz reliability growth curve using the results of the first 15 launches.

2. Find the reliability after Launch 22 as predicted by this curve.

3. Compare the reliability found in Case 2 and in Table 16.5 after Launch 22.

SOLUTIONS TO EXAMPLE 16–3

1. Prepare Table 16.5. Using this table, find the parameters of the Gompertz reliability growth curve, as follows:
Using Eq. (16.2),

TABLE 16.4 – Missile launch data for Example 16–3.

Launch number	Success	Failure	Cumulative reliability estimate, $\hat{\hat{R}}$, %
1		X	0.0
2		X	0.0
3		X	0.0
4	X		25.0
5		X	20.0
6		X	16.7
7	X		28.6
8	X		37.5
9	X		44.4
10	X		50.0
11	X		54.5
12	X		58.2
13	X		61.7
14	X		64.2
15	X		66.7
16	X		68.7
17		X	65.0
18	X		66.7
19		X	63.0
20	X		65.0
21	X		66.7
22	X		68.1

TABLE 16.5 – Missile launch data and their analysis for determining the parameters of the Gompertz reliability growth curve for Example 16–3.

Launch number, L_i		Cumulative number of successes, S_i	Cumulative reliability, $\hat{\bar{R}} = \frac{S_i}{L_i}$	$\log_{10} \hat{\bar{R}}$
	1	0	0	
	2	0	0	
	3	0	0	
Group 1	4	1	25.0	1.398
	5	1	20.0	1.301
	6	1	16.7	1.223
	7	2	28.6	<u>1.456</u>
				$S_1 = 5.378$
Group 2	8	3	37.5	1.574
	9	4	44.4	1.647
	10	5	50.0	1.699
	11	6	54.5	<u>1.736</u>
				$S_2 = 6.656$
Group 3	12	7	58.2	1.765
	13	8	61.7	1.790
	14	9	64.2	1.807
	15	10	66.7	<u>1.824</u>
				$S_3 = 7.186$
	16	11	68.7	1.837
	17	11	65.0	1.813
	18	12	66.7	1.824
	19	12	63.0	1.799
	20	13	65.0	1.813
	21	14	66.7	1.824
	22	15	68.1	1.833

$$c = \left(\frac{6.656 - 7.186}{5.378 - 6.656}\right)^{\frac{1}{4}},$$

or

$$c = 0.802. \tag{16.11}$$

Using Eq. (16.3),

$$a = \text{antilog}_{10}\left[\frac{1}{4}\left(5.378 - \frac{5.378 - 6.656}{1 - 0.802^4}\right)\right],$$

or

$a = 77.6\%$ (the upper limit for the reliability).

Using Eq. (16.4),

$$b = \text{antilog}_{10}\left[\frac{(5.378 - 6.656)(1 - 0.802)}{(1 - 0.802^4)^2}\right],$$

or

$b = 0.182.$

The Gompertz reliability growth curve may now be written as

$$R = 77.6(0.182)^{0.802^{L_G}}, \tag{16.12}$$

where L_G is the number of launches with the first successful launch being counted as $L_G = 0$.

2. The predicted reliability after Launch 22, based on Eq. (16.11), is

$$R = 77.6(0.182)^{0.802^{18}},$$

or

$$R = 75.15\%.$$

$L_G = 18$ is used because reliability growth starts with the fourth launch, which is taken to be $L_G = 0$; consequently, Launch 22 is $L_G = 18$.

3. The predicted reliability of 75.15% compares with the actual value of 68.1% given in Table 16.5 for Launch 22. The discrepancy may be attributed to the results of Launches 17 and 19 not being consistent with the previous results up to and including Launch 15, on which results the Gompertz model's parameters were obtained.

METHOD 2–FAILURE DISCOUNTING METHOD

During a reliability growth test, once a failure has been analyzed and corrective actions for that specific failure mode have been implemented, the probability of its recurrence is diminished. Then, for the succeeding attribute data, the value of the failure for which corrective actions have already been implemented should be subtracted from the total number of failures. But the questions are, "To what extent should the failure value be diminished or discounted?" and "How should the failure value be arrived at?" One answer would be to use engineering judgment; e.g., a panel of specialists would agree that the probability of failure has been reduced by 50% or 90% and therefore that failure should be given a value of 0.5 or 0.1. The obvious disadvantage of this approach is its arbitrariness and the difficulty of reaching an agreement. Therefore, a statistical basis was selected [7], one which is repeatable and less arbitrary. The value of the failure, f, is chosen to be the upper confidence limit on the probability of failure based on the number of successful tests following implementation of the corrective action. For example, after one successful test, following a corrective action, the failure is given a value of 0.9, based on a 90% confidence level; after two successful tests, the failure is given a value of 0.684, and so on. In general

$$f = 1 - (1 - CL)^{\frac{1}{S_n}}, \tag{16.13}$$

where

CL = confidence level,

and

S_n = number of successful tests after the first success following the corrective action.

The procedure for applying this method is illustrated in the next example.

EXAMPLE 16-4

Use Method 2 to answer the questions given in Example 16-3. Assume that during the 22 launches, the first failure is caused by Failure Mode 1, f_1, the second and fourth failures are caused by Failure Mode 2, f_2, the third and fifth failures are caused by Failure Mode 3, f_3, the sixth failure is caused by Failure Mode 4, f_4, and the seventh failure is caused by specification violation, f_5.

SOLUTIONS TO EXAMPLE 16-4

1. Prepare Table 16.6 as follows:

   (a) The value of failures calculated from Eq. (16.13) is based on $CL = 0.90$.

TABLE 16.6 – Launch sequence, its results, failure modes, and reliability computations for Example 16–4.

Launch, L number	Results*	Failure mode and value of failure					$\sum f_i$	$R = 1 - \dfrac{\sum f_i/L}{} \times 100,\%$	Remarks
		f_1	f_2	f_3	f_4	f_5			
1	F	1.000					1.000	0.00	Failure Mode f_1
2	F	1.000	1.000				2.000	0.00	Failure Mode f_2
3	F	0.900	1.000	1.000			2.900	3.33	Failure Mode f_3
4	S	0.684	0.900	1.000			2.584	35.40	
5	F	0.536	1.000	0.900			2.436	51.28	Failure Mode f_2 recurs
6	F	0.438	1.000	1.000			2.438	59.37	Failure Mode f_3 recurs
7	S	0.369	0.900	1.000			2.269	67.59	
8	S	0.319	0.684	0.900			1.903	76.21	
9	S	0.280	0.536	0.684			1.500	83.33	
10	S	0.250	0.438	0.536			1.224	87.76	
11	S	0.226	0.369	0.438			1.033	90.61	
12	S	0.206	0.319	0.369			0.894	92.55	
13	S	0.189	0.280	0.319			0.788	93.94	
14	S	0.175	0.250	0.280			0.705	94.96	
15	S	0.162	0.226	0.250			0.638	95.75	
16	S	0.152	0.206	0.226			0.584	96.35	
17	F	0.142	0.189	0.206	1.000		1.537	90.96	Failure Mode f_4
18	S	0.134	0.175	0.189	1.000		1.498	91.68	
19	F	0.127	0.162	0.175	0.900	1.000	2.364	87.56	Specification violation†
20	S	0.120	0.152	0.162	0.684	0.000	1.118	94.41	
21	S	0.114	0.142	0.152	0.536	0.000	0.944	95.50	
22	S	0.109	0.134	0.142	0.438	0.000	0.823	96.26	

*S = success; F = failure.

†Specification violation causes f_5. Specification change eliminates f_5. Specification change eliminates f_5.

(b) Failure Mode 1, f_1, is diminished on Launch 3, which is the second successful launch after the previous failure in the sense that Failure Mode 1 does not recur in the second and third launches, and f_1 continues to diminish in all subsequent tests since it does not recur.

(c) Failure Mode 2, f_2, is diminished on Launch 4. It recurs on Launch 5, and further corrective action is taken. Since, in Launch 6, f_2 does not recur, f_2 continues to diminish subsequently.

(d) Failure Mode 3 is similar to Failure Mode 2.

(e) Failure Mode 4 is similar to Failure Mode 1.

(f) On Launch 19, a small performance anomaly occurs, which is "being outside current specification limits." The corrective action for f_5 is to change the specification. With this change, f_5 becomes zero thereafter.

Prepare Table 16.7 to find the parameters of the Gompertz reliability growth curve, as follows: For Launch 3 and Failure Mode f_1, $S_n = 1$ in Eq. (16.13); then

$$f_{1/3} = 1 - (1 - 0.90)^{1/1} = 0.900.$$

And for Launch 4 and Failure Mode f_1, $S_n = 2$ in Eq. (16.13); then

$$f_{1/4} = 1 - (1 - 0.90)^{1/2} = 0.684,$$

etc. Using Eq. (16.2),

$$c = \left[\frac{7.7033 - 7.8979}{6.8624 - 7.7033}\right]^{\frac{1}{4}},$$

or

$$c = 0.6936.$$

Using Eq. (16.3),

$$a = \text{antilog}_{10}\left[\frac{1}{4}\left(6.8624 - \frac{6.8624 - 7.7033}{1 - 0.6936^4}\right)\right],$$

or

$$a = 97.53.$$

Using Eq. (16.4),

TABLE 16.7 – Missile launch data and their analysis to determine the parameters of the Gompertz reliability growth curve for Example 16–4.

Launch number, L		Modified cumulative number of failures, $\sum f$	Reliability, $R = (1 - \sum f/L)$ $\times 100, \%$	$\log_{10} R$
	1	1.000	0.0	
	2	2.000	0.0	
	3	2.900	3.33	
Group 1	4*	2.584	35.40	1.5490
	5	2.436	51.28	1.7099
	6	2.438	59.37	1.7736
	7	2.269	67.59	1.8299
				$S_1 = 6.8624$
Group 2	8	1.903	76.21	1.574
	9	1.500	83.33	1.9208
	10	1.224	87.76	1.9433
	11	1.033	90.61	1.9572
				$S_2 = 7.7033$
Group 3	12	0.894	92.55	1.9664
	13	0.788	93.94	1.9729
	14	0.705	94.96	1.9775
	15	0.638	95.75	1.9811
				$S_3 = 7.8979$
	16	0.584	96.35	
	17	1.537	90.96	
	18	1.498	91.68	
	19	2.364	87.56	
	20	1.118	94.41	
	21	0.944	95.50	
	22	0.823	96.26	

*It is assumed that there is no significant reliability growth until Launch 4.

$$b = \text{antilog}_{10} \frac{(6.8624 - 7.7033)(1 - 0.6936)}{(1 - 0.6936^4)^2},$$

or

$$b = 0.3663.$$

The Gompertz reliability growth curve may now be written as

$$R = 97.53(0.3663)^{0.6936^{L_G}}, \tag{16.14}$$

where L_G is the number of launches, with the first successful launch being counted as $L_G = 0$.

2. The predicted reliability after Launch 22, based on Eq. (16.14), is

$$R = 97.53(0.3663)^{0.6936^{18}},$$

or

$$R = 97.395\%.$$

$L_G = 18$ is used because reliability growth starts with Launch 4, which is taken to be $L_G = 0$; consequently, Launch 22 is $L_G = 18$.

3. In Table 16.8, the predicted reliability values, as calculated from Eq. (16.14), are compared with the reliabilities which are calculated from the raw data using Method 2 and given in Table 16.7. It may be seen in Table 16.8 that the Gompertz curve appears to provide a good fit to the actual data.

16.4 RELIABILITY GROWTH MODELS THAT GIVE S-SHAPED CURVES

The S-shaped reliability growth curve, such as shown in Figs. 16.3 and 16.4 (Type 2), is used to describe the reliability growth trend with a lower rate of debugging and growth at the early stage and a higher rate later on as good fixes are found and implemented. In some situations, it is unreasonable to set a high growth rate of improvement during the early stages of development for complex systems because of the difficulty of identifying the sources and causes of the failures, hence requiring the use of S–shaped reliability growth curves. In this section, models having S-shaped reliability growth curves are given and their use is illustrated by examples.

TABLE 16.8 – Comparison of the predicted reliability with the actual data for Example 16–4.

Launch number, L	$L_G = L - 4$, L_G	Reliability calculated from the Gompertz model, %	Reliability calculated from the available data*, %
1			
2			
3			
4	0	35.725	35.40
5	1	48.598	51.28
6	2	60.160	59.37
7	3	69.759	67.59
8	4	77.302	76.21
9	5	83.009	83.33
10	6	87.212	87.67
11	7	90.252	90.61
12	8	92.422	92.55
13	9	93.958	93.94
14	10	95.038	94.96
15	11	95.795	95.75
16	12	96.323	96.35
17	13	96.691	90.96
18	14	96.948	91.68
19	15	97.126	87.56
20	16	97.249	94.41
21	17	97.335	95.50
22	18	97.395	96.26

*Using failure discounting.

16.4.1   S-SHAPED RELIABILITY GROWTH CURVES

16.4.1.1   THE GOMPERTZ CURVE

The Gompertz curve given by Eq. (16.1) may be S-shaped if it has a point of inflection, T_{ai}, which is positive. This can be determined by setting the second derivative of Eq. (16.1) to zero; i.e.,

$$\frac{d^2R}{dT_a^2} = a \cdot \log b \cdot \log c \, [b^{c^{T_a}} \cdot c^{T_a} \cdot \log c + c^{T_a} \cdot b^{c^{T_a}} \cdot \log b \cdot c^{T_a} \cdot \log c] = 0.$$

$$(16.15)$$

Solving for T_a yields, T_a^*, or

$$T_a^* = \frac{\log[-(\log b)^{-1}]}{\log c},$$
$$(16.16)$$

where

T_a^* = growth time at the point of inflection.

Also, the growth rate

$$\frac{dR}{dT_a} = a \cdot \log b \cdot \log c \cdot b^{c^{T_a}} \cdot c^{T_a},$$
$$(16.17)$$

is positive for all values of T_a because $0 < b < 1$ and $0 < c < 1$. When $T_a < T_a^*$, then $d^2R / dT_a^2 < 0$; hence reliability increases at an increasing rate. And when $T_a > T_a^*$, then $d^2R / dT_a^2 > 0$; hence reliability increases at a decreasing rate. This means that the Gompertz curve may be S-shaped.

Since the domain of the Gompertz model is $T_a \geq 0$, it may be seen that if $T_a^* > 0$ the Gompertz curve is S-shaped, and if $T^* \leq 0$ it is not S-shaped. Setting $T_a^* = 0$ in Eq. (16.16) and solving for b yields

$$b = e^{-1} = 0.36788.$$
$$(16.18)$$

This means that if $b < 0.36788$ then $T_a^* > 0$ and there would be an S-shaped curve. If $b \geq 0.36788$, then $T_a^* \leq 0$ and the curve would not be S-shaped. Therefore, the following may be concluded:

1. When $0 < a \leq 1$, $0 < b < 0.36788a$, and $0 < c < 1$, the Gompertz curve is S-shaped.

2. When $0 < a \leq 1$, $b \geq 0.36788a$, and $0 < c < 1$, the Gompertz curve is not S-shaped, as may be seen in Fig. 16.6, where $b = 0.608 > 0.36788(0.920) = 0.338$.

3. At the point of inflection, the value of R is fixed and is equal to $0.36788 \cdot a$ for such S–shaped curves, as may be obtained from Eqs. (16.1) and (16.16).

The plots of some S-shaped Gompertz curves with $a = 1$, $b = 0.1$, and various values of c and their points of inflection are given in Fig. 16.8.

16.4.1.2   THE LOGISTIC RELIABILITY GROWTH CURVE

The logistic reliability growth curve [8 ; 9] is

$$R = \frac{1}{1 + be^{-kT_a}}, \quad b > 0, \quad k > 0, \quad T_a \geq 0, \qquad (16.19)$$

where b and k are parameters, and has an S-shaped curve. Similar to the analysis given for the Gompertz curve, the following may be concluded:

1. The point of inflection is given by

$$T_{ai} = \frac{\log b}{k}. \qquad (16.20)$$

2. When $b > 1$, then $T_{ai} > 0$ and an S-shaped curve results; when $0 < b \leq 1$, then $T_{ai} \leq 0$ and no S-shaped curve results. Therefore, the logistic reliability growth curve has an S-shape when $b > 1$ and $k > 0$.

3. At the points of inflection the value of R is fixed and is equal to 0.5 for such S-shaped curves.

The plots of some S-shaped logistic reliability growth curves with $b = 8$ and various values of k are given in Fig. 16.9.

EXAMPLE 16–5

Using the reliability growth data given in Table 16.9, do the following:

1. Plot the data points.

2. Find a Gompertz curve that represents the data well and plot the data comparatively with the raw data.

3. Find a logistic reliability growth curve which represents the data, and plot the data comparatively with the raw data.

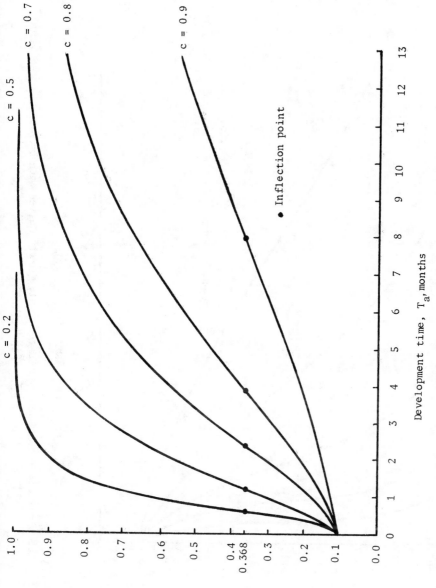

Fig. 16.8 – Plots of S-shaped Gompertz reliability growth curves with a=1, b=0.1 and various values of c, and their points of inflection.

427

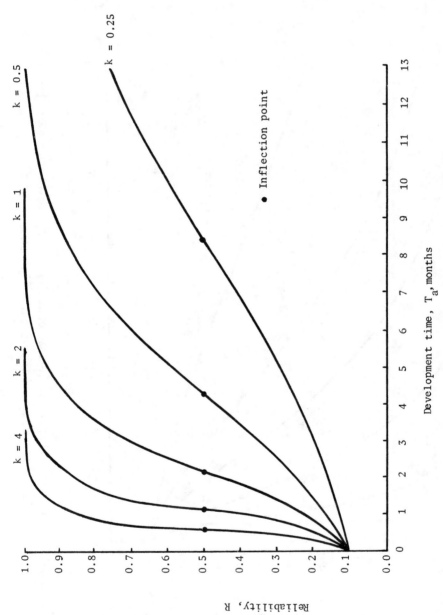

Fig. 16.9 – Plots of S-shaped logistic reliability growth curves with b=8 and various values of k, and their points of inflection.

TABLE 16.9 – The development time versus observed reliability data and predicted reliabilities for Example 16–5.

1	2	3	4	5
Time, months, T_a	Observed reliability, R_0, %	Predicted reliability by the Gompertz curve, R, %	Predicted reliability by the logistic curve, R, %	Predicted reliability by the modified Gompertz curve, R, %
0	31.0	23.1	2.60	31.0
1	35.5	40.6	9.25	34.8
2	49.3	57.8	28.01	49.8
3	70.1	72.0	59.77	69.4
4	83.0	82.7	85.01	83.9
5	92.2	90.1	95.59	92.2
6	96.4	95.1	98.80	96.4
7	98.6	98.4	99.68	98.3
8	99.0	100.5	99.92	99.2

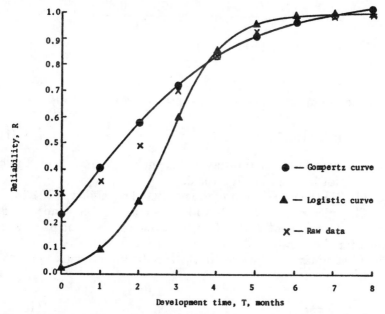

Fig. 16.10 – Plots of the raw data, the predicted reliabilities from the Gompertz curve, and the predicted reliabilities from the logistic curve for Example 16–5.

SOLUTIONS TO EXAMPLE 16–5

1. The plot of the raw data points is given in Fig. 16.10, where the trend of these points is S-shaped.

2. The Gompertz curve's parameters are determined as follows:

$$S_1 = \sum_{i=0}^{2} \log_{10} R_{oi} = 4.7344,$$

$$S_2 = \sum_{i=3}^{5} \log_{10} R_{oi} = 5.7295,$$

and

$$S_3 = \sum_{i=6}^{8} \log_{10} R_{oi} = 5.9736.$$

From Eqs. (16.2) through (16.4)

$$c = \left[\frac{5.7295 - 5.9736}{4.7344 - 5.7295}\right]^{\frac{1}{3}} = 0.6260,$$

$$a = \text{antilog}_{10}\left[\frac{1}{3}\left(4.7344 - \frac{4.7344 - 5.7295}{1 - 0.6260^3}\right)\right],$$

$$a = \text{antilog}_{10}(2.01765),$$

or

$$a = 104.149,$$

and

$$b = \text{antilog}_{10}\frac{(4.7344 - 5.7295)(1 - 0.6260)}{(1 - 0.6260^3)^2},$$

$$b = \text{antilog}_{10}(-0.65344),$$

or

$$b = 0.2221.$$

Therefore, the Gompertz curve which represents these data is

$$R = (104.149)(0.2221)^{0.6260^{T_a}}. \tag{16.21}$$

The values of predicted reliabilities are given in Column 3 of Table 16.9, and the plot is given in Fig. 16.10.

In Eq. (16.21) the upper limit of the predicted reliability is 104.149% > 100%. This means that $R = 104.149\%$ when $T \to \infty$, and this is unreasonable. Consequently, this Gompertz curve does not provide a good model for these data.

3. The least squares estimators of the logistic growth curve parameters are [9]

$$b = e^{\hat{b}_0}, \tag{16.22}$$

and

$$k = -\hat{b}_1, \tag{16.23}$$

where

$$\hat{b}_1 = \frac{\sum_{i=0}^{N-1} T_i Y_i - N\bar{T}\bar{Y}}{\sum_{i=0}^{N-1} T_i^2 - N\bar{T}^2}, \tag{16.24}$$

N = total number of available T_a values,

$$\hat{b}_0 = \bar{Y} - \hat{b}_1 \bar{T}, \tag{16.25}$$

$$\bar{T} = \frac{1}{N} \sum_{i=0}^{N-1} T_{ai}, \quad Y_i = \log\left(\frac{1}{R_i} - 1\right),$$

and

$$\bar{Y} = \frac{1}{N} \sum_{i=0}^{N-1} Y_i.$$

In this example $N=9$,

$$\bar{Y} = \frac{1}{9} \sum_{i=0}^{8} \log\left(\frac{1}{R_i} - 1\right) = -1.7355,$$

$$\bar{T} = \frac{1}{9} \sum_{i=0}^{8} T_{ai} = 4,$$

$$\sum T_{ai}^2 = 204 \text{ and } \sum_{i=0}^{8} T_{ai} Y_i = -142.86.$$

From Eqs. (16.24) and (16.25)

$$\hat{b}_1 = \frac{-142.86 - 9(4)(-1.7355)}{204 - 9(4)^2} = -1.3397,$$

$$\hat{b}_0 = -1.7355 - (-1.3397)(4) = 3.6233,$$

and from Eqs. (16.22) and (16.23)

$$b = e^{3.6233} = 37.461, \text{ and } k = -(-1.3397) = 1.3397.$$

Therefore, the logistic reliability growth curve which represents these data is

$$R = \frac{1}{1 + 37.461 \cdot e^{-1.3397 \cdot T_a}}. \tag{16.26}$$

The values of the predicted reliabilities are given in Column 4 of Table 16.9, and the plot is given in Fig. 16.10. From the plot it may be seen that these data cannot be represented well by the logistic reliability growth curve.

16.4.2 MODIFIED GOMPERTZ RELIABILITY GROWTH CURVE

From Cases 2 and 3 of Example 16–5, it may be seen that, in general, the reliability growth data with an S-shaped trend cannot be described well by the Gompertz or logistic curves, since these two models have fixed values of reliability at the inflection points. Thus only a few reliability growth data sets following an S-shaped reliability growth curve can be fitted to them. A modification of the Gompertz curve is given next which overcomes this shortcoming.

Consider a shift in the vertical coordinate, then the Gompertz model, or Eq. (16.1), becomes

$$R = d + ab^{c^{T_a}}, \tag{16.27}$$

where

$0 < a + d \le 1$, if R is in decimals,

$0\% < a + d \le 100\%$, if R is in percent,

$0 < b < 1 \ 0 < c < 1$, and $T_a \ge 0$,

R = system's or equipment's reliability at development time T_a,

d = shift parameter,

$d + a$ = upper limit that the reliability approaches asymptotically
     as $T_a \to \infty$,

$d + ab$ = original level of reliability at $T_a = 0$,

and

c = growth pattern parameter.

This modified Gompertz model is more flexible than the original, especially for fitting growth data with S-shaped trends.

The minimization technique may be used to estimate the parameters of the modified Gompertz reliability growth model. The method is illustrated in Example 16–6.

EXAMPLE 16–6

A reliability growth data set is given in Table 16.9. Find the modified Gompertz curve which represents these data, and plot it comparatively with the raw data.

SOLUTION TO EXAMPLE 16–6

Take the cumulative error between the predicted reliabilities and the observed reliabilities as the goal function; i.e.,

$$F = \sum |R_{oi} - R_{pi}|, \quad i = 0, 1, 2, ..., N - 1, \tag{16.28}$$

where

R_{oi} = observed reliability.

R_{pi} = predicted reliability given by the modified Gompertz curve,

or

$$R_{pi} = d + ab^{c^{T_i}}.$$

Since the R_{oi} and T_i are given, F is only a function of $a, b, c,$ and d; i.e.,

$$F = F(a, b, c, d). \tag{16.29}$$

Then the problem becomes one of finding the minimum point (a^*, b^*, c^*, d^*) such that the cumulative error is minimum; i.e.,

$$F(a^*, b^*, c^*, d^*) \to \text{minimum}, \tag{16.30}$$

where

$$0 < b^* < 1, \quad 0 < c^* < 1, \quad \text{and} \quad 0 < a^* + d^* \leq 1.$$

The computer program for determining $a^*, b^*, c^*,$ and d^* is given in Appendix 16A. Inputting the data given in Table 16.9, the outputs are

$$a^* = 69.107,$$
$$b^* = 0.00176,$$
$$c^* = 0.45203,$$

and

$$d^* = 30.891.$$

Therefore, the modified Gompertz model is

$$R = 30.891 + (69.107)(0.00176)^{0.45203^{T_a}}. \tag{16.31}$$

The predicted reliability, using Eq. (16.31), is given in Column 5 of Table 16.9 and the plot in Fig. 16.11. From the plot it may be seen that the modified Gompertz curve of Eq. (16.31) represents the data very well.

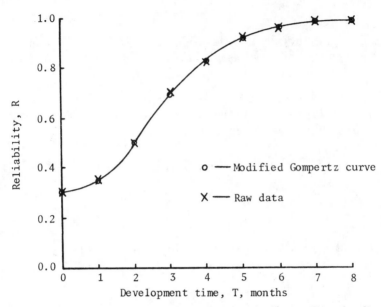

Fig. 16.11 – Plots of the raw data and the predicted reliabilities from the modified Gompertz curve for
Example 16–6.

16.5   MTBF GROWTH AND FAILURE RATE IMPROVEMENT MODELS

If the MTBF growth, or the failure rate improvement, trend with development or TAAF time of a device, equipment, or system is like that
shown in Fig. 16.5, then a good math model to use is that postulated
by Duane [4], thus called the Duane growth curve. The development
or TAAF time considered in reliability growth in terms of the MTBF
or failure rate or reliability improvement for continuously operating
units is the unit hours accumulated by one or more units involved in
the growth process. For one-shot items the TAAF time considered
in reliability growth is the accumulated number of missions, events,
launches, or discrete periods of operation, e.g., month of TAAF or cycles of operations to a failure. The Duane failure rate improvement
model, also see Appendix 16C, is

$$\hat{\bar{\lambda}} = \frac{r}{T_a} = aT_a^{-\alpha},\qquad(16.32)$$

where

$\hat{\bar{\lambda}}$ = average estimate of the cumulative failure rate, fr/hr,

r = total failures occurring in time T_a,

T_a = total accumulated unit hours of test and/or development time,

a = cumulative failure rate at $T_a = 1$, or at the beginning of the test, or the earliest time at which the first $\hat{\lambda}$ is predicted, or the $\hat{\lambda}$ for the equipment at the start of the design and development process $(a > 0)$,

and

α = improvement rate in the $\hat{\lambda}$, $0 \leq \alpha \leq 1$.

The corresponding MTBF, $\hat{m}$, improvement or growth model is

$$\hat{m} = \frac{T_a}{r} = bT_a^{\alpha}, \tag{16.33}$$

where
   $b = 1/a$ = cumulative MTBF at $T_a = 1$ or at the beginning of the test, or the earliest time at which the first $\hat{m}$ can be determined, or the $\hat{m}$ predicted at the start of the design and development process $(b > 0)$.

It may be seen that Eq. (16.32) may be linearized by taking the logs of both sides, or

$$\log \hat{\lambda} = \log a - \alpha \ \log T_a. \tag{16.34}$$

Consequently, plotting $\hat{\lambda}$ versus T_a on log-log paper will result in a straight line with a negative slope, such that $\log a$ is the y intercept when $T_a = 1$, or a is the cumulative failure rate at $T_a = 1$, and $-\alpha$ is the slope of the straight line on the log-log plot.
   Similarly, Eq. (16.33) may be linearized by taking the logs of both sides, or

$$\log \hat{m} = \log b + \alpha \ \log T_a. \tag{16.35}$$

Consequently, plotting $\hat{m}$ versus T_a on log-log paper will result in a straight line with a positive slope such that $\log b$ is the y intercept when $T_a = 1$, or b is the cumulative mean-time-between-failures at $T_a = 1$ or essentially at the beginning of the design and development period, and α is the slope of the straight line on the log-log plot.
   Two ways of determining these curves may be the following:

1. Predict the $\hat{\lambda}_0$ and $\hat{m} = 1/\hat{\lambda}_0$ of the device, equipment, or system from its reliability block diagram and available component failure

rates. Plot this value on log-log plotting paper at $T_a = 1$. From past experience and from past data available for similar equipment, find values of α_1, the slope of the improvement lines for $\hat{\bar{\lambda}}$ or m. Modify this α as necessary. If a better design effort is expected and a more intensive research, test, and development, or a TAAF, program is to be implemented, then, say, a 15% improvement in the growth rate may be attainable; consequently, the available value for α, α_1, should be adjusted by this amount. The value to be used will then be $\alpha = 1.15\alpha_1$. A line is then drawn through point $\hat{\bar{\lambda}}_0$ and $T_a = 1$ with the just determined slope α, keeping in mind that α is negative for the $\hat{\bar{\lambda}}$ curve. This line should be extended to the design, development, and test time scheduled to be expended to achieve the failure rate goal to see if this goal will indeed be achieved on schedule. It is also possible to find the design, development, and test time to achieve the goal which may be earlier than the delivery date or later. If earlier, then either the reliability program effort can be judiciously and appropriately trimmed; or if it is an incentive contract, full advantage is taken of the fact that the failure rate goal can be exceeded, with the associated increased profits to the company.

A similar approach may be used for the MTBF growth model, where $\hat{m}_0 = 1/\lambda_0$ is plotted at $T_a = 1$, and a line is drawn through the point $\hat{m}_0$ and $T_a = 1$ with slope α to obtain the MTBF growth line.

If α values are not available, consult Table 16.10, which gives actual α values for various types of equipment. These have to be obtained from the literature or by MTBF growth tests. It may be seen from Table 16.10 that α values range between 0.24 and 0.65. The lower values reflect slow early growth and the higher values fast early growth.

2. During the design, development and test phase and at specific milestones, the $\hat{\bar{\lambda}} = 1/\hat{m}$ are calculated from generated r and T_a values. These values of $\hat{\bar{\lambda}}$ or $\hat{m}$ are plotted above the corresponding T_a values on log-log plotting paper. A straight line is drawn favoring these points so as to minimize the distance between these points and the line, thus establishing the improvement or growth model and its parameters a or b and α graphically. If needed, linear regression analysis techniques can be used to determine these parameters.

16.5.1   CURRENT OR INSTANTANEOUS $\hat{\bar{\lambda}}$ AND $\hat{m}$

The cumulative MTBF, $\hat{m}_c$, and $\hat{\bar{\lambda}}_c$ do tell us whether m is increasing or λ is decreasing with time, utilizing all data up to that time. We

TABLE 16.10 – Values for the slope, α, of the $\hat{\bar{\lambda}}$ and $\hat{m}$ improvement curves for various equipment.

Equipment		Slope, α
Computer system:	Actual	0.24
	When easy to find failures were eliminated	0.26
	When all known failure causes were eliminated	0.36
Mainframe computer		0.50
Aerospace electronics:	All malfunctions	0.57
	"Relevant" failures only	0.65
Attack radar		0.60
Rocket engine		0.46
Afterburning turbojet		0.35
Complex hydromechanical system		0.60
Aircraft generator		0.38
Modern dry turbojet		0.48

may want to know, however, the current or instantaneous $\hat{m}_i$ or $\hat{\lambda}_i$ to see what we are doing at a specific instant or after a specific test and development time. This is obtained by differentiating Eq. (16.32) with respect to T_a, or

$$\lambda_i(T_a) = \lim_{\Delta T_a \to 0} \left(\frac{\Delta r}{\Delta T_a}\right) = \frac{\partial r}{\partial T_a},$$

where from

$$\lambda = \frac{r}{T_a} = aT_a^{-\alpha}, \quad r = aT_a^{1-\alpha}.$$

Then

$$\lambda_i(T_a) = \frac{\partial(aT_a^{1-\alpha})}{\partial T_a}.$$

Therefore,

$$\lambda_i(T_a) = (1 - \alpha)aT_a^{-\alpha}, \tag{16.36}$$

or

$$\lambda_i(T_a) = (1 - \alpha)\hat{\lambda}_c. \tag{16.37}$$

Similarly, for Eq. (16.33) this procedure yields

$$m_i(T_a) = \frac{1}{1 - \alpha}bT_a^{\alpha},$$

or

$$m_i(T_a) = \frac{1}{1 - \alpha}\hat{m}_c, \quad \alpha \neq 1, \tag{16.38}$$

where $\alpha = 1$ implies infinite MTBF growth.

It may be seen from Eq. (16.37) that the current or instantaneous failure rate improvement line is obtained by shifting the cumulative failure rate line down, parallel to itself, by a distance of $(1 - \alpha)$. Similarly, it may be seen from Eq. (16.38) that the current, or instantaneous, MTBF growth line is obtained by shifting the cumulative MTBF line up, parallel to itself, by a distance of $1/(1 - \alpha)$, as illustrated in Fig. 16.12. The instantaneous MTBF is the MTBF the product will exibit in the field if the design is frozen at the TAAF time at which the product is fielded, assuming the instantaneous MTBF shall not be altered during manufacturing and thereafter.

EXAMPLE 16-7

It is required to design, develop, test, and deliver a new avionics equipment within 36 months with an MTBF of at least 150 hr. The reliability program is to be in accordance with MIL-STD-785, with testing per MIL-STD-781 Test Level F. First article configuration inspection (FACI) and configuration control are required on the first production item.

SOLUTION TO EXAMPLE 16-7

Using MIL-HDBK-217, it is predicted that an MTBF of 220 hr is attainable if screened, high-reliability, MIL-STD parts and components

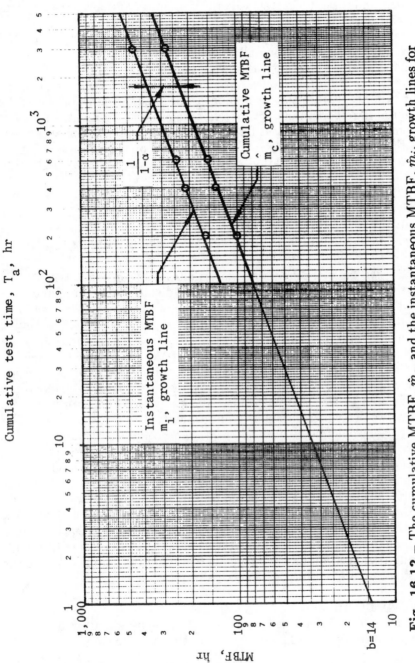

Fig. 16.12 – The cumulative MTBF, $\hat{m}_c$, and the instantaneous MTBF, $\hat{m}_i$, growth lines for Example 16–8 and the data given in Table 16.11.

439

are used under an integrated reliability program and exacting application and derating criteria. This prediction meets the good reliability program management criterion by exceeding the MTBF goal of 150 hr by more than 46%, thus allowing for potential growth in product complexity to meet its performance goals.

Next, schedule milestones are established, based on past development experience, resulting in 15 months for design, 6 additional months for initial hardware manufacture and ambient test, 12 months for evaluation testing, and 3 months for the documentation of the final changes and their incorporation into the hardware prior to production FACI, and to meet all configuration control requirements, for a total of 36 months.

It is concluded that, for the new design that this is, the initial hardware MTBF will be 10% of that predicted or $\hat{m}_0 = 22$ hr, after about 100 hr of test time, with a growth rate of 0.50 based on the diligent implementation of an integrated reliability program plan per MIL-STD-785 and of testing per MIL-STD-781.

Now we have the necessary information to construct the MTBF growth curve given in Fig. 16.13, which indicates that compliance can be achieved at $T_a = 4,800$ hr of test time. Prior test experience indicates that for such a product 200 hr of testing time can be achieved per system per month. Now several plans can be devised to accumulate the 4,800 hr of test time required to meet the MTBF goal of 150 hr. The first plan would be to test one system continuously for 24 months. This would require the minimum assets, but the longest calendar time, as indicated at the bottom of Fig. 16.13. The second plan would be to test two systems continuously and concurrently for 12 months. The third plan would be to test three systems continuously and concurrently for 8 months, accommodating additional time for reaction to contingency, including growth of up to 25% in product complexity as the predicted MTBF of 220 hr was based on an 11,000 parts count for this product, which can grow to 14,000 parts and still meet the MTBF goal of 150 hr. The third plan is the one with the least risk and the least calendar time, but it is the costliest in assets. These trade-offs, past experience, and the capability of the reliability engineering organization and management provide top management clear decision-making options for a successful program.

EXAMPLE 16–8

A complex system's reliability growth is being monitored and the data given in Table 16.11 are obtained. Do the following:

1. Plot the cumulative MTBF growth curve.

2. Write the equation of this growth curve.

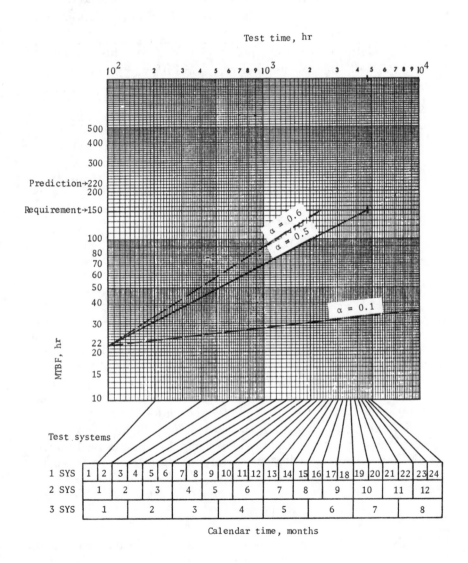

Fig. 16.13 – MTBF growth planning for Example 16–7.

TABLE 16.11 – Cumulative test hours and the corresponding observed failures for the complex system of Example 16-8. The cumulative and the instantaneous MTBFs are also given.

Point number	Cumulative test time, T_a, hr	Cumulative number of observed failures, r	Cumulative MTBF= $\hat{m}_c = \frac{T_a}{r}$, hr	Instantaneous MTBF= $\hat{m}_i = \frac{\hat{m}_c}{(1-\alpha)}$, hr
1	200	2	100	158.9
2	400	3	133	211.3
3	600	4	150	238.3
4	3,000	11	273	433.7

3. Write the equation of the "instantaneous" (current) MTBF growth model.

4. Plot the "instantaneous" (current) MTBF growth curve.

SOLUTIONS TO EXAMPLE 16–8

1. Given the data in the second and third columns of Table 16.11, the cumulative MTBF, $\hat{m}_c$, values are calculated. These are given in the fourth column. The information in the second and fourth columns is then plotted as shown in Fig. 16.12. It may be seen that a straight line represents the MTBF growth very well on log-log scales.

2. Extending the line to $T_a = 1$ yields $b = 14$ hr. Another way of determining b is to calculate α by using two points on the fitted straight line and substituting the corresponding $\hat{m}_c$ and T_a values into

$$\alpha = \frac{\log \hat{m}_{c_2} - \log \hat{m}_{c_1}}{\log T_{a_2} - \log T_{a_1}}. \tag{16.39}$$

Then substitute this α value and the chosen set of values for $\hat{m}_{c_1}$ and T_{a_1} into Eq. (16.33) and solve for b.

The slope, α, of the line may also be found from Eq. (16.35), knowing that at point (1; 14) log 1 = 0, or from

$$\alpha = \frac{\log \hat{m}_c - \log b}{\log T_a - \log 1}. \tag{16.40}$$

Read off the straight line in Fig. 16.12 the values $T_{a_1} = 200$ hr, $\hat{m}_{c_1} = 100$ hr, and $T_{a_2} = 3,500$ hr, $\hat{m}_{c_2} = 300$ hr. Substituting the first set of values and $b = 14$ hr into Eq. (16.40) yields

$$\alpha_1 = \frac{\log(100) - \log(14)}{\log(200) - \log 1},$$

or

$$\alpha_1 = 0.371.$$

Substituting the second set of values and $b = 14$ hr into Eq. (16.40) yields

$$\alpha_2 = \frac{\log(300) - \log(14)}{\log(3,500) - \log 1},$$

or

$$\alpha_2 = 0.376.$$

Averaging these two values yields a better estimate of $\alpha = 0.3735$. Now we can write the equation for the cumulative MTBF growth curve as

$$\hat{m}_c = 14T_a^{0.3735}. \tag{16.41}$$

3. The equation for the instantaneous MTBF growth curve, using Eq. (16.38), is

$$\hat{m}_i = \frac{1}{1 - 0.3735}(14)T_a^{0.3735}. \tag{16.42}$$

4. Equation (16.42) is shown plotted in Fig. 16.12 where it may be seen that a parallel shift upward of the cumulative MTBF, $\hat{m}_c$, line by a distance of $1/(1 - \alpha)$ gives the instantaneous MTBF, or the $\hat{m}_i$, line.

16.6   THE AMSAA RELIABILITY GROWTH MODEL

16.6.1   INTRODUCTION

The AMSAA (U.S. Army Material Systems Analysis Activity) reliability growth model assumes that, within a test phase, failures are occurring according to a nonhomogeneous Poisson process. The development or TAAF time considered in reliability growth in terms of the MTBF, or failure rate, or reliability improvement, for continuously operating units is the unit hours accumulated by one or more units involved in the growth process. For one-shot items the TAAF time considered in reliability growth is the accumulated number of missions, launches, or discrete periods of operation, e.g., month of TAAF or cycles of operations to a failure. It is further assumed that the instantaneous failure rate can be represented by the Weibull failure rate function

$$\lambda_i(T_a) = \frac{\beta}{\eta}(\frac{T_a}{\eta})^{\beta-1}, \tag{16.43}$$

where $\beta > 0$ and $\eta > 0$ are parameters, and T_a is the cumulative test time [10; 11; 12].

Here a single repairable equipment is considered which operates from age zero, fails at age T_{a1}, is restored to satisfactory function by a corrective action which involves design improvement, operates again, fails later at age T_{a2} measured from age zero, gets restored again to satisfactory function by a design improvement, operates again, and the process gets repeated. If more than one unit is involved in the reliability growth process, then T_a^* is the unit hours of operation accumulated by all units involved in the process by the time the ith failure occurs.

Under this model the instantaneous MTBF function is

$$m_i(T_a) = \frac{1}{\lambda_i(T_a)} = \frac{\eta}{\beta}(\frac{T_a}{\eta})^{1-\beta}. \tag{16.44}$$

The total number of failures, $N(T^*)$, accumulated on all test items in a specific cumulative test time T^* is a random variable which is Poisson distributed. The probability that exactly n failures occur between the initiation of testing and the total test time T^* is

$$P\{N(T^*) = n\} = \frac{[\bar{N}(T^*)]^n e^{-\bar{N}(T^*)}}{n!}, \tag{16.45}$$

where the mean number of failures in test period $0 \to T^*$, $\bar{N}(T^*)$, is given by

$$\bar{N}(T^*) = \int_0^{T^*} \lambda_i(T_a)\, dT_a,$$

or

$$\bar{N}(T^*) = (\frac{T^*}{\eta})^{\beta}. \tag{16.46}$$

From Eq. (16.46) the expected number of failures occurring in the interval from the cumulative test time T_1 until the cumulative test time T_2 is

$$\bar{N}(T_2) - \bar{N}(T_1) = \frac{1}{\eta^{\beta}}(T_2^{\beta} - T_1^{\beta}). \tag{16.47}$$

During the developmental program, the failure rate of a system decreases as test time T_a increases, and at time T_0 the failure rate is

$$\lambda_i(T_0) = \frac{\beta}{\eta}(\frac{T_0}{\eta})^{\beta-1}. \tag{16.48}$$

In practice, it is generally assumed that, if no improvements are incorporated into the system after time T_0, then failures would continue at the constant rate $\lambda_i(T_0)$ with further testing. That is, if no additional modifications are made on the system after time T_0, then future failures would follow an exponential distribution with an MTBF of

$$m_i(T_0) = \frac{\eta}{\beta}(\frac{T_0}{\eta})^{1-\beta}. \tag{16.49}$$

16.6.2 GRAPHICAL ESTIMATION OF PARAMETERS

A graph of the observed cumulative number of failures plotted against cumulative test time on log-log paper furnishes crude estimates of the parameters which describe the failure rate function given by Eq. (16.43). Taking logarithms of the expression for the mean number of failures, Eq. (16.46), yields

$$\log[\bar{N}(T^*)] = \beta \cdot \log(T^*) - \beta \cdot \log(\eta). \tag{16.50}$$

Therefore, the expression for the mean number of failures is represented by a straight line on log-log paper. A line drawn to fit the data points representing the cumulative number of failures at the time each failure occurs is a suitable approximation of the true line. Taking an arbitrary set of two points on the fitted straight line, $[T_{a1}, \bar{N}(T_{a1})]$ and $[T_{a2}, \bar{N}(T_{a2})]$, and substituting into Eq. (16.50) yields

$$\log[\bar{N}(T_{a1})] = \beta \cdot \log(T_{a1}) - \beta \cdot \log(\eta),$$

and

$$\log[\bar{N}(T_{a2})] = \beta \cdot \log(T_{a2}) - \beta \cdot \log(\eta).$$

Solving for β yields

$$\beta = \frac{\log[\bar{N}(T_{a2})] - \log[\bar{N}(T_{a1})]}{\log(T_{a2}) - \log(T_{a1})}, \tag{16.51}$$

and

$$\eta = \frac{T_{a1}}{[\bar{N}(T_{a1})]^{\frac{1}{\beta}}}, \tag{16.52}$$

or

$$\eta = \frac{T_{2a}}{[\bar{N}(T_{a2})]^{\frac{1}{\beta}}}.$$

EXAMPLE 16-9

A prototype of a system is tested with the incorporation of design changes. A total of 23 failures occurs during a cumulative test time of 22,000 hr. The data are given in Table 16.12. Determine the AMSAA reliability growth model that represents the data using a graphical estimate.

SOLUTION TO EXAMPLE 16-9

The plot of the observed cumulative number of failures versus the cumulative test time is given in Fig. 16.14. It may be seen from the plot that the observed data tend to a straight line. A straight line is drawn to fit these data points, and two points on the straight line are chosen, say (5.4, 1) and (2,160, 10).

From Eqs. (16.51) and (16.52)

$$\hat{\beta} = \frac{\log_{10}(10) - \log_{10}(1)}{\log_{10}(2,160) - \log_{10}(5.4)},$$

or

$$\hat{\beta} = 0.3843,$$

and

$$\hat{\eta} = \frac{5.4}{1^{\frac{1}{0.3843}}} = 5.4,$$

or

$$\hat{\eta} = \frac{2,160}{10^{\frac{1}{0.3843}}} = 5.4.$$

The failure rate improvement function representing these data can be written as

$$\lambda_i(T_a) = \frac{0.3843}{5.4} \left(\frac{T_a}{5.4}\right)^{0.3843-1},$$

or

$$\lambda_i(T_a) = 0.2010 \cdot T_a^{-0.6157}. \tag{16.53}$$

16.6.3   STATISTICAL ESTIMATION OF PARAMETERS

Modeling reliability growth as a nonhomogeneous Poisson process permits the assessment of the demonstrated reliability performance by statistical procedures. The method of maximum likelihood provides estimates of the scale parameter, η, and the shape parameter, β, and then of the failure rate and MTBF functions [9 ; 10].

TABLE 16.12 – The development test data of a prototype of a system for Example 16–9.

Number of failures, N	Cumulative test time, T_{ai}, hr	$\log_e T_{ai}$
1	9.2	2.2192
2	25.0	3.2189
3	61.5	4.1190
4	260.0	5.5609
5	300.0	5.7038
6	710.0	6.5653
7	916.0	6.8189
8	1,010.0	6.9177
9	1,220.0	7.1066
10	2,530.0	7.8360
11	3,350.0	8.1167
12	4,200.0	8.3428
13	4,410.0	8.3916
14	4,990.0	8.5152
15	5,570.0	8.6252
16	8,310.0	9.0252
17	8,530.0	9.0513
18	9,200.0	9.1270
19	10,500.0	9.2591
20	12,100.0	9.4010
21	13,400.0	9.5030
22	14,600.0	9.5888
23	22,000.0	9.9988

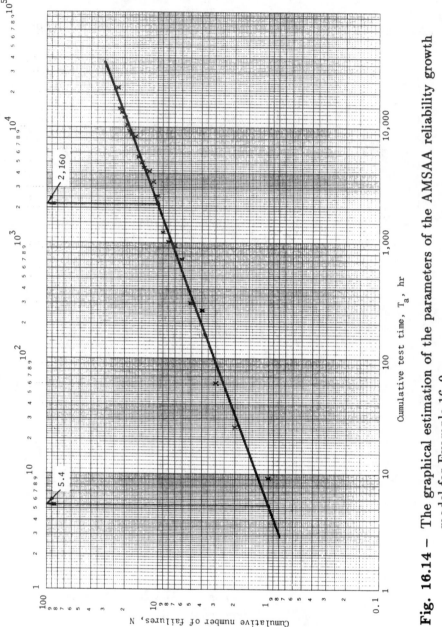

Fig. 16.14 – The graphical estimation of the parameters of the AMSAA reliability growth model for Example 16–9.

448

16.6.3.1 TIME TERMINATED TEST

For the time terminated test the maximum likelihood estimator of the shape parameter, β, is

$$\hat{\beta} = \frac{N}{N \cdot \log_e(T_a^*) - \sum\limits_{i=1}^{N} \log_e(T_{ai})}, \qquad (16.54)$$

and of the scale parameter, η, is

$$\hat{\eta} = \frac{T^*}{N^{1/\hat{\beta}}}, \qquad (16.55)$$

where

$$
\begin{aligned}
T_a^* &= \text{total accumulated specific test time,} \\
N &= \text{total number of failures from time 0 to } T^*, \\
T_{ai} &= \text{time-to-failure data}, i = 1, 2, ..., N.
\end{aligned}
$$

16.6.3.2 FAILURE TERMINATED TEST

For the failure terminated test the maximum likelihood estimator of the shape parameter, β, is

$$\hat{\beta} = \frac{N}{(N-1) \cdot \log_e(T_{aN}) - \sum\limits_{i=1}^{N-1} \log_e(T_{ai})}, \qquad (16.56)$$

and of the scale parameter, η, is

$$\hat{\eta} = \frac{T_{aN}}{N^{1/\hat{\beta}}}, \qquad (16.57)$$

where

T_{aN} = test time accumulated by the Nth time to failure,

and

N and T_{ai} have the same definitions as in Eqs. (16.54) and (16.55).

It has to be observed that the maximum likelihood estimators of the shape parameter given by Eqs. (16.54) and (16.56) are not unbiased [10; 13]. The unbiased estimate of the shape parameter, β', is

$$\beta' = \frac{N-1}{N}\hat{\beta}. \qquad (16.58)$$

Hence the scale parameter for the time terminated test is

$$\eta' = \frac{T^*}{N^{1/\beta'}},$$
(16.59)

and for the failure terminated test is

$$\eta' = \frac{T_{aN}}{N^{1/\beta'}}.$$
(16.60)

Then the estimate of the failure rate function is

$$\hat{\lambda}_i(T_a) = \frac{\beta'}{\eta'}\left(\frac{T_a}{\eta'}\right)^{\beta'-1},$$
(16.61)

and the estimate of the MTBF function is

$$\hat{m}_i(T_a) = \frac{\eta'}{\beta'}\left(\frac{T_a}{\eta'}\right)^{1-\beta'}.$$
(16.62)

EXAMPLE 16-10

Using the data given in Table 16.9 of Example 16-9, find the AMSAA reliability growth model that represents the data using the statistical estimates and assuming the following:

1. The test is time terminated.

2. The test is failure terminated.

SOLUTIONS TO EXAMPLE 16-10

1. The values of the $\log_e(T_{ai})$ are listed in Table 16.12. For the time terminated test, from Eq. (16.54) the maximum likelihood estimate of β is

$$\hat{\beta} = \frac{23}{(23)\log_e(22,000) - \sum_{i=1}^{23}\log_e(T_{ai})},$$

where

$$\sum_{i=1}^{n}\log_e(T_{ai}) = \sum_{i=1}^{23}\log_e(T_{ai}) = 173.012.$$

Then

$$\hat{\beta} = \frac{23}{(23)\log_e(22,000) - 173.012},$$

or

$$\hat{\beta} = 0.40456.$$

From Eq. (16.58) the unbiased estimate of β is

$$\beta' = \frac{23 - 1}{23}(0.40456),$$

or

$$\beta' = 0.3870,$$

and from Eq. (16.59)

$$\eta' = \frac{22,000}{23^{\frac{1}{0.3870}}},$$

or

$$\eta' = 6.6642 \text{ hr.}$$

Then the failure rate function is

$$\hat{\lambda}_i(T_a) = \frac{0.3870}{6.6642}\left(\frac{T_a}{6.6642}\right)^{0.3870-1},$$

or

$$\hat{\lambda}_i(T_a) = 0.1857 \cdot T_a^{-0.6130}. \tag{16.63}$$

The MTBF growth function is

$$\hat{m}_i(T_a) = 5.3837 \cdot T_a^{0.6130}. \tag{16.64}$$

2. For the failure terminated test, from Eq. (16.56), the maximum likelihood estimate of β is

$$\hat{\beta} = \frac{23}{(22)\log_e(22,000) - \sum_{i=1}^{22}\log_e(T_{ai})},$$

where

$$\sum_{i=1}^{n-1}\log_e(T_{ai}) = \sum_{i=1}^{22}\log_e(T_{ai}) = 163.0132.$$

Then

$$\hat{\beta} = \frac{23}{(22)\log_e(22,000) - 163.0132},$$

or

$$\hat{\beta} = 0.4038.$$

From Eq. (16.58) the unbiased estimate of β is

$$\beta' = \frac{23 - 1}{23}(0.4038),$$

or

$$\beta' = 0.3862,$$

and from Eq. (16.59)

$$\eta' = \frac{22,000}{23^{\frac{1}{0.3862}}}$$

or

$$\eta' = 6.5533 \ \text{hr.}$$

Then the failure rate function is

$$\hat{\lambda}_i(T_a) = \frac{0.3862}{6.5533}\left(\frac{T_a}{6.5533}\right)^{0.3862-1},$$

or

$$\hat{\lambda}_i(T_a) = 0.1869 \cdot T_a^{-0.6138}. \tag{16.65}$$

The MTBF growth function is

$$\hat{m}_i(T_a) = 5.3505 \cdot T_a^{0.6138}. \tag{16.66}$$

From Eqs. (16.53), (16.63), and (16.65), it may be seen that there is not much difference among these estimates.

EXAMPLE 16–11

Two identical systems are tested. Any design changes made to improve the reliability of these systems are incorporated into both systems when any one of the two systems fails. A total of 29 failures occur during the test. The data are given in Table 16.13. Determine the AMSAA failure rate improvement and MTBF growth models that represent the data using (1) the graphical estimate and (2) the maximum likelihood estimate. (3) Compare the results.

SOLUTIONS TO EXAMPLE 16–11

1. The plot of the observed cumulative number of failures versus the cumulative test time listed in Column 4 of Table 16.13, is given in Fig. 16.15. It may be seen from the plot that the observed data tend to a straight line. A straight line is drawn to fit these data points, and two points on the straight line are chosen, say (1.7, 1) and (155.0, 10). From Eqs. (16.51) and (16.52)

$$\hat{\beta} = \frac{\log_{10}(10) - \log_{10}(1)}{\log_{10}(155.0) - \log_{10}(1.7)},$$

or

$$\hat{\beta} = 0.5102,$$

and

TABLE 16.13 – The development test data of two identical systems for Example 16–11.

Number of failures, N	Accumulated test time of: System 1, hr	System 2, hr	Cumulative test time, T_{ai}, hr	$\log_e T_{ai}$
1	0.2	(2.0)*	2.2	0.788
2	(1.7)	2.9	4.6	1.526
3	(4.5)	5.2	9.7	2.272
4	(5.8)	9.0	14.8	2.695
5	(17.3)	9.2	26.5	3.277
6	(29.3)	24.2	53.5	3.980
7	36.5	(61.1)	97.6	4.581
8	(46.3)	69.6	116.9	4.753
9	63.6	(78.1)	141.7	4.954
10	(64.4)	85.5	149.9	5.010
11	74.3	(93.6)	167.9	5.123
12	106.6	(103.1)	209.7	5.346
13	(195.2)	117.0	312.2	5.744
14	(235.1)	134.4	369.5	5.912
15	248.7	(150.2)	398.9	5.989
16	(256.8)	164.6	421.4	6.044
17	(261.1)	174.3	435.4	6.076
18	(299.4)	193.2	492.6	6.200
19	305.3	(234.2)	539.5	6.291
20	326.9	(257.3)	584.2	6.370
21	339.2	(290.3)	629.5	6.445
22	366.1	(293.1)	659.2	6.491
23	(466.4)	316.4	782.8	6.663
24	504.0	(373.2)	877.2	6.777
25	510.0	(375.1)	885.1	6.786
26	(543.2)	386.1	929.3	6.834
27	(635.4)	453.3	1,088.7	6.993
28	641.2	(485.8)	1,127.0	7.027
29	(755.8)	573.6	1,329.4	7.192

* The values in parentheses are the accumulated test times for System 1 (or System 2) when System 2 (or System 1) fails.

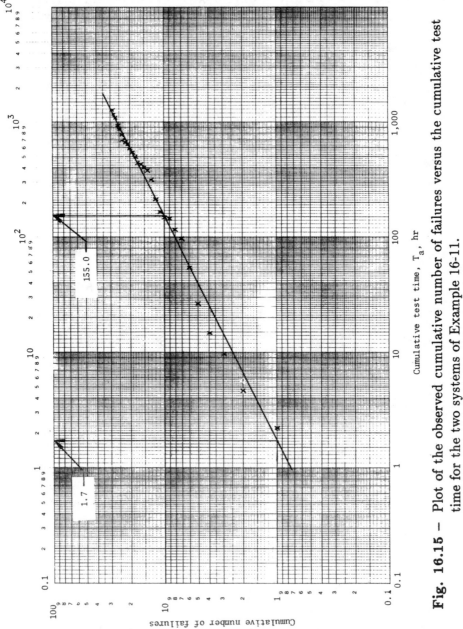

Cumulative test time, T_a, hr

Fig. 16.15 – Plot of the observed cumulative number of failures versus the cumulative test time for the two systems of Example 16-11.

$$\hat{\eta} = \frac{1.7}{10^{\frac{1}{0.5102}}} = 1.7,$$

or

$$\hat{\eta} = \frac{155.0}{10^{\frac{1}{0.5102}}} = 1.7.$$

The failure rate improvement function representing these data can be written as

$$\lambda_i(T_a) = \frac{0.5102}{1.7}(\frac{T_a}{1.7})^{0.5102-1},$$

or

$$\lambda_i(T_a) = 0.3892 \cdot T_a^{-0.4898}.$$

The MTBF growth function is

$$m_i(T_a) = 2.570 T_a^{0.4898}.$$

2. For the failure terminated test, from Eq. (16.56), the maximum likelihood estimate of β is

$$\hat{\beta} = \frac{29}{(28)\log_e(1,329.4) - \sum\limits_{i=1}^{28} \log_e(T_{ai})},$$

where

$$\sum_{i=1}^{n-1} \log_e(T_{ai}) = \sum_{i=1}^{28} \log_e(T_{ai}) = 146.947.$$

Then

$$\hat{\beta} = \frac{29}{(28)\log_e(1,329.4) - 146.947},$$

or

$$\hat{\beta} = 0.5327,$$

From Eq. (16.58) the unbiased estimate of β is

$$\beta' = \frac{29-1}{29}(0.5327),$$

or

$$\beta' = 0.5143,$$

and from Eq. (16.59)

$$\eta' = \frac{1,329.4}{29^{\frac{1}{0.5143}}}$$

or

$$\eta' = 1.906 \text{ hr.}$$

Then the failure rate improvement function is

$$\lambda_i(T_a) = \frac{0.5143}{1.906}(\frac{T_a}{1.906})^{0.5143-1},$$

TABLE 16.14 – Spacecraft reliability growth data for Problem 16-1.

Group number	Time, T_a	Reliability, %
1	0-June 19XX	51.0
	1-July	59.4
2	2-August	63.8
	3-September	66.6
3	4-October	68.8
	5-November	65.5

or
$$\lambda_i(T_a) = 0.3691 \cdot T_a^{-0.4857}.$$
The MTBF growth function is
$$m_i(T_a) = 2.709 \cdot T_a^{0.4857}.$$

3. The results using the graphical and the maximum likelihood approaches are quite close, preference going to the latter.

PROBLEMS

16-1. Reliability growth data from a spacecraft are given in Table 16.14. These data are based on successive design analysis only, because no test or use data were available at the time the predictions were made. The increase in the predicted reliability reflected the increase in the detailed knowledge about the design, its components, their failure rates, and the mission duration. Do the following:

(1) Find the math model for the reliability growth curve, using the Gompertz model.

(2) Find the reliability that may be attained at the end of January of the following year, i.e., two months after November when the last reliability prediction was made, or in the seventh month ($T_a = 7$).

(3) Find the upper reliability limit that may be achieved if the current effort continues.

Observe that for $T_a = 0$ the Gompertz model gives $R(T_a = 0)$ when growth is about to start, in this case the month of June

TABLE 16.15 – Reliability growth data for the solid propellant rocket of Problem 16-2.

Month	Reliability, %
0	41
1	54
2	63
3	75
4	83
5	89
6	91
7	92
8	93

19XX. The first month's reliability growth is obtained by substituting $T_a = 1$, the second month's by substituting $T_a = 2$, etc.

16-2. Given the reliability growth data of Table 16.15, do the following:

(1) Find the parameters of the Gompertz reliability growth curve.

(2) Write the Gompertz reliability growth equation.

(3) Plot the Gompertz curve and the data points on linear-linear scale.

(4) Discuss the results of Case 3 comparatively and describe the shape of the reliability growth curve and the raw data.

16-3. Given the reliability growth data of Table 16.16, do the following:

(1) Find the parameters of the Gompertz reliability growth curve.

(2) Write the Gompertz reliability growth equation.

(3) Plot the Gompertz curve and the data points on linear-linear scale.

(4) Discuss the results of Case 3 comparatively and describe the shape of the reliability growth curve and the raw data.

16-4. A 16-stage reliability development test program was performed. Fifteen groups of success-failure data were obtained and are given in Table 16.17. Do the following:

TABLE 16.16 – **Reliability growth data for the space satellite of Problem 16-3.**

Month	Reliability, %
0	22
1	25
2	29
3	33
4	46
5	59
6	68
7	75
8	81
9	86
10	90
11	93

(1) Fit these data to the Lloyd-Lipow model.

(2) Plot the reliabilities predicted by the Lloyd-Lipow model and the observed reliabilities and discuss comparatively the results.

16-5. A test-fix-test development program of a missile was performed. Out of 31 missiles launched, 9 failed. To increase their reliability, corrective actions were taken after each failure. The test data are given in Table 16.18. Assume that during the 31 launches failure 1 is caused by Failure Mode 1, f_1; failures 2 and 5 are caused by Failure Mode 2, f_2; failures 3 and 4 are caused by Failure Mode 3, f_3; failure 6 is caused by Failure Mode 4, f_4; failure 7 is caused by Failure Mode 5, f_5; failure 8 is caused by Failure Mode 6, f_6; and failure 9 is caused by Failure Mode 7, f_7. Do the following:

(1) Find the Gompertz reliability growth model to fit the cumulative reliabilities calculated directly from the raw data.

(2) Find the Gompertz reliability growth model to fit the reliabilities obtained by failure discounting.

(3) Comparatively discuss the fitted models from the raw data and the discounted data.

16-6. A reliability growth data set is given in Table 16.19. Do the following:

TABLE 16.17 – The reliability growth test results for Problem 16-4.

Number of test stage, k	Number of tests in kth stage, N_k	Number of successful tests in kth stage, S_k
1	5	2
2	8	5
3	9	6
4	9	7
5	8	5
6	10	7
7	10	8
8	10	7
9	10	6
10	11	7
11	15	11
12	15	13
13	15	12
14	14	12
15	15	13

TABLE 16.18 – Missile launch data for Problem 16-5.

Launch number	Success	Failure	Failure mode
1		X	Mode 1, f_1
2		X	Mode 2, f_2
3		X	Mode 3, f_3
4	X		
5		X	Mode 3, f_4
6		X	Mode 2, f_2
7	X		
8	X		
9	X		
10	X		
11		X	Mode 4, f_4
12	X		
13	X		
14		X	Mode 5, f_5
15	X		
16	X		
17	X		
18	X		
19		X	Mode 6, f_6
20	X		
21	X		
22	X		
23	X		
24	X		
25	X		
26		X	Mode 7, f_7
27	X		
28	X		
29	X		
30	X		
31	X		

* Failure Mode 3 reoccurred at failure 4.

** Failure Mode 2 reoccurred at failure 5.

TABLE 16.19 – The observed reliability data for Problem 16-6.

Time, months	Observed reliability, R_0, %
1	25.0
2	28.0
3	31.0
4	37.0
5	45.0
6	59.0
7	65.0
8	77.0
9	82.0
10	87.0
11	93.0
12	95.0
13	95.0

(1) Plot the data points.

(2) Find a Gompertz curve that represent the data well and plot it comparatively with the raw data.

(3) Find a logistic reliability growth curve which represents the data and plot it comparatively with the raw data.

(4) Find the modified Gompertz curve which represents the data and plot it comparatively with the raw data.

16-7. An MTBF growth curve for a pump has been determined to have the following Duane model parameters:

$$b = 30 \text{ hr}, \ \alpha = 0.55.$$

(1) Write the Duane cumulative MTBF growth curve's equation.

(2) Plot the equation found in Case 1.

(3) Write the Duane instantaneous MTBF growth curve's equation.

(4) Plot the equation found in Case 3.

(5) An instantaneous MTBF goal of 2,100 hr is desired to be attained after 500 cumulative hours of test and development time. Can that goal be attained?

(6) What instantaneous MTBF goal may be attained after 500 cumulative hours of test and development time?

(7) After how many cumulative test and development hours can the instantaneous MTBF goal of 2,100 hr be met?

16-8. The following parameters of an MTBF growth curve have been determined for a Duane model:

$$b = 20 \text{ hr}, \ \alpha = 0.40.$$

(1) Write the equation of the cumulative MTBF growth curve.

(2) Write the equation of the instantaneous MTBF growth curve.

(3) Will an instantaneous MTBF goal of 200 hr be attained after 800 cumulative test and development hours?

(4) What instantaneous MTBF may be attained after 800 cumulative hours of test and development time?

16-9. A prototype of a system is tested with the incorporation of design changes. A total of 15 failures occur. Failure 15 occurs at the cumulative test time of 3,200.0 hr. The data are given in Table 16.20. Determine the AMSAA reliability growth model that represents the data using:

(1) A graphical estimation method.

(2) The maximum likelihood estimation method.

16-10. Two identical systems are tested. Any design changes made to improve the reliability of these systems are incorporated into both systems when any one of the two systems fails. A total of 22 failures occur during the test. The data are given in Table 16.21. Determine the AMSAA failure rate improvement and MTBF growth models that represent the data using (1) the graphical estimate and (2) the maximum likelihood estimate. (3) Compare the results.

TABLE 16.20 – **The development test data of a prototype system for Problem 16-9.**

Number of failures, N	Cumulative test time, T_{ai}, hr
1	35.0
2	140.0
3	210.0
4	300.0
5	600.0
6	640.0
7	710.0
8	1,210.0
9	1,530.0
10	1,690.0
11	1,750.0
12	2,050.0
13	2,180.0
14	2,500.0
15	3,200.0

TABLE 16.21 – The development test data of two identical systems for Problem 16-10.

Number of failures, N	Accumulated test time of:	
	System 1, hr	System 2, hr
1	1.0	(1.7)*
2	7.3	(3.0)
3	(8.7)	3.8
4	(23.3)	7.3
5	(46.4)	10.6
6	50.1	(11.2)
7	57.8	(22.2)
8	(72.1)	37.4
9	(86.6)	38.4
10	87.0	(41.6)
11	(98.7)	45.1
12	102.2	(65.7)
13	149.2	(122.0)
14	166.6	(130.1)
15	(180.8)	139.8
16	181.3	(146.9)
17	(207.9)	158.3
18	(209.8)	186.9
19	(226.9)	194.2
20	232.2	(206.0)
21	(257.5)	223.7
22	(330.4)	285.9

* The values in parentheses are the accumulated test times for System 1 (or System 2) when System 2 (or System 1) fails.

REFERENCES

1. *Reliability Growth-Methods and Management,* United Army Management Engineering Training Agency, Rock Island Arsenal, Ill., 129 pp., May 1972.

2. Virene, E. P.,"Reliability Growth and the Upper Limit," *Proc. 1968 Annual Symposium on Reliability,* Boston, Mass., pp. 265-270, Jan. 16-18, 1968.

3. Codier, Ernest O., "Reliability Growth in Real Life," *Proc. 1968 Annual Symposium on Reliability,* Boston, Mass., pp. 458-469, Jan. 16-18, 1968.

4. Duane, J. T., "Learning Curve Approach to Reliability Monitoring," *IEEE Transactions on Aerospace,* Vol. 2, No. 2, 1964.

5. Crow, Larry H., *Reliability Growth Modeling,* U.S. Army Material Systems Analysis Agency, Aberdeen Proving Ground, Md., Technical Report No. 55, 45 pp., Aug. 1972.

6. Lloyd, David K., and Lipow, Myron, *Reliability: Management, Methods and Mathematics,* 201 Calle Miramar, Redondo Beach, Calif. 90277, 589 pp., 1977.

7. Lloyd, David K., "Forecasting Reliability Growth," *Quality and Reliability Engineering International,* Vol. 2, pp. 19-23, 1986.

8. Weiss, H. K., *An Analytical Model of Reliability Growth through Testing,* Report No. 54304, AD-035767, Northrop Aircraft Inc., Hawthorne, Calif., 17 pp., May 1954.

9. Cheng, Hanchang, *Reliability Growth Monitoring Techniques,* Master's Report, The University of Arizona, Tucson, 161 pp., 1982.

10. Crow, Larry H., *Confidence Interval Procedures for Reliability Growth Analysis,* AD-A044788, 30 pp., 1977.

11. MIL-HDBK-189, 13 Feb. 1981, *Reliability Growth Management.*

12. MIL-STD-1635(EC), *Reliability Growth Testing,* 3 Feb. 1978.

13. Lin, Tsung-Ming T., "A New Method for Estimating Duane Growth Model Parameters," *Annual Reliability and Maintainability Symposium,* Philadelphia, Penn., pp. 389-393, Jan. 22-24, 1985.

APPENDIX 16A

DERIVATION OF EQUATIONS (16.2), (16.3), AND (16.4)

Consider the case where each group has n items, the increment of time is I, and the T_a with the corresponding R_i have been determined. Then we have

T_a	R_i
T_{a_0}	R_0
T_{a_1}	R_1
T_{a_2}	R_2
.	.
.	.
.	.
$T_{a_{m-1}}$	R_{m-1}

with $m = 3 \times n$, $T_{a_i} - T_{a_1-1} = I$, and $i = 1, 2, ..., m - 1$.

From the Gompertz equation

$$R = ab^{c^{T_a}}, \tag{16A.1}$$

$$\log R = \log a + c^{T_a} \log b, \tag{16A.2}$$

and

$$S_1 = \sum_{i=0}^{n-1} \log R_i = n \log a + \log b \sum_{i=0}^{n-1} c^{T_{a_i}},$$

$$S_2 = \sum_{i=n}^{2n-1} \log R_i = n \log a + \log b \sum_{i=n}^{2n-1} c^{T_{a_i}}, \tag{16A.3}$$

and

$$S_3 = \sum_{i=2n}^{m-1} \log R_i = n \log a + \log b \sum_{i=2n}^{m-1} c^{T_{a_i}}.$$

Then

$$\frac{S_3 - S_2}{S_2 - S_1} = \frac{\sum_{i=2n}^{m-1} c^{T_{a_i}} - \sum_{i=n}^{2n-1} c^{T_{a_i}}}{\sum_{i=n}^{2n-1} c^{T_{a_i}} - \sum_{i=0}^{n-1} c^{T_{a_i}}},$$

$$\frac{S_3 - S_2}{S_2 - S_1} = \frac{c^{T_{a_{2n}}} \sum_{i=0}^{n-1} c^{T_{a_i}} - c^{T_{a_n}} \sum_{i=0}^{n-1} c^{T_{a_i}}}{c^{T_{a_n}} \sum_{i=0}^{n-1} c^{T_{a_i}} - \sum_{i=0}^{n-1} c^{T_{a_i}}},$$

or

$$\frac{S_3 - S_2}{S_2 - S_1} = \frac{c^{T_{a2n}} - c^{T_{an}}}{c^{T_{an}} - 1} = c^{T_{an}} = c^{n \cdot I + T_{a0}}.$$

Without loss of generality, take $T_{a0} = 0$; then

$$\frac{S_3 - S_2}{S_2 - S_1} = c^{n \cdot I}.$$

Solving for c yields

$$c = \left(\frac{S_3 - S_2}{S_2 - S_1}\right)^{\frac{1}{n \cdot I}}. \tag{16A.4}$$

Consider Eq. (16A.3) again; then

$$S_1 - n \cdot \log a = \log b \sum_{i=0}^{n-1} c^{T_{ai}},$$

$$S_2 - n \cdot \log a = \log b \sum_{i=n}^{2n-1} c^{T_{ai}},$$

or

$$\frac{S_1 - n \cdot \log a}{S_2 - n \cdot \log a} = \frac{1}{c^{n \cdot I}}.$$

Reordering the equation yields

$$\log a = \frac{1}{n}\left(S_1 + \frac{S_2 - S_1}{1 - c^{n \cdot I}}\right),$$

or

$$a = \text{anti} \log \left[\frac{1}{n}\left[S_1 + \frac{S_2 - S_1}{1 - c^{n \cdot I}}\right]\right]. \tag{16A.5}$$

Consider Eq. (16A.3) where

$$S_1 - \log b \sum_{i=0}^{n-1} c^{T_{ai}} = n \cdot \log a,$$

$$S_2 - \log b \sum_{i=n}^{2n-1} c^{T_{ai}} = n \cdot \log a,$$

$$\frac{S_1 - \log b \sum_{i=0}^{n-1} c^{T_{ai}}}{S_2 - \log b \sum_{i=n}^{2n-1} c^{T_{ai}}} = 1,$$

or

$$S_1 - \log b \sum_{i=0}^{n-1} c^{T_{ai}} = S_2 - \log b \sum_{i=n}^{2n-1} c^{T_{ai}}. \tag{16A.6}$$

Reordering Eq. (16A.6) yields

$$\log b = \frac{(S_2 - S_1)(c^I - 1)}{(1 - c^{n \cdot I})^2},$$

or

$$b = \text{anti} \log \left[\frac{(S_2 - S_1)(c^I - 1)}{(1 - c^{n \cdot I})^2}\right]. \tag{16A.7}$$

For the special case where $I = 1$, from Eqs. (16A.4), (16A.5), and (16A.7)

$$c = \left(\frac{S_3 - S_2}{S_2 - S_1}\right)^{\frac{1}{n}},$$

$$a = \text{anti}\log\left[\frac{1}{n}[S_1 + \frac{S_2 - S_1}{1 - c^n}]\right],$$

and

$$b = \text{anti}\log\left[\frac{(S_2 - S_1)(c - 1)}{(1 - c^n)^2}\right].$$

These are Eqs. (16.2), (16.3), and (16.4), respectively.

APPENDIX 16B
COMPUTER PROGRAM AND
OUTPUT FOR EXAMPLE 16–6

```
C        THIS PRCGRAM IS TO ESTIMATE THE PARAMETERS OF MODIFIEC
C        GOMPERTZ RELIABILITY GROWTH MODEL.

         PROGRAM JIANG(INPUT,OUTPUT,TAPE5=INPUT,TAPE6=OUTPUT)
         EXTERNAL FCN
         REAL A(4),B(4),X(4),WORK(54),R(9),T(9),R1(9)
         INTEGER N,NSIG,NSRCH,IWORK(4),IER
         COMMON /B/M,T,R,R1

C        M--INPUT, INTEGEP, THE NUMBER OF OBSERVED RELIABILITY.
C        P--INPUT, A ARRAY, THE OBSERVED RELIABILITY.
         READ(5,15)M
15       FORMAT(I5)
         READ(5,16)R
16       FORMAT(F1C.6)
         DO 130 I=1,M
         T(I)=I-1.0
130      CONTINUE
C        N--THE NUMBER OF VARIABLE
         N=4
C        NSIG--EXPECTED ERROR
         NSIG=4
         NSRCH=12
C        A--ARRAY, THE LCWER BOUND OF VARIABLE X
C        B--ARRAY, THE UPPER BCUND OF VARIABLE X
         A(1)=0.C
         A(2)=0.C
         A(3)=0.C
         A(4)=C.0
         B(1)=1.C
         B(2)=1.0
         B(3)=1.0
         B(4)=R(1)
         CALL ZXMWD (FCN,N,NSIG,A,B,NSRCH,X,F,WORK,IWORK,IER)
         WRITE(6,21C)X
210      FORMAT(10X,'A=',F10.5,5X,'B=',F1C.5,5X,'C=',F1C.5
     *   ,5X,'D=',F1C.5)
         WRITE(6,220)F
220      FORMAT(1CX,'F=',E15.5)
         WRITE(6,245)
245      FORMAT(15X,'TIME',9X,'OBSERVED R',11X,'PREDICTED R')
         DO 250 I=1,M
         WRITE(6,24C)T(I),R(I),R1(I)
240      FORMAT(1CX,F1C.5,5X,F10.5,10X,F10.5)
250      CONTINUE
         STOP
         END
```

APPENDIX 16B - (continued)

```
        SUBROUTINE FCN(N,X,F)
        INTEGER I,N,M
        REAL X(N),F,R(9),T(9),R1(9)
        COMMON /B/M,T,R,R1
        IF(X(1)+X(4)-1.C) 3,3,5
5       F=SUM+SUM
        GOTO 11
3       SUM=0.0
        DO 10 I=1,M
        R1(I)=X(4)+X(1)*X(2)**(X(3)**T(I))
        SUM=SUM+(ABS(R1(I)-R(I)))
10      CONTINUE
        F=SUM
11      RETURN
        END
```

A =   .69107   B =   .00176   C =   .45203   D =   .30891
F =   .34049E-01

TIME	OBSERVED R	PREDICTED R
.00CCC	.31000	.31013
1.00000	.35500	.34817
2.00000	.49300	.49793
3.00000	.70100	.69352
4.00C00	.83000	.83916
5.00000	.92200	.92200
6.000CC	.96400	.96357
7.0000C	.98600	.98328
8.00CC0	.99000	.99239

APPENDIX 16C

RELATIONSHIP OF EQUATION (16.32) AND THE WEIBULL FAILURE RATE

If in the two-parameter Weibull failure rate equation

$$\lambda = \frac{\beta}{\eta}(\frac{T}{\eta})^{\beta-1} = \frac{\beta}{\eta \cdot \eta^{\beta-1}} \cdot T^{\beta-1} \qquad (16C.1)$$

the substitutions

$$a = \frac{\beta}{\eta \cdot \eta^{\beta-1}} \text{ and } -\alpha = \beta - 1$$

are made, the result is

$$\lambda = aT^{-\alpha}. \qquad (16C.2)$$

This is Eq. (16.32), indicating that the Duane failure rate improvement model is a Weibull failure rate model.

Chapter 17

FAILURE MODES, EFFECTS, AND CRITICALITY ANALYSIS

17.1 INTRODUCTION

Two methods of failure modes, effects, and criticality analysis
(FAMECA) are covered in this chapter and illustrated by two examples. Such an analysis identifies those components in an equipment or
system whose design needs to be changed or improved upon to increase
their reliability and safety of operation. In a complex equipment or system, all components cannot be redesigned! There just is not enough
time, engineers, and money to do this! Consequently, the components
that are the most critical, i.e., if they fail the equipment or system will
fail, need to be scientifically singled out. A thorough FAMECA done
by experienced design and reliability Engineers, together, will accomplish this task and find which components and which failure modes
should be tackled first and improved on.

17.2 METHOD 1

This method is covered in two parts. Part 1 covers the purpose and
outlines the manner in which the FAMECA is applied. Part 2 describes
the procedures required to conduct a FAMECA and the sequence of responsibilities required at the component, subsystem, and system levels
of reliability analysis. The format used in illustrating the application
of the technique is precisely the same format that should be used when
this technique is incorporated into a design assurance manual.

17.2.1 SYSTEMATIC TECHNIQUE

In the design of complex systems, in general, the critical components in the design cannot be identified through a simple inspection of the design. Consequently, a systematic approach to identify and rank the criticality and the corresponding increases in system reliability is deemed necessary. When the critical components have been identified and ranked, a concerted effort can be placed on reducing the criticality of the most critical components in the system. The result is that greater returns, in terms of system reliability improvement, are realizable for a given engineering effort.

To insure maximum accuracy of the FAMECA, the component design engineer, the system design engineer, and the reliability engineer must integrate their efforts. This technique delineates the inputs required from each discipline, as well as the technical approach to be used in conducting the FAMECA.

Overall responsibility for implementing and controlling the execution of FAMECA is placed with the project reliability engineer. The specific FAMECA's to be conducted at the component and subsystem design levels are the responsibilities of the component design engineer and the system design engineer, respectively. The sequence of activities and the groups responsible for performing these activities are summarized in Table 17.1.

The sequence of analyses and the associated forms have been designed in such a manner as to result in the output of a given analysis being the input for the next level of analysis.

17.2.2 COMPONENT FAILURE MODES ANALYSIS

Here a systematic technique for identifying the critical components in a design is provided, as well as possible courses of action for reducing component criticality. A FAMECA shall be conducted for each mission phase. Mission phases are defined as marked changes in environment, such as count-down, lift-off, etc., and/or changes in the functional complexity, such as booster engine staging, programmed pitch-over, etc. The failure modes analysis must also be continually updated to insure compatibility of the criticality ranking list with the latest design changes.

The following documents may be used as supplements:

1. Design assurance manual for engineers, which may contain system models, system reliability apportionment, system reliability prediction, stress analysis, and component reliability sensitivity analysis.

TABLE 17.1– Sequence of FAMECA activities and responsibilities.

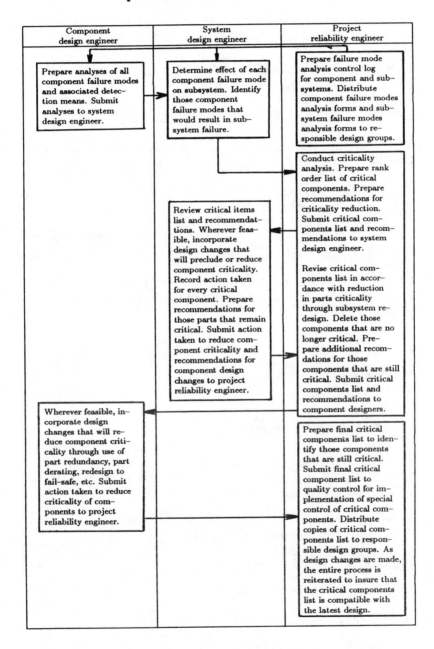

Component design engineer	System design engineer	Project reliability engineer
		Prepare failure mode analysis control log for component and subsystems. Distribute component failure modes analysis forms and subsystem failure analysis forms to responsible design groups.
Prepare analyses of all component failure modes and associated detection means. Submit analyses to system design engineer.	Determine effect of each component failure mode on subsystem. Identify those component failure modes that would result in subsystem failure.	
		Conduct criticality analysis. Prepare rank order list of critical components. Prepare recommendations for criticality reduction. Submit critical components list and recommendations to system design engineer.
	Review critical items list and recommendations. Wherever feasible, incorporate design changes that will preclude or reduce component criticality. Record action taken for every critical component. Prepare recommendations for those parts that remain critical. Submit action taken to reduce component criticality and recommendations for component design changes to project reliability engineer.	Revise critical components list in accordance with reduction in parts criticality through subsystem redesign. Delete those components that are no longer critical. Prepare additional recomdations for those components that are still critical. Submit critical components list and recommendations to component designers.
Wherever feasible, incorporate design changes that will reduce component criticality through use of part redundancy, part derating, redesign to fail–safe, etc. Submit action taken to reduce criticality of components to project reliability engineer.		Prepare final critical components list to identify those components that are still critical. Submit final critical component list to quality control for implementation of special control of critical components. Distribute copies of critical components list to responsible design groups. As design changes are made, the entire process is reiterated to insure that the critical components list is compatible with the latest design.

2. MIL–STD–1629A (24 November 1980), Procedures for Performing a Failure Modes, Effects, and Criticality Analysis.

3. MIL–STD–2070 (AS), Procedures for Performing a Failure Modes, Effects and Criticality Analysis for Aeronautical Equipment.

17.2.2.1–RESPONSIBILITIES OF VARIOUS ENGINEERS IN CONDUCTING A FAMECA

PROJECT RELIABILITY ENGINEER SHALL:

1. Initiate requirements for component failure modes analysis by formal announcement to department concerned.

2. Create and distribute the following required failure modes analysis forms to responsible design groups:

   2.1 Component FAMECA form to component design engineer.

   2.2 Subsystem FAMECA form to the subsystem design engineer.

3. Implement and maintain a failure modes analysis log. Table 17.2 is an example of a failure modes and effects analysis log. Log entries include:

   3.1 Analysis to be performed at both system and component levels.

   3.2 Analysis due dates.

   3.3 Analysis completion dates.

COMPONENT DESIGN ENGINEER SHALL:

1. Perform a component FAMECA for each component within responsibility. Table 17.3 is an example of a component failure modes and effects analysis form.

2. Enter the appropriate information in the blocks on the component failure modes effects analysis form.

3. In Column 1, list all pieces or parts that comprise the component.

4. In Column 2, briefly describe the function of each part.

5. In Column 3, list all ways each part can fail.

6. In Column 4, list the effect that each possible part failure will have on component operation.

TABLE 17.2–Example of a failure modes and effects analysis log.

		Failure modes and effects analysis log			Project: FUMTU	
System	Component analysis required	Component analysis due date	Component analysis date	Subsystem analysis due date	Subsystem analysis completion date	Criticality analysis completion date
Hydraulics	Hydraulic control valve XX-66912	5/12/19XX	5/8/19XX	5/30/19XX		
	Hydraulic pump XX-66842		5/6/19XX			
	Accumulator XX-66321					
	Bypass valve XX-66333					
	Reservoir XX-66382					
	Actuator XX-68721					
	Pressure sensor XX-35421					
Propulsion	Flex level sensor XX-21320	5/27/19XX		5/30/19XX		
	Pressure regulator XX-21433					
	Flow control valve XX-21444					

TABLE 17.3–Example of component failure modes and effects form and analysis.

Component failure modes and effects analysis

Component name: Hydraulic control valve		Operating time: 76 min	Subsystem: Hydraulics		Due data: 5/12/19XX
Part no: XX-66912	Predicted reliability: 0.99996		Project: FUMTU		Date:5/6/19XX
Part	Part function	Part failure	Failure effect	Detection means	Remarks
1	2	3	4	5	6
1. Spool centering spring	Maintains control spool in "neutral" position.	a. Centering spring breaks or becomes disengaged.	a. Valve will remain in "pressure" condition when electric power is removed.	a. Visual indicator will show "pressure" when electric power is removed.	a. Spool is by–passed by 300–psi relief valve.
		b. Centering spring loses resilience.	b. Valve will slowly return to pressure condition when electric power is removed.	b. More than 5 seconds will be required for visual indica–tor to return to "pressure" position.	b. Time required to return to "pressure" is dependent on magnitude of resiliency loss.
2. Valve spool	Controls flow to pressure and return parts.	Galling of spool or spool cylinder.	a. Control of hydraulic pres–sure will be erratic.	a. Erratic flow of hydraulic fluid.	a. Reaction time is dependent on severity of galling.
			b. Valve will jam closed.	b. Valve will not respond to input signals. Hydraulic fluid will continue to flow when electric power is removed.	b. Spool jamming will normally be preceded by erratic hydraulic flow.
			c. Valve will jam open.	c. Hydraulic fluid will not flow when electric power is applied.	
3. Valve input lever	Positions control spool.	Disengagement of input lever.	Valve will remain closed when input lever is moved to open position.	No hydraulic flow when input lever is moved to open position.	If lever dis–engages while valve is open, valve will close.

7. In Column 5, list the ways in which each component failure can be detected. If a part failure can occur without a resultant component failure, as would be the case in parallel redundancy, this fact shall be so noted in the remarks column.

8. In Column 6, enter any clarifying or qualifying statements deemed necessary or any other additional information on the component's failure modes and/or effects of failures on component operation.

9. Submit a copy of the completed component failure modes analysis to the responsible system design engineer.

SYSTEM DESIGN ENGINEER SHALL:

1. Perform a subsystem failure modes effects analysis. Table 17.4 is an example of a subsystem failure modes and effects analysis form.

2. Enter the appropriate information in the blocks on the subsystem failure modes effects analysis form.

3. In Column 1, list all of the components that comprise the subsystem.

4. In Column 2, briefly describe the component's function in the subsystem.

5. In Column 3, list all modes of failure for each component. The failure modes for each component are obtained from the component FAMECA previously discussed.

6. In Column 4, enter the effect each component failure would have on the subsystem's operation. To determine this, consider the effect that failure would have if the failure resulted in premature operation of the component, in failure of the component to operate at the selected time, in failure of the component to cease operation at the selected time, or in failure of the component during operation.

7. In Column 5, enter the estimated time that would elapse from component failure to the time of loss of vehicle or mission.

8. In Column 6, enter the time of operation of each component as determined by the system-model-sequence-time bar-graphs in the project system model.

TABLE 17.4—Subsystem failure modes and effects analysis.

Project: FUMTU		Mission Phase: Flight			Date: 5/30/19XX	
Component name and part number	Component function	Failure modes	Failure effect on subsystem	Reaction time, min	Component operating time, min	Probability of loss, P_L, %
1	2	3	4	5	6	7
Hydraulic control valve XX-66912.	Control hydraulic flow to XX-68721 actuator.	1. Valve remains in "pressure" condition when electrical power is removed.	XX-68721 Actuator will not release the payload latching mechanism.	0.5	10	100
		2. Valve slowly returns to neutral when electrical power is removed.	No effect on subsystem during flight phase.	1.5	10	0
		3. Control of hydraulic pressure is erratic.	Erratic hydraulic flow could prevent operation of XX-68721 actuator due to resultant vibration.	1.5	10	10
		4. Valve jams closed.	XX-68721 Actuator will not release the payload latching mechanism.	0.5	10	50

480

9. In Column 7, enter the probability of loss of the vehicle or mission for each possible failure. The percentage probability of loss is to be assigned in the following increments:

Effect of item failure	Probability of loss of mission or vehicle, %
Actual loss	100
Probable loss	50
Possible loss	10
No loss	0

10. Submit completed copy of the subsystem failure modes and effects analysis to the project reliability engineer.

PROJECT RELIABILITY ENGINEER SHALL:

1. Compute component criticality ranking and prepare critical components ranking list as follows:

2. Enter the appropriate information in the blocks at the top of the component criticality analysis form. An example is illustrated in Table 17.5. The predicted reliability number is obtained from the component's reliability prediction.

3. If the failure history of the component is known, enter the number of times the component failed in each mode in Column 5. No entries are made in Columns 1 through 4.

4. If the failure history of the component is not known, enter the environments to which the component will be exposed at the tops of Columns 1 through 4. Environmental conditions for the mission phase should be defined in the project system model. For each environmental condition, evaluate the effect that condition would have on each mode of failure. A number from 1 to 10 is selected to indicate the estimated effect that the environmental condition will have on each failure mode. Where it is judged to contribute significantly to the possibility of failure in a given mode, the number 10 is used. Intermediate numbers are used to express the possible conditions lying between the extremes. The product of the estimates for each failure mode is then computed and entered in Column 5.

5. Column 5 is summed and the ratio of the number of failures in each mode to the total number of failures is determined. In the

TABLE 17.5–Component criticality analysis.

Component criticality analysis

Component: Hydraulic control valve				Mission phase: Flight			Predicted reliability = 0.99960	
Part no. XX-66912				Subsystem: Hydraulic	Project: FUMTU		Date: 5 November 19XX	
Component failure modes	Environment				Number of failures or product, P	Failure mode frequency ratio, $P/\Sigma P_i$ = FMFR	Probability of loss of mission or vehicle, P_L,%	Unreliability= 1 - reliability, Q	Criticality,CR, (FMFR) $(P_L)(Q)$
	Temp.	Shock	Vib.	Humid.					
	1	2	3	4	5	6	7	8	9
Valve remains in "pressure" condition when electrical power is removed.	6	8	1	2	96	0.59	100	0.00040	0.000236
Control of hydraulic pressure is erratic.	4	4	1	1	16	0.10	10	0.00040	0.000004
Valve jams closed.	10	5	1	1	50	0.31	50	0.00040	0.000062

$\Sigma P = 162$

case where products are used, the ratio of the product for each mode to the sum of the products is computed. The resulting ratios are the failure mode frequency ratios, FMFR, estimated for each failure mode. This FMFR is entered in Column 6.

6. In Column 7, enter the estimated probability of loss, P_L, of the mission or vehicle if the failure in a given mode should occur. This probability of loss figure is obtained from the subsystem FAMECA previously discussed.

7. In Column 8, enter the component's unreliability, Q. The Q for a component is determined by subtracting the component's predicted reliability from 1.

8. In Column 9, enter the product of the failure mode frequency ratio, the probability of loss, and the component unreliability. These products are the criticality ranking numbers, CR, for each component failure mode, or

$$CR = (FMFR)(F_L)(Q).$$

9. When the CR for each failure mode has been determined, rank the component failure modes by CR. The ranked critical components are entered on the critical components ranking list, as shown in Table 17.6.

10. Recommend, whenever possible, subsystem redesign to reduce or preclude component criticality in Column 4. Subsystem criticality reduction can be achieved by:

    10.1 The use of redundancy.

    10.2 The realignment of component/subsystem functional interdependence such that the subsystem will fail-safe.

    10.3 The incorporation of instrumentation for monitoring critical component operation so that incipient failures can be detected by the operator in time to preclude system failure.

    10.4 The redesign of the subsystem to take advantage of a lower failure mode frequency ratio for a component. For example, if a check-valve fails open more often than it fails closed, the subsystem might be redesigned such that a failure in the open condition does not result in subsystem loss.

11. Distribute critical component list to system design engineers for corrective action on the subsystems for which they are responsible.

TABLE 17.6–Critical components ranking list.

Critical component ranking list				
Project: FUMTU				
Date: 3 Dec. 19XX			Date revised	
Component part number and name	Failure mode	Criticality ranking ($\times 10^6$)	Recommended action	Action taken and drawing change
1	2	3	4	5
XX-66912– Hydraulic control valve.	Valve remains in "pressure" condition when electrical power is removed.	236	Use redundant x x-66912 hydraulic control valve. Redesign.	Redundant hydraulic control valve would result in system exceeding weight limitations. Not accepted. Redesign initiated.
XX-52120– Guidance signal discriminator.	Fails to generate pitch signal during pitchover.	221	Derate parts used in discriminator.	Part derating investigation initiated.
XX-22138– Electric actuator.	Actuator does not open XX-33221 valve.	210	Install actuator on opposite side of valve so that valve will "fail safe."	Actuator moved to opposite side of valve. See Dwg. Chg. XX-22100E.

484

SUBSYSTEM DESIGN ENGINEER SHALL:

1. Review critical component list.

2. Evaluate impact of component redesign on subsystem cost, performance, weight, space, etc.

3. Redesign subsystem if criticality can be reduced more easily then by component redesign.

4. Note in the "Action Taken" column those cases in which criticality of a component cannot be reduced through practical subsystem redesign.

5. Recommend possible component redesign that will improve component reliability, result in fail-safe operation of the component, or otherwise reduce component criticality.

6. Return critical component list to project reliability engineer for revision and updating.

PROJECT RELIABILITY ENGINEER SHALL:

1. Conduct second component criticality analysis.

2. Remove components from critical component list that are no longer considered critical.

3. Revise rank order of remaining critical components to accommodate changes due to subsystem redesign.

4. Recommend component modifications to reduce or preclude component criticality. Component criticality reduction can be achieved by:

   4.1 Derating of parts. A technique for determining which parts are most sensitive to derating changes should be developed. Part stress level through stress analysis should be developed.

   4.2 Selection of high-reliability parts.

   4.3 Use of part redundancy and incorporation of independent instrumentation and test points for monitoring each leg of redundant circuits.

COMPONENT DESIGN ENGINEER SHALL:

1. Analyze feasibility of recommended component modifications.

2. Redesign components, if practical.

3. Note redesign action taken on critical components list along with the drawing changes that implement the design changes.

4. Return list to project reliability engineer.

PROJECT RELIABILITY ENGINEER SHALL:

1. Change criticality ranking of components on critical components list to reflect design changes.

2. Update list periodically to reflect subsequent changes in component criticality resulting from component or subsystem design changes, state-of-the-art improvements, etc.

17.2.3   REQUIREMENTS FOR SPECIAL HANDLING AND TESTING OF CRITICAL COMPONENTS

COMPONENT DESIGN ENGINEER SHALL:

1. Determine the requirements for special testing, handling, and monitoring of critical components.

2. Incorporate the requirements for special testing of components in procurement specification or drawing, if part is in-house manufactured, to insure that the critical mode of failure is examined.

3. Incorporate requirements for special marking, handling, and transportation as required in the procurement specification or applicable drawings if in-house manufactured.

4. Request that vendor data control establish vendor data requirements to insure identification and receipt of special test reports on critical components.

5. Design special instrumentation or monitoring devices, wherever feasible, to enable the operator to detect an incipient failure and preclude a serious subsystem failure.

QUALITY ASSURANCE ENGINEER SHALL:

1. Design special receiving inspection tests to ensure that handling and testing of vendor parts is commensurate with their critical nature.

2. Design tests that specifically test components at the assembly and system levels to ensure that the critical failure mode is exercised.

3. Process data accumulated from testing of individual components and testing and/or operation of systems containing critical components.

4. Distribute above to responsible design, reliability, and quality assurance departments.

UPDATING THE CRITICAL COMPONENTS RANKING LIST

Design changes will, in many cases, result in changes in the criticality of individual components. For this reason, a FAMECA must be accomplished each time a component or subsystem redesign is executed. The criticality ranking list is to be correspondingly revised to reflect changes in component criticality.

17.3   METHOD 2

This method is illustrated by applying it to aircraft splines. To evaluate the designs of aircraft splines and recommend areas of design, material, lubrication, treatment, and environment changes to achieve reliability and maintainability improvement, it is necessary to identify the critical failure modes and design factors involved. The FAMECA is one method that can be used to determine the spline failure modes involved in the prediction of a spline's reliability, the critical effect of each failure mode on the operation of the equipment in which the spline is operating, and the design factors which should be considered to improve the reliability and maintainability of splines.

Through the evaluation of all failure modes, their effects, causes and criticality, appropriate design changes can be selectively applied to obtain the most cost effective reliability and maintainability benefits.

17.3.1   THE FAILURE MODES AND EFFECTS ANALYSIS

The failure modes and effects analysis for aircraft splines [1; 2] is given in Table 17.7. In the analysis, the spline assembly is broken down into interlocking splines and the spline shaft. The tooth wear of the interlocking splines is considered first because it is the most predominant failure mode. Tooth wear in splines is comprised primarily of fretting wear, fretting corrosion wear, and abrasion wear.

Table 17.7—Failure modes and effects analysis

Failure mode	Failure effect on assembly function	Causes (in decreasing order of importance)	Factors which reduce failures	Critical failure criterion	Range of time to critical failure	Remarks
Tooth wear.	Vibration.	Misalignment.	Alignment guides; muffs; crowning.	Wear from a few thousandths of an inch to a knife edge tooth land.	300-2,500 hr spline life.	Misalignment will increase the vibration, fretting action, sinusoidal contact stress, and temperature which in turn will increase the wear in the tooth. Experimental data show that wear rate as a function of misalignment is approximated by an exponential function $y = e^{az}$, which reduces to $\log_e y = 0.5329 + 16.4234x$, where y = wear rate/hour and x = degrees of misalignment.
	Noisy operation.					
	Rough operation.					
	Loss of operation.	Inadequate lubrication.	Better greases or oils; favorable wet spline environment.		120-3,000 hr to an overhaul.	Grease in splines tends to produce a blanketing lubricating effect which reduces wear and a debris-retaining effect which tends to promote wear due to abrasion. Of seven greases evaluated by SWRI, grease B (MIL–G–21164B) provided the longest no–wear induction period. Grease C (MIL–G–27617) provided no induction period but provided very high performance in a jet fuel environment, possibly due to low solubility. Mineral oil provided the most favorable wet spline environment.

488

Table 17.7–(Cont'd) Failure modes and effects analysis

Failure mode	Failure effect on assembly function	Causes (in decreasing order of importance)	Factors which reduce failures	Critical failure criterion	Range of time to critical failure	Remarks
Tooth wear.	- - - - - - - - - - - - - - - →	Excessive contact stresses.	Reduced stresses; reduction in number of teeth.			Reduction in the number of spline teeth results in improved load sharing, which reduces the contact load on the more heavily loaded teeth. Concurrently, the contact stresses may be reduced by increasing tooth height.
		Soft material characteristics.	Surface treatment; harder materials.			Materials with higher Young's modulus are preferred. AISI 9310 steel carburized splines give significantly longer wear life than the commonly used AISI 4120 steel splines. Surface treatment with a 0.001 in. coating of Electroless Nickel II increases wear life at 0.004 in. total wear from about 50 hr to about 960 hr.

Table 17.7–(Cont'd) Failure modes and effects analysis

Failure mode	Failure effect on assembly function	Causes (in decreasing order of importance)	Factors which reduce failures	Critical failure criterion	Range of time to critical failure	Remarks
Tooth wear.	----→	Vibration.	Alignment; balance; speed control; appropriate stiffness of spline assembly; avoidance of resonance.			Vibration increases the fretting corrosion, hence the wear.
		Temperature too high.	Better lubrication.			In general, there is a small positive correlation between temperature and tooth wear which is more significant for softer spline materials, indicating a possible temperature–chemical interaction.

Table 17.7–(Cont'd) Failure modes and effects analysis

Failure mode	Failure effect on assembly function	Causes (in decreasing order of importance)	Factors which reduce failures	Critical failure criterion	Range of time to critical failure	Remarks
Tooth wear.	---------->	Hostile environment; corrosive fluids; dry air.	Wet environment with 0-rings.			Jet fuels JP-5 and JP-6 combine a lubricating effect with the exclusion of oxygen to inhibit corrosion and mitigate wear. Mineral oil with additives provides the most favorable environment.
		Inadequate maintenance.	Improved preventive maintenance schedules; effective corrective maintenance procedures.			Preventive maintenance such as thorough cleaning and re-greasing and prevention of misalignment will reduce the wear rate. Corrective maintenance procedures should include the prevention of mismatching of used splines or of misalignment to reduce spline tooth wear.
		Pump cavitation.	Check valve.			

491

Table 17.7–(Cont'd) Failure modes and effects analysis

Failure mode	Failure effect on assembly function	Causes (in decreasing order of importance)	Factors which reduce failures	Critical failure criterion	Range of time to critical failure	Remarks
Tooth wear.	— — — →	Crowning radius too small.				Crowning which provides tolerance for misalignment results in increased wear. Wear decreases as the crown radius of curvature increases. Within the tooth wear range of 0.004 to 0.012 in., wear life is increased from 57% to 72% as the crown radius is increased from 11.25 in. to ∞.

492

Table 17.7–(Cont'd) Failure modes and effects analysis

Failure mode	Failure effect on assembly function	Causes (in decreasing order of importance)	Factors which reduce failures	Critical failure criterion	Range of time to critical failure	Remarks
Tooth fracture.	Vibration. Noisy operation. Erratic operation.	Excessive shear stresses.	Reduced loads.	Fracture.		In experimental tests, tooth fracture occurred frequently when stresses exceeded 4,620 psi.
	Loss of operation.	Tooth surface too rough.	Optimum roughness.			Roughness of the tooth surface increases the no–wear induction period, but provides stress risers for early microscopic crack initiation.

493

Table 17.7–(Cont'd) Failure modes and effects analysis

Failure mode	Failure effect on assembly function	Causes (in decreasing order of importance)	Factors which reduce failures	Critical failure criterion	Range of time to critical failure	Remarks
Tooth fracture.	- - - - - - - - - - -→	Vibration.	Alignment; balance.			Vibration adds additional stresses which decrease endurance life and result in fracture.
		Material too weak or too brittle.	Stronger materials.			A hard case with a softer, more ductile core is desirable.

Table 17.7–(Cont'd) Failure modes and effects analysis

Interlocking splines

Failure mode	Failure effect on assembly function	Causes (in decreasing order of importance)	Factors which reduce failures	Critical failure criterion	Range of time to critical failure	Remarks
Tooth fracture.	⟶	Excessive strain hardening.	Reduced strain hardening.			Select materials which do not tend to become brittle or are fracture prone through strain hardening.
		Inadequate surface hardening.	Improved quality of hardening.			Case hardening may decrease the toughness of the material and increase the crack initiation and propagation tendency.

Table 17.7–(Cont'd) Failure modes and effects analysis

Failure mode	Failure effect on assembly function	Causes (in decreasing order of importance)	Factors which reduce failures	Critical failure criterion	Range of time to critical failure	Remarks
Spalling.	Vibration. Accelerated wear. Noisy and rough operation. Loss of operation.	Below surface tensile stresses; vibratory motion of the teeth in contact.	Reduction of contact stress and vibration.			Spalling is the result of tensile stresses present at the point of maximum stress a few thousandths of an inch below the contact surfaces. Due to vibratory motion of the teeth in contact, the loading and unloading of stresses initiates subsurface cracks which propagate to the surface, causing a layer of material to separate from the contacting surfaces. Spalling provides debris which contributes to abrasion wear. Also, sufficient metal may be removed by spalling to eventually lead to tooth weakening and subsequent fracture.

Table 17.7-(Cont'd) Failure modes and effects analysis

Failure mode	Failure effect on assembly function	Causes (in decreasing order of importance)	Factors which reduce failures	Critical failure criterion	Range of time to critical failure	Remarks
Bending deformation.	Vibration.	Force causing excessive bending moment.	Redesign to reduce forces.	Deformation of the shaft.		Bending increases the misalignment of the spline teeth, the vibration, the fretting corrosion, and the wear.
	Erratic operation.					
	Binding.	Inadequate shaft diameter.	Increased shaft diameter.			Misalignment which exceeds the clearance between the interlocking splines will cause the spline assembly to bind.
		Inadequate material characteristics.	Improved material selection.			The ductility of materials must be considered in design to provide adequate shaft stiffness with a minimum diameter.

Table 17.7.–(Cont'd) Failure modes and effects analysis

Failure mode	Failure effect on assembly function	Causes (in decreasing order of importance)	Factors which reduce failures	Critical failure criterion	Range of time to critical failure	Remarks
Shear fracture.	Loss of power transmission, therefore loss of operation.	High shear stresses.	Reduced load in torque.	Fracture.		When the operating conditions dictate a high torque load, the shear stresses must be reduced through design control of shaft dimensions.
		Excessive stress concentration at critical locations.	Increased radius of shoulders and fillets.			High-stress concentration most likely occurs at the neck where the spline is mounted on the shaft.
		Shaft cross section too small.	Increased shaft diameter.			
		Material too ductile or too weak.	Stronger materials.			

498

Table 17.7–(Cont'd) Failure modes and effects analysis

Interlocking splines

Failure mode	Failure effect on assembly function	Causes (in decreasing order of importance)	Factors which reduce failures	Critical failure criterion	Range of time to critical failure	Remarks
Bursting of the spline shell.	Loss of operation.	Excessive radial force at pitch line.	Reduced loads; increased face width.	Fracture.		All three stresses are tensile stresses and may add up to cause failure.
	------------→	Excessive stress due to beam loading of the teeth.	Reduced loads; increased face width.	Fracture.		
	------------→	Excessive stress caused by centrifugal force.	Reduced rpm.	Fracture.		

499

1. Fretting wear is caused by the mechanical welding (bonding) and tearing apart of the asperities on the contacting surfaces under contact pressure, which are subjected to relative oscillatory or vibratory motion.

2. Fretting corrosion wear is the removal of the corrosion products resulting from chemical reaction between the pure nascent surfaces generated by fretting wear and a corrosive atmosphere. Weak oxides which are formed on the fretted surfaces break off to form debris, which adds to abrasion wear.

3. Abrasion wear is caused by the abrasive action of the debris from fretting wear and fretting corrosion, which generate additional debris-caused wear.

The failure effect of tooth wear on the function of the next higher assembly is interactive and progressive until all of the teeth are worn away and there is loss of function. The causes of tooth wear are considered in the decreasing order of their importance. For each failure mode and its cause, factors are identified which can be addressed in design to reduce tooth wear. Column 5 of Table 17.7 shows that much progress is needed in establishing critical wear failure criteria to design for a specified reliability and life in the required operating environment. Column 6 of Table 17.7, based on depot overhaul records, shows that ranges of operating lives of aircraft splines are extremely broad, and data on which those ranges are based do not appear to follow any well-behaved pattern.

The remaining failure modes of tooth fracture, spalling, shaft bending deformation, shear fracture, and bursting of the spline shell are analyzed next. The primary causes of failure, the factors which reduce such failures, the critical failure criteria, the range of time to critical failure, and remarks on reducing these failures are given.The results are recorded in the same manner as those for tooth wear.

17.3.2 CRITICALITY ANALYSIS AND RANKING

The results of the failure modes and effects are incorporated into the criticality analysis to provide a ranking, or order of priority, for all significant failure modes and are given in Tables 17.8 through 17.11.

A systematic approach is used to determine a criticality ranking for the spline failure modes. Since the failure history for splines is not known, the approach uses quantitative estimates based on the FAMECA results to evaluate the effect potential environment conditions would have on each one of the failure modes . A number from 1 to

Table 17.8–Spline (including shaft) criticality analysis.

| Failure mode | Environments | | | | | Product of environment effect ratings, P | Failure mode frequency ratio, $P/\Sigma P$ = FMFR | Probability of loss of mission, P_L | Unreliability weighting factor, Q | Criticality, (FMFR)$(P_L)(Q)$ |
	Temp.	Vib.	Lub.	Mis–align–ment	Load					
Tooth wear										
Fretting wear	3	5.25	5.75	9.75	8	7,063.87	0.1753	0.119	0.41	0.00856
Fretting corrosion	4.75	4.75	9	4.75	7.25	6,992.96	0.1736	0.119	0.41	0.00846
Abrasion wear	4.5	6.25	8	7.75	7.5	13,078.13	0.3241	0.119	0.41	0.01580
Spalling wear	2.75	7	6.25	7.75	8.25	7,692.48	0.1910	0.119	0.41	0.00931
Tooth fracture	2.25	7	2.5	6	9.75	2,303.44	0.0572	0.600	0.13	0.00446
Tooth deformation	3	5.25	2.5	7	9.5	2,618.44	0.0649	0.250	0.22	0.00357
Shaft fracture	2.25	3	1	8.75	9.75	575.86	0.0143	1.000	0.21	0.00300

$$\sum P = 40,325.18$$

501

Table 17.9–Spline (including shaft) criticality analysis.

Failure mode	Environments					Product of environment effect ratings, P	Failure mode frequency ratio, $P/\Sigma P$ = FMFR	Probability of loss of mission, P_L	Unreliability weighting factor, Q	Criticality, (FMFR)$(P_L)(Q)$
	Temp.	Vib.	Lub.	Mis-align-ment	Load					
Tooth wear										
Fretting wear	3	5.25	5.75	9.75	8	7,063.87	0.1741	0.119	0.41	0.00849
Fretting corrosion	4.75	4.75	9	4.75	7.25	6,992.96	0.1723	0.119	0.41	0.00841
Abrasion wear	4.5	6.25	8	7.75	7.5	13,078.13	0.3223	0.119	0.41	0.01750
Spalling wear	2.75	7	6.25	7.75	8.25	7,692.48	0.1896	0.119	0.41	0.00925
Tooth fracture	2.25	7	2.5	6	9.75	2,303.44	0.0568	0.600	0.13	0.00443
Tooth deformation	3	5.25	2.5	7	9.5	2,618.44	0.0645	0.250	0.22	0.00354
Shaft fracture	2.25	4	1.5	7	8.75	826.88	0.0204	0.988	0.11	0.00222

$$\sum P = 40,576.20$$

TABLE 17.10–Values of probability of loss of mission.

Effect of item failure	Probability of loss of mission
Actual loss	1
Probable loss	↑
Possible loss	↓
No loss	0

10 is assigned, using the Delphi technique, to indicate the estimated effect the environmental condition will have on each failure mode. Where it is judged that a particular environment will have a negligible effect in a failure mode, the number 1 is assigned. The number 10 is used where the environmental condition is judged to contribute significantly to the possibility of spline failure from that particular failure mode. Conditions between these two extremes are assigned numbers between 1 and 10. The results appear in the five columns under "Environments" in Tables 17.8 and 17.9. Based on these ratings, the columns under "Product of Environment Effect Rating, P," and "Failure Mode Frequency Ratio" are calculated as shown. The "Probability of Loss of Mission , P_L, " is based on a range of values given in Table 17.10. The specific values listed are arrived at by using the Delphi technique. The " Unreliability Weighing Factor, Q, " for each failure mode is determined by calculating the reliability of each failure mode using the methodology discussed in [2], as follows:

$$Q = \frac{25 + m}{50},$$

where m is the standardized normal variate used in calculating the reliability from the normal distribution area tables.

The last column in Tables 17.8 and 17.9 gives the final results of the criticality analysis in terms of a criticality value which is the product of the previous three columns. A criticality ranking list for the spline system is given in Table 17.11, with the failure mode having the highest criticality value listed on the top. The failure modes with the highest criticality values should be the ones that receive the highest priority on all design improvement efforts for higher product reliability and maintainability. It is of interest to note that the most critical failure mode is wear, in all of its three modes, whereas spline design usually concentrates on designing against tooth fracture!

TABLE 17.11-Spline criticality ranking.

Failure mode	Criticality ranking $\times 10^4$	Recommendations
Abrasion wear.	158	Improve lubrication to guard against debris from fretting wear and fretting corrosion which generate additional debris-caused wear; minimize misalignment.
Spalling wear.	93	Improve design through better material and sizing; minimize misalignment.
Fretting wear.	85	Improve design through better material and sizing; minimize misalignment.
Fretting corrosion.	84	Improve lubrication and operating environment.
Tooth fracture.	44	Improve design through better material and sizing; minimize vibration.
Tooth deformation.	35	Improve design through better material and sizing.
Shaft fracture.	30	Improve design through better material and sizing.
Bursting of spline shell.	22	Improve design through better material and sizing; reduce rpm, minimize misalignment.

PROBLEMS

17-1. Conduct a comprehensive FAMECA for the accessory power unit system of Example 12-3 determining the criticality ranking of all units in this power unit using your own inputs following the format of Table 17.5 of the text. Explain why you chose the "Environment factor," the "Probability of loss of mission," and the "Unreliability weighting factor" values you used to determine the criticality ranking.

17-2. Same as Problem 17-1, but for Problem 12-8.

17-3. Same as Problem 17-1, but for Problem 12-9.

17-4. Conduct a FAMECA on one of the following units:

(1) Your car's water pump.

(2) Your car's fuel pump.

You may consider going to a car repair shop or a supplier of the unit you choose to work on and get exploded views of that unit which shows all the components. All drawings, pictures, sketches, etc., should be sumitted with your FAMECA.

17-5. Conduct a FAMECA on one of the following units:

(1) Room window air conditioner.

(2) Electric chain saw.

You may consider going to a supplier of the unit you choose to work on to get exploded views, drawings and pictures. All drawings, pictures, etc., should be submitted with your FAMECA.

REFERENCES

1. Kececioglu, D., Chester, L.B., Koharchek, A., and Shehata, M., *Aircraft Spline Reliability Predictive Technique Development and Design Improvement Methodology,* First Progress Report submitted to the Weapons Engineering Support Activity, Department of the Navy, Washington Navy Yard, Washington, D.C, under Contract NOO156–75–C–0944, 31 Jan. 1975, 202 pp.

2. Kececioglu, D., L. B. Chester, Koharchek, A., and Shehata, M.,*Aircraft Spline Reliability Predictive Technique Development and Design Improvement Methodology,* Second Progress Report submitted to the Weapons Engineering Support Activity, Department of the Navy, Washington Navy Yard, Washington, D.C, under Contract NOO156–75–C–0944, 31 Jan. 1975, 202 pp.

3. Heather, B.D., *The Evaluation and Practical Applications of FAMECA*, RADC–TR– 83–72, Rome Air Development Center, Griffiss Air Force Base, N.Y. 13341, AD/A131 358, Mar. 1983, 94 pp.

4. MIL–STD–1629A, *Procedures for Performing a Failure Modes, Effects and Criticality Analysis*, 24 Nov. 1980.

5. MIL–STD–2070 (AS), *Procedures for Performing a Failure Modes, Effects and Criticality Analysis for Aeronautical Equipment.*

APPENDICES

APPENDIX A
RANK TABLES

SAMPLE SIZE = 1

ORDER NUMBER	99.9	99.5	99.0	97.5	95.0	90.0	75.0	50.0	25.0	10.0	5.0	2.5	1.0	0.5	0.1	ORDER NUMBER
1	99.900	99.500	99.000	97.500	95.000	90.000	75.000	50.000	25.000	10.000	5.000	2.500	1.000	.500	.100	1
	0.1	0.5	1.0	2.5	5.0	10.0	25.0	50.0	75.0	90.0	95.0	97.5	99.0	99.5	99.9	

SAMPLE SIZE = 2

ORDER NUMBER	99.9	99.5	99.0	97.5	95.0	90.0	75.0	50.0	25.0	10.0	5.0	2.5	1.0	0.5	0.1	ORDER NUMBER
1	96.838	92.929	90.000	84.189	77.639	68.377	50.000	29.289	13.397	5.132	2.532	1.258	.501	.250	.050	2
2	99.950	99.750	99.499	98.742	97.468	94.868	86.603	70.711	50.000	31.623	22.361	15.811	10.000	7.071	3.162	1
	0.1	0.5	1.0	2.5	5.0	10.0	25.0	50.0	75.0	90.0	95.0	97.5	99.0	99.5	99.9	

SAMPLE SIZE = 3

ORDER NUMBER	99.9	99.5	99.0	97.5	95.0	90.0	75.0	50.0	25.0	10.0	5.0	2.5	1.0	0.5	0.1	ORDER NUMBER
1	90.000	82.900	78.456	70.760	63.160	53.584	37.004	20.630	9.144	3.451	1.695	.840	.334	.167	.033	3
2	98.164	95.860	94.109	90.569	86.465	80.421	67.365	50.001	32.635	19.579	13.535	9.431	5.891	4.140	1.836	2
3	99.967	99.833	99.666	99.160	98.305	96.549	90.856	79.370	62.996	46.416	36.840	29.240	21.544	17.100	10.000	1
	0.1	0.5	1.0	2.5	5.0	10.0	25.0	50.0	75.0	90.0	95.0	97.5	99.0	99.5	99.9	

SAMPLE SIZE = 4

ORDER NUMBER	99.9	99.5	99.0	97.5	95.0	90.0	75.0	50.0	25.0	10.0	5.0	2.5	1.0	0.5	0.1	ORDER NUMBER
1	82.217	73.409	68.377	60.236	52.713	43.766	29.289	15.910	6.940	2.600	1.274	.631	.251	.125	.025	4
2	93.597	88.912	85.914	80.589	75.140	67.954	54.368	38.573	24.302	14.256	9.762	6.759	4.200	2.944	1.302	3
3	98.698	97.056	95.800	93.241	90.238	85.744	75.698	61.427	45.632	32.046	24.860	19.411	14.086	11.088	6.403	2
4	99.975	99.875	99.749	99.369	98.726	97.400	93.060	84.090	70.711	56.234	47.287	39.764	31.623	26.591	17.783	1
	0.1	0.5	1.0	2.5	5.0	10.0	25.0	50.0	75.0	90.0	95.0	97.5	99.0	99.5	99.9	

SAMPLE SIZE = 5

ORDER NO.	99.9	99.5	99.0	97.5	95.0	90.0	75.0	50.0	25.0	10.0	5.0	2.5	1.0	0.5	0.1	ORDER NO.
5	74.881	65.343	60.189	52.182	45.072	36.904	24.214	12.945	5.591	2.085	1.021	0.505	0.201	0.100	0.020	5
4	87.798	81.490	77.793	71.642	65.740	58.389	45.419	31.380	19.376	11.224	7.644	5.274	3.268	2.288	1.011	4
3	95.245	91.717	89.436	85.337	81.074	75.336	64.056	50.001	35.944	24.664	18.926	14.663	10.564	8.283	4.755	3
2	98.989	97.712	96.732	94.726	92.356	88.776	80.624	68.620	54.581	41.611	34.260	28.358	22.207	18.510	12.202	2
1	99.980	99.900	99.799	99.495	98.979	97.915	94.409	87.055	75.786	63.096	54.928	47.818	39.811	34.657	25.119	1
	0.1	0.5	1.0	2.5	5.0	10.0	25.0	50.0	75.0	90.0	95.0	97.5	99.0	99.5	99.9	

SAMPLE SIZE = 6

ORDER NO.	99.9	99.5	99.0	97.5	95.0	90.0	75.0	50.0	25.0	10.0	5.0	2.5	1.0	0.5	0.1	ORDER NO.
6	68.377	58.648	53.584	45.926	39.304	31.871	20.630	10.910	4.682	1.741	0.851	0.421	0.167	0.084	0.017	6
5	81.861	74.601	70.568	64.123	58.181	51.032	38.947	26.444	16.116	9.260	6.284	4.327	2.676	1.871	0.826	5
4	90.604	85.641	82.693	77.723	72.866	66.680	55.320	42.141	29.691	20.090	15.316	11.811	8.472	6.628	3.791	4
3	96.209	93.372	91.528	88.189	84.684	79.910	70.309	57.859	44.680	33.320	27.134	22.277	17.307	14.359	9.396	3
2	99.174	98.129	97.324	95.673	93.716	90.740	83.884	73.556	61.053	48.968	41.819	35.877	29.432	25.399	18.139	2
1	99.983	99.916	99.833	99.579	99.149	98.259	95.318	89.090	79.370	68.129	60.696	54.074	46.416	41.352	31.623	1
	0.1	0.5	1.0	2.5	5.0	10.0	25.0	50.0	75.0	90.0	95.0	97.5	99.0	99.5	99.9	

SAMPLE SIZE = 7

ORDER NO.	99.9	99.5	99.0	97.5	95.0	90.0	75.0	50.0	25.0	10.0	5.0	2.5	1.0	0.5	0.1	ORDER NO.
7	62.724	53.088	48.205	40.962	34.816	28.031	17.966	9.428	4.026	1.494	0.730	0.361	0.143	0.072	0.014	7
6	76.252	68.491	64.336	57.873	52.070	45.257	34.071	22.849	13.798	7.882	5.338	3.669	2.267	1.585	0.698	6
5	85.622	79.704	76.368	70.957	65.874	59.618	48.609	36.411	25.307	16.964	12.876	9.899	7.081	5.531	3.156	5
4	92.335	88.230	85.773	81.594	77.468	72.140	62.115	49.999	37.885	27.860	22.532	18.406	14.227	11.770	7.665	4
3	96.844	94.469	92.919	90.101	87.124	83.036	74.693	63.589	51.391	40.382	34.126	29.043	23.632	20.296	14.378	3
2	99.302	98.415	97.733	96.331	94.662	92.118	86.202	77.151	65.929	54.743	47.930	42.127	35.664	31.509	23.748	2
1	99.986	99.928	99.857	99.639	99.270	98.506	95.974	90.572	82.034	71.969	65.184	59.038	51.795	46.912	37.276	1
	0.1	0.5	1.0	2.5	5.0	10.0	25.0	50.0	75.0	90.0	95.0	97.5	99.0	99.5	99.9	

SAMPLE SIZE = 8

ORDER NO.	0.1	0.5	1.0	2.5	5.0	10.0	25.0	50.0	75.0	90.0	95.0	97.5	99.0	99.5	99.9	ORDER NO.
1	0.013	0.063	0.126	0.316	0.639	1.308	3.532	8.300	15.910	25.011	31.234	36.942	43.766	48.433	57.830	8
2	0.605	1.374	1.966	3.185	4.639	6.863	12.063	20.113	30.270	40.624	47.068	52.651	58.994	63.152	71.128	7
3	2.705	4.746	6.084	8.523	11.111	14.686	22.057	32.052	43.320	53.822	59.969	65.086	70.676	74.216	80.730	6
4	6.483	9.986	12.095	15.701	19.291	23.966	32.908	44.016	55.549	65.537	71.076	75.513	80.180	83.030	88.042	5
5	11.958	16.970	19.820	24.487	28.924	34.463	44.451	55.984	67.092	76.034	80.709	84.299	87.905	90.014	93.517	4
6	19.270	25.784	29.324	34.914	40.031	46.178	56.680	67.948	77.943	85.314	88.889	91.477	93.916	95.254	97.295	3
7	28.872	36.848	41.006	47.349	52.932	59.376	69.730	79.887	87.937	93.137	95.361	96.815	98.034	98.626	99.395	2
8	42.170	51.567	56.234	63.058	68.766	74.989	84.090	91.700	96.468	98.692	99.361	99.684	99.874	99.937	99.987	1
	99.9	99.5	99.0	97.5	95.0	90.0	75.0	50.0	25.0	10.0	5.0	2.5	1.0	0.5	0.1	ORDER NO.

SAMPLE SIZE = 9

ORDER NO.	0.1	0.5	1.0	2.5	5.0	10.0	25.0	50.0	75.0	90.0	95.0	97.5	99.0	99.5	99.9	ORDER NO.
1	0.011	0.056	0.112	0.281	0.568	1.164	3.146	7.413	14.276	22.574	28.313	33.627	40.052	44.495	53.584	9
2	0.533	1.212	1.736	2.814	4.102	6.077	10.717	17.962	27.227	36.835	42.913	48.249	54.403	58.497	66.511	8
3	2.366	4.159	5.335	7.485	9.774	12.949	19.550	28.623	39.054	49.007	54.964	60.009	65.632	69.260	76.115	7
4	5.621	8.678	10.526	13.700	16.875	21.040	29.099	39.309	50.199	59.942	65.505	70.071	74.998	78.086	83.714	6
5	10.252	14.605	17.097	21.201	25.137	30.097	39.196	49.999	60.804	70.903	74.863	78.799	82.903	85.395	89.748	5
6	16.286	21.914	25.002	29.929	34.495	40.058	49.801	60.691	70.901	78.960	83.125	86.300	89.474	91.322	94.379	4
7	23.885	30.740	34.368	39.991	45.036	50.993	60.946	71.377	80.450	87.051	90.226	92.515	94.665	95.841	97.634	3
8	33.489	41.503	45.597	51.751	57.087	63.165	72.773	82.038	89.283	93.923	95.898	97.186	98.264	98.788	99.467	2
9	46.416	55.505	59.948	66.373	71.687	77.426	85.724	92.587	96.854	98.836	99.432	99.719	99.888	99.944	99.989	1
	99.9	99.5	99.0	97.5	95.0	90.0	75.0	50.0	25.0	10.0	5.0	2.5	1.0	0.5	0.1	ORDER NO.

SAMPLE SIZE = 10

ORDER NO.	0.1	0.5	1.0	2.5	5.0	10.0	25.0	50.0	75.0	90.0	95.0	97.5	99.0	99.5	99.9	ORDER
1	0.010	0.050	0.100	0.253	0.512	1.048	2.836	6.697	12.945	20.567	25.887	30.850	36.904	41.130	49.881	10
2	0.477	1.086	1.554	2.522	3.677	5.453	9.640	16.227	24.737	33.685	39.416	44.501	50.436	54.429	62.373	9
3	2.103	3.701	4.751	6.673	8.726	11.582	17.558	25.857	35.545	44.961	50.690	55.610	61.175	64.820	71.845	8
4	4.963	7.677	9.321	12.156	15.003	18.757	26.086	35.509	45.770	55.173	60.662	65.246	70.289	73.512	79.536	7
5	8.981	12.830	15.044	18.708	22.243	26.731	35.069	45.170	55.549	64.579	69.646	73.762	78.165	80.909	85.870	6
6	14.130	19.091	21.835	26.238	30.354	35.421	44.451	54.830	64.931	73.269	77.757	81.292	84.956	87.170	91.019	5
7	20.464	26.488	29.711	34.754	39.338	44.827	54.230	64.491	73.914	81.243	84.997	87.844	90.679	92.323	95.037	4
8	28.155	35.180	38.825	44.390	49.310	55.039	64.455	74.143	82.442	88.418	91.274	93.327	95.249	96.299	97.897	3
9	37.627	45.571	49.564	55.499	60.584	66.315	75.263	83.773	90.360	94.547	96.323	97.478	98.446	98.914	99.523	2
10	50.119	58.870	63.096	69.150	74.113	79.433	87.055	93.303	97.164	98.952	99.488	99.747	99.900	99.950	99.990	1
	99.9	99.5	99.0	97.5	95.0	90.0	75.0	50.0	25.0	10.0	5.0	2.5	1.0	0.5	0.1	ORDER

SAMPLE SIZE = 11

ORDER NO.	0.1	0.5	1.0	2.5	5.0	10.0	25.0	50.0	75.0	90.0	95.0	97.5	99.0	99.5	99.9	ORDER NO.
1	0.009	0.046	0.091	0.230	0.465	0.953	2.581	6.107	11.841	18.887	23.840	28.491	34.207	38.225	46.633	11
2	0.433	0.982	1.406	2.283	3.332	4.945	8.761	14.796	22.663	31.025	36.436	41.277	46.981	50.857	58.665	10
3	1.894	3.335	4.282	6.022	7.882	10.477	15.934	23.579	32.609	41.515	47.009	51.775	57.232	60.850	67.931	9
4	4.444	6.884	8.366	10.926	13.508	16.923	23.640	32.380	42.046	51.077	56.437	60.975	66.042	69.328	75.594	8
5	7.995	11.446	13.439	16.749	19.958	24.053	31.734	41.189	51.107	59.948	65.018	69.210	73.780	76.679	82.061	7
6	12.493	16.932	19.398	23.379	27.125	31.773	40.157	50.001	59.843	68.227	72.875	76.621	80.602	83.068	87.507	6
7	17.939	23.321	26.220	30.790	34.982	40.052	48.893	58.811	68.266	75.947	80.042	83.251	86.561	88.554	92.005	5
8	24.406	30.672	33.958	39.025	43.563	48.923	57.954	67.620	76.360	83.077	86.492	89.074	91.634	93.116	95.556	4
9	32.069	39.150	42.768	48.225	52.991	58.485	67.391	76.421	84.066	89.523	92.118	93.978	95.718	96.665	98.106	3
10	41.335	49.143	53.019	58.723	63.564	68.975	77.337	85.204	91.239	95.055	96.668	97.717	98.594	99.018	99.567	2
11	53.367	61.775	65.793	71.509	76.160	81.113	88.159	93.893	97.419	99.047	99.535	99.770	99.909	99.954	99.991	1
	99.9	99.5	99.0	97.5	95.0	90.0	75.0	50.0	25.0	10.0	5.0	2.5	1.0	0.5	0.1	

SAMPLE SIZE = 12

ORDER NO.	0.1	0.5	1.0	2.5	5.0	10.0	25.0	50.0	75.0	90.0	95.0	97.5	99.0	99.5	99.9	ORDER NO.
1	0.008	0.042	0.084	0.211	0.427	0.874	2.369	5.613	10.910	17.460	22.092	26.465	31.871	35.695	43.766	12
2	0.394	0.896	1.284	2.087	3.046	4.523	8.028	13.598	20.908	28.750	33.868	38.480	43.954	47.703	55.335	11
3	1.722	3.034	3.898	5.486	7.188	9.565	14.585	21.668	30.119	38.552	43.810	48.414	53.735	57.294	64.356	10
4	4.023	6.240	7.589	9.925	12.286	15.418	21.616	29.757	38.877	47.527	52.732	57.186	62.220	65.522	71.920	9
5	7.206	10.336	12.147	15.165	18.102	21.868	28.985	37.853	47.309	55.900	60.914	65.113	69.759	72.752	78.413	8
6	11.202	15.220	17.461	21.095	24.529	28.817	36.633	45.951	55.466	63.772	68.476	72.334	76.511	79.147	84.011	7
7	15.989	20.853	23.489	27.666	31.524	36.228	44.534	54.049	63.367	71.183	75.471	78.905	82.539	84.780	88.798	6
8	21.587	27.248	30.241	34.887	39.086	44.100	52.691	62.147	71.015	78.132	81.898	84.835	87.853	89.664	92.794	5
9	28.080	34.478	37.780	42.814	47.268	52.473	61.123	70.243	78.332	84.582	87.714	90.075	92.411	93.760	95.977	4
10	35.644	42.706	46.265	51.586	56.190	61.448	69.881	78.332	85.415	90.435	92.812	94.514	96.102	96.966	98.278	3
11	44.665	52.297	56.046	61.520	66.132	71.250	79.092	86.402	91.972	95.477	96.954	97.913	98.716	99.104	99.606	2
12	56.234	64.305	68.129	73.535	77.908	82.540	89.090	94.387	97.631	99.126	99.573	99.789	99.916	99.958	99.992	1
	99.9	99.5	99.0	97.5	95.0	90.0	75.0	50.0	25.0	10.0	5.0	2.5	1.0	0.5	0.1	

SAMPLE SIZE = 13

ORDER NO.	0.1	0.5	1.0	2.5	5.0	10.0	25.0	50.0	75.0	90.0	95.0	97.5	99.0	99.5	99.9	ORDER NO.
1	0.008	0.039	0.077	0.195	0.394	0.807	2.189	5.192	10.115	16.232	20.582	24.705	29.830	33.473	41.220	13
2	0.362	0.825	1.182	1.920	2.805	4.169	7.409	12.579	19.405	26.783	31.634	36.030	41.282	44.903	52.340	12
3	1.579	2.782	3.577	5.038	6.605	8.799	13.448	20.045	27.979	35.978	41.010	45.448	50.617	54.104	61.095	11
4	3.677	5.708	6.945	9.092	11.266	14.161	19.913	27.528	36.149	44.427	49.465	53.812	58.776	62.063	68.514	10
5	6.561	9.423	11.083	13.857	16.566	20.051	26.676	35.017	44.028	52.343	57.262	61.427	66.090	69.128	74.966	9
6	10.157	13.827	15.882	19.224	22.396	26.373	33.681	42.507	51.670	59.824	64.520	68.423	72.712	75.457	80.622	8
7	14.431	18.870	21.288	25.135	28.706	33.086	40.902	49.999	59.098	66.914	71.294	74.865	78.712	81.130	85.569	7
8	19.378	24.543	27.288	31.577	35.480	40.176	48.330	57.493	66.319	73.627	77.604	80.776	84.118	86.173	89.843	6
9	25.034	30.872	33.910	38.573	42.738	47.657	55.972	64.983	73.324	79.949	83.434	86.143	88.917	90.577	93.439	5
10	31.486	37.937	41.224	46.188	50.535	55.573	63.851	72.472	80.087	85.839	88.734	90.908	93.055	94.292	96.323	4
11	38.905	45.896	49.383	54.552	58.990	64.022	72.021	79.955	86.552	91.201	93.395	94.962	96.423	97.218	98.421	3
12	47.660	55.097	58.718	63.970	68.366	73.217	80.595	87.421	92.591	95.831	97.195	98.080	98.818	99.175	99.638	2
13	58.780	66.527	70.170	75.295	79.418	83.768	89.885	94.808	97.811	99.193	99.606	99.805	99.923	99.961	99.992	1
	99.9	99.5	99.0	97.5	95.0	90.0	75.0	50.0	25.0	10.0	5.0	2.5	1.0	0.5	0.1	ORDER NO.

SAMPLE SIZE = 14

ORDER NO.	0.1	0.5	1.0	2.5	5.0	10.0	25.0	50.0	75.0	90.0	95.0	97.5	99.0	99.5	99.9	ORDER NO.
1	0.007	0.036	0.072	0.181	0.366	0.750	2.034	4.830	9.428	15.166	19.264	23.164	28.031	31.508	38.946	14
2	0.336	0.764	1.095	1.780	2.599	3.866	6.879	11.703	18.104	25.068	29.673	33.869	38.909	42.402	49.635	13
3	1.458	2.570	3.306	4.658	6.110	8.147	12.475	18.647	26.122	33.721	38.538	42.812	47.826	51.231	58.117	12
4	3.385	5.259	6.403	8.389	10.404	13.094	18.459	25.608	33.775	41.698	46.566	50.797	55.666	58.918	65.366	11
5	6.022	8.660	10.192	12.760	15.272	18.513	24.709	32.575	41.167	49.197	54.000	58.103	62.743	65.795	71.732	10
6	9.293	12.672	14.568	17.661	20.608	24.316	31.174	39.544	48.351	56.310	60.958	64.861	69.202	72.015	77.384	9
7	13.155	17.240	19.473	23.035	26.359	30.456	37.824	46.514	55.350	63.087	67.496	71.139	75.120	77.657	82.407	8
8	17.593	22.343	24.880	28.861	32.504	36.913	44.650	53.486	62.176	69.544	73.641	76.965	80.527	82.760	86.845	7
9	22.616	27.985	30.798	35.139	39.042	43.690	51.649	60.456	68.826	75.684	79.392	82.339	85.432	87.328	90.707	6
10	28.268	34.205	37.257	41.897	46.000	50.803	58.833	67.425	75.291	81.487	84.728	87.240	89.808	91.340	93.978	5
11	34.634	41.082	44.334	49.203	53.434	58.302	66.225	74.392	81.541	86.906	89.596	91.611	93.597	94.741	96.615	4
12	41.883	48.769	52.174	57.188	61.462	66.279	73.878	81.353	87.525	91.853	93.890	95.342	96.694	97.430	98.542	3
13	50.365	57.598	61.091	66.131	70.327	74.932	81.896	88.297	93.121	96.134	97.401	98.220	98.905	99.236	99.664	2
14	61.054	68.492	71.969	76.836	80.736	84.834	90.572	95.170	97.966	99.250	99.634	99.819	99.928	99.964	99.993	1
	99.9	99.5	99.0	97.5	95.0	90.0	75.0	50.0	25.0	10.0	5.0	2.5	1.0	0.5	0.1	ORDER NO.

SAMPLE SIZE = 15

ORDER NO.	0.1	0.5	1.0	2.5	5.0	10.0	25.0	50.0	75.0	90.0	95.0	97.5	99.0	99.5	99.9	ORDER NO.
1	0.007	0.033	0.067	0.169	0.341	0.700	1.900	4.516	8.828	14.230	18.104	21.802	26.436	29.758	36.904	15
2	0.314	0.712	1.020	1.658	2.422	3.603	6.420	10.940	16.965	23.557	27.940	31.948	36.790	40.159	47.182	14
3	1.354	2.389	3.072	4.331	5.685	7.586	11.634	17.432	24.496	31.728	36.344	40.461	45.316	48.634	55.393	13
4	3.136	4.876	5.939	7.787	9.666	12.177	17.204	23.939	31.692	39.280	43.978	48.089	52.851	56.052	62.460	12
5	5.566	8.012	9.435	11.825	14.167	17.197	23.016	30.453	38.653	46.397	51.075	55.100	59.689	62.731	68.713	11
6	8.566	11.697	13.457	16.337	19.086	22.559	29.017	36.967	45.426	53.170	57.745	61.620	65.971	68.816	74.317	10
7	12.090	15.873	17.947	21.267	24.372	28.219	35.183	43.484	52.039	59.647	64.043	67.713	71.770	74.387	79.359	9
8	16.117	20.515	22.872	26.586	29.998	34.151	41.499	50.001	58.501	65.849	70.002	73.414	77.128	79.485	83.883	8
9	20.641	25.613	28.230	32.287	35.957	40.353	47.961	56.516	64.817	71.781	75.628	78.733	82.053	84.127	87.910	7
10	25.683	31.184	34.029	38.380	42.255	46.830	54.574	63.033	70.983	77.441	80.914	83.663	86.543	88.303	91.434	6
11	31.287	37.269	40.311	44.900	48.925	53.603	61.347	69.547	76.984	82.803	85.833	88.175	90.565	91.988	94.434	5
12	37.540	43.948	47.149	51.911	56.022	60.720	68.308	76.061	82.796	87.823	90.334	92.213	94.061	95.124	96.864	4
13	44.607	51.366	54.684	59.539	63.656	68.272	75.504	82.568	88.366	92.414	94.315	95.669	96.928	97.611	98.646	3
14	52.818	59.841	63.210	68.052	72.060	76.443	83.035	89.060	93.580	96.397	97.578	98.342	98.980	99.288	99.686	2
15	63.096	70.242	73.564	78.198	81.896	85.770	91.172	95.484	98.100	99.300	99.659	99.831	99.933	99.967	99.993	1
ORDER NO.	99.9	99.5	99.0	97.5	95.0	90.0	75.0	50.0	25.0	10.0	5.0	2.5	1.0	0.5	0.1	ORDER NO.

SAMPLE SIZE = 16

ORDER NO.	0.1	0.5	1.0	2.5	5.0	10.0	25.0	50.0	75.0	90.0	95.0	97.5	99.0	99.5	99.9	ORDER NO.
1	0.006	0.031	0.063	0.158	0.320	0.656	1.782	4.240	8.300	13.404	17.075	20.591	25.011	28.190	35.062	16
2	0.292	0.666	0.954	1.551	2.268	3.374	6.017	10.270	15.961	22.218	26.395	30.231	34.884	38.136	44.952	15
3	1.264	2.230	2.869	4.047	5.314	7.096	10.899	16.366	23.058	29.957	34.382	38.348	43.049	46.276	52.897	14
4	2.921	4.545	5.538	7.265	9.025	11.381	16.108	22.474	29.848	37.122	41.657	45.646	50.294	53.436	59.772	13
5	5.173	7.455	8.784	11.018	13.212	16.056	21.539	28.588	36.423	43.893	48.440	52.377	56.898	59.913	65.899	12
6	7.944	10.862	12.505	15.199	17.777	21.041	27.140	34.705	42.831	50.352	54.835	58.662	62.995	65.849	71.429	11
7	11.189	14.710	16.647	19.753	22.669	26.292	32.890	40.823	49.093	56.544	60.899	64.565	68.659	71.323	76.449	10
8	14.874	18.969	21.172	24.651	27.860	31.783	38.769	46.941	55.225	62.496	66.662	70.123	73.931	76.377	81.007	9
9	18.993	23.623	26.069	29.877	33.338	37.504	44.775	53.059	61.231	68.217	72.140	75.349	78.828	81.031	85.126	8
10	23.551	28.677	31.341	35.435	39.101	43.456	50.907	59.177	67.110	73.708	77.331	80.247	83.353	85.290	88.811	7
11	28.571	34.151	37.005	41.338	45.165	49.648	57.169	65.295	72.860	78.959	82.223	84.801	87.495	89.138	92.056	6
12	34.101	40.087	43.102	47.623	51.560	56.107	63.577	71.412	78.461	83.944	86.788	88.982	91.216	92.545	94.827	5
13	40.228	46.564	49.706	54.354	58.343	62.878	70.152	77.526	83.892	88.619	90.975	92.735	94.462	95.455	97.079	4
14	47.103	53.724	56.951	61.652	65.618	70.043	76.942	83.634	89.101	92.904	94.686	95.953	97.131	97.770	98.736	3
15	55.048	61.864	65.116	69.769	73.605	77.782	84.039	89.730	93.983	96.626	97.732	98.449	99.046	99.334	99.708	2
16	64.938	71.810	74.989	79.409	82.925	86.596	91.700	95.760	98.218	99.344	99.680	99.842	99.937	99.969	99.994	1
ORDER NO.	99.9	99.5	99.0	97.5	95.0	90.0	75.0	50.0	25.0	10.0	5.0	2.5	1.0	0.5	0.1	ORDER NO.

SAMPLE SIZE = 17

ORDER NO.	0.1	0.5	1.0	2.5	5.0	10.0	25.0	50.0	75.0	90.0	95.0	97.5	99.0	99.5	99.9	ORDER NO.
1	0.006	0.029	0.059	0.149	0.301	0.618	1.678	3.995	7.831	12.667	16.157	19.506	23.730	26.777	33.392	17
2	0.275	0.625	0.896	1.458	2.132	3.173	5.663	9.678	15.069	21.021	25.013	28.689	33.163	36.303	42.916	16
3	1.185	2.093	2.692	3.799	4.990	6.667	10.252	15.421	21.781	28.370	32.619	36.440	40.992	44.129	50.603	15
4	2.735	4.256	5.189	6.811	8.465	10.682	15.144	21.178	28.208	35.188	39.564	43.432	47.962	51.040	57.290	14
5	4.833	6.969	8.216	10.314	12.377	15.058	20.240	26.940	34.437	41.639	46.055	49.900	54.339	57.317	63.277	13
6	7.409	10.139	11.681	14.210	16.636	19.717	25.494	32.703	40.513	47.807	52.192	55.958	60.252	63.099	68.716	12
7	10.412	13.708	15.524	18.444	21.191	24.613	30.878	38.468	46.459	53.736	58.030	61.672	65.771	68.459	73.687	11
8	13.813	17.645	19.711	22.984	26.011	29.726	36.381	44.234	52.288	59.449	63.599	67.075	70.939	73.443	78.242	10
9	17.596	21.928	24.224	27.811	31.083	35.040	41.993	50.001	58.007	64.960	68.917	72.189	75.776	78.072	82.404	9
10	21.758	26.557	29.061	32.925	36.401	40.551	47.712	55.766	63.619	70.274	73.989	77.016	80.289	82.355	86.187	8
11	26.313	31.541	34.229	38.328	41.970	46.264	53.541	61.532	69.122	75.387	78.809	81.556	84.476	86.292	89.588	7
12	31.284	36.901	39.748	44.042	47.808	52.193	59.487	67.297	74.506	80.283	83.364	85.790	88.319	89.861	92.591	6
13	36.723	42.683	45.661	50.100	53.945	58.361	65.563	73.060	79.760	84.942	87.623	89.686	91.784	93.031	95.167	5
14	42.710	48.960	52.038	56.568	60.436	64.812	71.792	78.822	84.856	89.318	91.535	93.189	94.811	95.744	97.265	4
15	49.397	55.871	59.008	63.560	67.381	71.630	78.219	84.579	89.748	93.333	95.010	96.201	97.308	97.907	98.815	3
16	57.084	63.697	66.837	71.311	74.987	78.979	84.931	90.322	94.337	96.827	97.868	98.542	99.104	99.375	99.725	2
17	66.608	73.223	76.270	80.494	83.843	87.333	92.169	96.005	98.322	99.382	99.699	99.851	99.941	99.971	99.994	1
	99.9	99.5	99.0	97.5	95.0	90.0	75.0	50.0	25.0	10.0	5.0	2.5	1.0	0.5	0.1	ORDER NO.

SAMPLE SIZE = 18

ORDER NO.	0.1	0.5	1.0	2.5	5.0	10.0	25.0	50.0	75.0	90.0	95.0	97.5	99.0	99.5	99.9	ORDER NO.
1	0.006	0.028	0.056	0.141	0.285	0.584	1.586	3.778	7.413	12.008	15.332	18.530	22.574	25.499	31.871	18
2	0.259	0.590	0.846	1.376	2.010	2.995	5.347	9.150	14.271	19.947	23.766	27.294	31.602	34.635	41.051	17
3	1.116	1.971	2.535	3.579	4.702	6.286	9.676	14.581	20.637	26.942	31.026	34.711	39.118	42.167	48.492	16
4	2.570	4.002	4.879	6.409	7.969	10.064	14.289	20.023	26.737	33.442	37.669	41.418	45.831	48.841	54.990	15
5	4.536	6.544	7.719	9.695	11.643	14.176	19.091	25.471	32.655	39.602	43.888	47.637	51.989	54.923	60.835	14
6	6.940	9.507	10.958	13.343	15.635	18.549	24.035	30.921	38.430	45.502	49.783	53.480	57.720	60.548	66.169	13
7	9.737	12.835	14.544	17.298	19.895	23.139	29.101	36.372	44.088	51.183	55.404	59.008	63.091	65.786	71.075	12
8	12.894	16.494	18.441	21.529	24.397	27.923	34.270	41.822	49.642	56.672	60.784	64.256	68.142	70.683	75.600	11
9	16.393	20.466	22.630	26.019	29.121	32.885	39.539	47.274	55.099	61.980	65.940	69.242	72.898	75.260	79.772	10
10	20.228	24.740	27.102	30.758	34.060	38.020	44.901	52.726	60.461	67.115	70.879	73.981	77.370	79.534	83.607	9
11	24.400	29.317	31.858	35.744	39.216	43.328	50.358	58.178	65.730	72.077	75.603	78.471	81.559	83.506	87.106	8
12	28.925	34.214	36.909	40.992	44.596	48.817	55.912	63.628	70.899	76.861	80.105	82.702	85.456	87.165	90.263	7
13	33.831	39.452	42.280	46.520	50.217	54.498	61.570	69.079	75.965	81.451	84.365	86.657	89.042	90.493	93.060	6
14	39.165	45.077	48.011	52.363	56.112	60.398	67.345	74.529	80.909	85.824	88.357	90.305	92.281	93.456	95.464	5
15	45.010	51.159	54.169	58.582	62.331	66.558	73.263	79.977	85.711	89.936	92.031	93.591	95.121	95.998	97.430	4
16	51.508	57.833	60.882	65.289	68.974	73.058	79.363	85.419	90.324	93.714	95.298	96.421	97.465	98.029	98.884	3
17	58.949	65.365	68.398	72.706	76.234	80.053	85.729	90.850	94.653	97.005	97.990	98.624	99.121	99.154	99.741	2
18	68.129	74.501	77.426	81.470	84.668	87.992	92.587	96.222	98.414	99.416	99.715	99.859	99.944	99.972	99.994	1
	99.9	99.5	99.0	97.5	95.0	90.0	75.0	50.0	25.0	10.0	5.0	2.5	1.0	0.5	0.1	ORDER NO.

SAMPLE SIZE = 19

ORDER NO.	0.1	0.5	1.0	2.5	5.0	10.0	25.0	50.0	75.0	90.0	95.0	97.5	99.0	99.5	99.9	ORDER NO.
1	0.005	0.026	0.053	0.133	0.270	0.553	1.503	3.582	7.036	11.413	14.587	17.647	21.524	24.335	30.481	19
2	0.245	0.558	0.800	1.301	1.904	2.834	5.065	8.677	13.554	18.977	22.637	26.028	30.180	33.111	39.338	18
3	1.054	1.861	2.396	3.382	4.447	5.947	9.164	13.827	19.607	25.651	29.580	33.138	37.406	40.368	46.542	17
4	2.424	3.777	4.606	6.052	7.529	9.515	13.526	18.989	25.411	31.860	35.943	39.579	43.873	46.816	52.856	16
5	4.272	6.168	7.278	9.147	10.990	13.393	18.064	24.154	31.046	37.753	41.912	45.565	49.825	52.711	58.556	15
6	6.528	8.950	10.322	12.576	14.747	17.513	22.736	29.322	36.550	43.404	47.579	51.203	55.379	58.179	63.781	14
7	9.147	12.067	13.682	16.288	18.751	21.833	27.517	34.490	41.946	48.856	52.996	56.550	60.601	63.291	68.609	13
8	12.093	15.488	17.327	20.252	22.971	26.327	32.395	39.660	47.247	54.133	58.193	61.642	65.531	68.090	73.087	12
9	15.348	19.189	21.235	24.447	27.395	30.984	37.360	44.830	52.460	59.246	63.189	66.500	70.196	72.602	77.244	11
10	18.903	23.161	25.396	28.864	32.009	35.793	42.408	50.001	57.592	64.207	67.991	71.136	74.604	76.839	81.097	10
11	22.756	27.398	29.804	33.500	36.811	40.754	47.540	55.170	62.640	69.016	72.605	75.553	78.765	80.811	84.652	9
12	26.913	31.910	34.469	38.358	41.807	45.867	52.753	60.340	67.605	73.673	77.029	79.748	82.673	84.512	87.907	8
13	31.391	36.709	39.399	43.450	47.004	51.144	58.054	65.510	72.483	78.167	81.249	83.712	86.318	87.933	90.853	7
14	36.219	41.821	44.621	48.797	52.421	56.596	63.450	70.678	77.264	82.487	85.253	87.424	89.678	91.050	93.472	6
15	41.444	47.289	50.175	54.435	58.088	62.247	68.954	75.846	81.936	86.607	89.010	90.853	92.722	93.832	95.728	5
16	47.144	53.184	56.127	60.421	64.057	68.140	74.589	81.011	86.474	90.485	92.471	93.948	95.394	96.223	97.576	4
17	53.458	59.632	62.594	66.862	70.420	74.349	80.393	86.173	90.836	94.053	95.553	96.618	97.604	98.139	98.946	3
18	60.662	66.889	69.820	73.972	77.363	81.023	86.446	91.323	94.935	97.166	98.096	98.699	99.200	99.442	99.755	2
19	69.519	75.665	78.476	82.353	85.413	88.587	92.964	96.418	98.497	99.447	99.730	99.867	99.947	99.974	99.995	1
	99.9	99.5	99.0	97.5	95.0	90.0	75.0	50.0	25.0	10.0	5.0	2.5	1.0	0.5	0.1	ORDER NO.

ORDER NO.	99.9	99.5	99.0	97.5	95.0	90.0	75.0	50.0	25.0	10.0	5.0	2.5	1.0	0.5	0.1	ORDER NO.
1	29.205	23.273	20.567	16.843	13.911	10.875	6.697	3.406	1.428	0.525	0.256	0.127	0.050	0.025	0.005	20
2	37.759	31.715	28.880	24.873	21.610	18.096	12.905	8.251	4.812	2.691	1.806	1.235	0.759	0.530	0.233	19
3	44.738	38.712	35.833	31.698	28.262	24.476	18.674	13.148	8.701	5.642	4.217	3.207	2.271	1.765	0.999	18
4	50.874	44.947	42.072	37.893	34.367	30.419	24.210	18.055	12.840	9.022	7.136	5.733	4.362	3.576	2.294	17
5	56.431	50.661	47.828	43.661	40.102	36.066	29.588	22.967	17.142	12.693	10.409	8.657	6.884	5.833	4.038	16
6	61.543	55.976	53.211	49.105	45.559	41.489	34.844	27.880	21.571	16.587	13.955	11.893	9.754	8.456	6.164	15
7	66.283	60.961	58.286	54.279	50.782	46.726	40.000	32.795	26.098	20.666	17.731	15.391	12.917	11.388	8.624	14
8	70.702	65.656	63.094	59.219	55.804	51.803	45.069	37.710	30.715	24.906	21.706	19.118	16.341	14.599	11.387	13
9	74.827	70.090	67.658	63.946	60.642	56.733	50.060	42.626	35.410	29.293	25.864	23.058	20.005	18.066	14.431	12
10	78.674	74.277	71.992	68.472	65.307	61.524	54.975	47.543	40.182	33.817	30.195	27.197	23.896	21.775	17.747	11
11	82.253	78.225	76.104	72.803	69.805	66.183	59.818	52.457	45.025	38.476	34.693	31.528	28.008	25.723	21.326	10
12	85.569	81.934	79.995	76.942	74.136	70.707	64.590	57.374	49.940	43.267	39.358	36.054	32.342	29.910	25.173	9
13	88.613	85.401	83.659	80.882	78.294	75.094	69.285	62.290	54.931	48.197	44.196	40.781	36.906	34.344	29.298	8
14	91.376	88.612	87.063	84.609	82.269	79.334	73.902	67.205	60.000	53.274	49.218	45.721	41.714	39.039	33.717	7
15	93.836	91.544	90.246	88.107	86.045	83.413	78.429	72.120	65.156	58.511	54.441	50.895	46.789	44.024	38.457	6
16	95.962	94.167	93.116	91.343	89.591	87.307	82.858	77.033	70.412	63.934	59.898	56.339	52.172	49.339	43.569	5
17	97.706	96.424	95.638	94.267	92.864	90.978	87.160	81.945	75.790	69.581	65.633	62.107	57.928	55.053	49.126	4
18	99.001	98.235	97.729	96.793	95.783	94.358	91.299	86.852	81.326	75.524	71.738	68.302	64.167	61.288	55.262	3
19	99.767	99.470	99.241	98.765	98.194	97.309	95.188	91.749	87.095	81.904	78.390	75.127	71.120	68.285	62.241	2
20	99.995	99.975	99.950	99.873	99.744	99.475	98.572	96.594	93.303	89.125	86.089	83.157	79.433	76.727	70.795	1
ORDER NO.	0.1	0.5	1.0	2.5	5.0	10.0	25.0	50.0	75.0	90.0	95.0	97.5	99.0	99.5	99.9	ORDER NO.

SAMPLE SIZE = 21

ORDER NO.	99.9	99.5	99.0	97.5	95.0	90.0	75.0	50.0	25.0	10.0	5.0	2.5	1.0	0.5	0.1	ORDER NO.
1	28.031	22.299	19.691	16.110	13.295	10.385	6.388	3.247	1.361	0.500	0.244	0.120	0.048	0.024	0.005	21
2	36.298	30.429	27.684	23.816	20.673	17.294	12.315	7.864	4.583	2.562	1.719	1.175	0.722	0.504	0.221	20
3	43.063	37.185	34.386	30.378	27.055	23.405	17.828	12.531	8.283	5.367	4.009	3.049	2.158	1.676	0.948	19
4	49.026	43.217	40.411	36.343	32.922	29.102	23.118	17.210	12.220	8.578	6.780	5.447	4.142	3.394	2.178	18
5	54.443	48.757	45.979	41.906	38.441	34.522	28.260	21.891	16.311	12.061	9.884	8.218	6.532	5.534	3.828	17
6	59.440	53.924	51.199	47.166	43.697	39.733	33.289	26.574	20.518	15.755	13.245	11.282	9.246	8.012	5.837	16
7	64.091	58.782	56.130	52.175	48.739	44.772	38.226	31.258	24.819	19.619	16.817	14.588	12.235	10.781	8.157	15
8	68.443	63.374	60.816	56.968	53.594	49.660	43.082	35.943	29.202	23.632	20.576	18.107	15.464	13.807	10.758	14
9	72.521	67.722	65.276	61.564	58.280	54.417	47.866	40.629	33.656	27.779	24.499	21.819	18.913	17.068	13.618	13
10	76.347	71.847	69.527	65.980	62.811	59.046	52.583	45.315	38.178	32.052	28.580	25.713	22.567	20.548	16.724	12
11	79.929	75.755	73.579	70.219	67.190	63.557	57.233	50.001	42.767	36.443	32.810	29.781	26.421	24.245	20.071	11
12	83.276	79.452	77.433	74.287	71.420	67.948	61.822	54.685	47.417	40.954	37.189	34.020	30.473	28.153	23.653	10
13	86.382	82.932	81.087	78.181	75.501	72.221	66.344	59.371	52.134	45.583	41.720	38.436	34.724	32.278	27.479	9
14	89.242	86.193	84.536	81.893	79.424	76.368	70.798	64.057	56.918	50.340	46.406	43.032	39.184	36.626	31.557	8
15	91.843	89.219	87.765	85.412	83.183	80.381	75.181	68.742	61.774	55.228	51.261	47.825	43.870	41.218	35.909	7
16	94.163	91.988	90.754	88.718	86.755	84.245	79.482	73.426	66.711	60.267	56.303	52.834	48.801	46.076	40.560	6
17	96.172	94.466	93.468	91.782	90.116	87.939	83.689	78.109	71.740	65.478	61.559	58.094	54.021	51.243	45.557	5
18	97.822	96.606	95.858	94.553	93.220	91.422	87.780	82.790	76.882	70.898	67.078	63.657	59.589	56.783	50.974	4
19	99.052	98.324	97.842	96.951	95.991	94.633	91.717	87.469	82.172	76.595	72.945	69.622	65.614	62.815	56.937	3
20	99.779	99.496	99.278	98.825	98.281	97.438	95.417	92.136	87.685	82.706	79.327	76.184	72.316	69.571	63.702	2
21	99.995	99.976	99.952	99.880	99.756	99.500	98.639	96.753	93.612	89.615	86.705	83.890	80.309	77.701	71.969	1
	0.1	0.5	1.0	2.5	5.0	10.0	25.0	50.0	75.0	90.0	95.0	97.5	99.0	99.5	99.9	ORDER NO.

517

SAMPLE SIZE = 22

ORDER NO.	0.1	0.5	1.0	2.5	5.0	10.0	25.0	50.0	75.0	90.0	95.0	97.5	99.0	99.5	99.9	ORDER NO.
1	0.005	0.023	0.046	0.115	0.233	0.478	1.299	3.102	6.107	9.937	12.731	15.437	18.887	21.403	26.947	22
2	0.211	0.480	0.689	1.121	1.640	2.444	4.374	7.512	11.777	16.559	19.812	22.844	26.584	29.243	34.945	21
3	0.904	1.597	2.056	2.906	3.823	5.117	7.905	11.970	17.052	22.422	25.947	29.162	33.050	35.772	41.506	20
4	2.071	3.231	3.944	5.187	6.460	8.175	11.658	16.439	22.120	27.894	31.591	34.913	38.873	41.611	47.305	19
5	3.640	5.262	6.214	7.821	9.411	11.491	15.557	20.911	27.045	33.105	36.909	40.284	44.263	46.987	52.583	18
6	5.543	7.613	8.790	10.729	12.603	15.002	19.564	25.384	31.866	38.117	41.981	45.370	49.326	52.010	57.467	17
7	7.740	10.236	11.620	13.865	15.993	18.674	23.660	29.859	36.600	42.970	46.850	50.222	54.121	56.744	62.025	16
8	10.197	13.097	14.677	17.197	19.556	22.483	27.831	34.335	41.261	47.684	51.545	54.873	58.685	61.228	66.302	15
9	12.893	16.175	17.934	20.710	23.272	26.417	32.069	38.810	45.853	52.274	56.087	59.342	63.041	65.490	70.327	14
10	15.816	19.456	21.380	24.386	27.131	30.463	36.369	43.285	50.385	56.751	60.485	63.645	67.205	69.544	74.119	13
11	18.957	22.932	25.008	28.220	31.126	34.618	40.726	47.762	54.858	61.120	64.745	67.789	71.188	73.402	77.688	12
12	22.312	26.598	28.812	32.211	35.255	38.880	45.142	52.238	59.274	65.382	68.874	71.780	74.992	77.068	81.043	11
13	25.881	30.456	32.795	36.355	39.515	43.249	49.615	56.715	63.631	69.537	72.869	75.614	78.620	80.544	84.184	10
14	29.673	34.510	36.959	40.658	43.913	47.726	54.147	61.190	67.931	73.583	76.728	79.290	82.066	83.825	87.107	9
15	33.698	38.772	41.315	45.127	48.455	52.316	58.739	65.665	72.169	77.517	80.444	82.803	85.323	86.903	89.803	8
16	37.975	43.256	45.879	49.778	53.150	57.030	63.400	70.141	76.340	81.326	84.007	86.135	88.380	89.764	92.260	7
17	42.533	47.990	50.674	54.630	58.019	61.883	68.134	74.616	80.436	84.998	87.397	89.271	91.210	92.387	94.457	6
18	47.417	53.013	55.737	59.716	63.091	66.895	72.955	79.089	84.443	88.509	90.589	92.179	93.786	94.738	96.360	5
19	52.695	58.389	61.127	65.087	68.409	72.106	77.880	83.561	88.342	91.825	93.540	94.813	96.056	96.769	97.929	4
20	58.494	64.228	66.950	70.838	74.053	77.578	82.948	88.030	92.095	94.883	96.177	97.094	97.944	98.403	99.096	3
21	65.055	70.757	73.416	77.156	80.188	83.441	88.223	92.488	95.626	97.556	98.360	98.879	99.311	99.520	99.789	2
22	73.053	78.597	81.113	84.563	87.269	90.063	93.893	96.898	98.701	99.522	99.767	99.885	99.954	99.977	99.995	1
	99.9	99.5	99.0	97.5	95.0	90.0	75.0	50.0	25.0	10.0	5.0	2.5	1.0	0.5	0.1	ORDER NO.

SAMPLE SIZE = 23

ORDER NO.	0.1	0.5	1.0	2.5	5.0	10.0	25.0	50.0	75.0	90.0	95.0	97.5	99.0	99.5	99.9	ORDER NO.
1	0.004	0.022	0.044	0.110	0.223	0.457	1.243	2.969	5.849	9.526	12.212	14.819	18.145	20.575	25.943	23
2	0.202	0.459	0.658	1.071	1.567	2.337	4.184	7.191	11.284	15.884	19.020	21.949	25.567	28.144	33.687	22
3	0.863	1.525	1.965	2.775	3.652	4.890	7.558	11.457	16.343	21.519	24.924	28.037	31.812	34.460	40.055	21
4	1.977	3.083	3.762	4.951	6.167	7.809	11.144	15.734	21.203	26.781	30.364	33.588	37.446	40.118	45.693	20
5	3.468	5.016	5.926	7.461	8.981	10.972	14.869	20.014	25.932	31.797	35.493	38.781	42.667	45.336	50.840	19
6	5.277	7.253	8.375	10.229	12.022	14.318	18.696	24.297	30.560	36.626	40.389	43.703	47.581	50.220	55.610	18
7	7.363	9.743	11.066	13.210	15.247	17.815	22.605	28.580	35.107	41.305	45.098	48.405	52.242	54.833	60.075	17
8	9.692	12.457	13.966	16.376	18.635	21.441	26.583	32.864	39.585	45.855	49.644	52.920	56.687	59.214	64.274	16
9	12.243	15.372	17.052	19.707	22.164	25.182	30.625	37.147	44.001	50.291	54.046	57.265	60.940	63.384	68.240	15
10	15.003	18.475	20.315	23.191	25.825	29.028	34.722	41.431	48.362	54.621	58.315	61.459	65.015	67.365	71.989	14
11	17.963	21.755	23.742	26.820	29.609	32.972	38.874	45.716	52.668	58.852	62.461	65.505	68.923	71.165	75.533	13
12	21.117	25.210	27.329	30.589	33.515	37.011	43.078	50.001	56.922	62.989	66.485	69.411	72.671	74.790	78.883	12
13	24.467	28.835	31.077	34.495	37.539	41.148	47.332	54.284	61.126	67.028	70.391	73.180	76.258	78.245	82.037	11
14	28.011	32.635	34.985	38.541	41.685	45.379	51.638	58.569	65.278	70.972	74.175	76.809	79.685	81.525	84.997	10
15	31.760	36.616	39.060	42.735	45.954	49.709	55.999	62.853	69.375	74.818	77.836	80.293	82.948	84.628	87.757	9
16	35.726	40.786	43.313	47.080	50.356	54.145	60.415	67.136	73.417	78.559	81.365	83.624	86.034	87.543	90.308	8
17	39.925	45.167	47.758	51.595	54.902	58.695	64.893	71.420	77.395	82.185	84.753	86.790	88.934	90.257	92.637	7
18	44.390	49.780	52.419	56.297	59.611	63.374	69.440	75.703	81.304	85.682	87.978	89.771	91.625	92.747	94.723	6
19	49.160	54.664	57.333	61.219	64.507	68.203	74.068	79.986	85.131	89.028	91.019	92.539	94.074	94.984	96.532	5
20	54.307	59.882	62.554	66.412	69.636	73.219	78.797	84.266	88.856	92.191	93.833	95.049	96.238	96.917	98.023	4
21	59.945	65.540	68.188	71.963	75.076	78.481	83.657	88.543	92.442	95.110	96.348	97.225	98.035	98.475	99.137	3
22	66.313	71.856	74.433	78.051	80.980	84.116	88.716	92.809	95.816	97.663	98.433	98.929	99.342	99.541	99.798	2
23	74.057	79.425	81.855	85.181	87.788	90.474	94.151	97.031	98.757	99.543	99.777	99.890	99.956	99.978	99.996	1
	99.9	99.5	99.0	97.5	95.0	90.0	75.0	50.0	25.0	10.0	5.0	2.5	1.0	0.5	0.1	ORDER NO.

SAMPLE SIZE = 24

ORDER NO.	0.1	0.5	1.0	2.5	5.0	10.0	25.0	50.0	75.0	90.0	95.0	97.5	99.0	99.5	99.9	ORDER NO.
1	0.004	0.021	0.042	0.105	0.213	0.438	1.192	2.847	5.613	9.148	11.735	14.247	17.460	19.809	25.011	24
2	0.193	0.440	0.630	1.026	1.501	2.238	4.009	6.895	10.830	15.262	18.289	21.120	24.625	27.125	32.516	23
3	0.826	1.460	1.879	2.656	3.495	4.682	7.241	10.987	15.690	20.686	23.980	26.997	30.663	33.239	38.700	22
4	1.890	2.947	3.597	4.736	5.901	7.473	10.676	15.087	20.361	25.755	29.227	32.362	36.117	38.726	44.186	21
5	3.312	4.794	5.662	7.131	8.588	10.497	14.240	19.192	24.905	30.589	34.180	37.385	41.181	43.795	49.203	20
6	5.038	6.925	7.999	9.772	11.491	13.694	17.901	23.298	29.356	35.247	38.914	42.152	45.953	48.547	53.863	19
7	7.021	9.296	10.561	12.615	14.568	17.034	21.639	27.406	33.730	39.764	43.468	46.711	50.484	53.042	58.233	18
8	9.234	11.878	13.320	15.630	17.796	20.493	25.443	31.513	38.039	44.160	47.872	51.095	54.815	57.317	62.354	17
9	11.655	14.646	16.254	18.800	21.157	24.058	29.305	35.621	42.292	48.449	52.142	55.322	58.965	61.399	66.256	16
10	14.271	17.590	19.350	22.109	24.639	27.721	33.221	39.729	46.493	52.641	56.289	59.406	62.951	65.302	69.956	15
11	17.069	20.696	22.599	25.553	28.236	31.475	37.185	43.838	50.643	56.742	60.322	63.357	66.782	69.039	73.467	14
12	20.048	23.962	25.994	29.124	31.942	35.317	41.196	47.945	54.746	60.755	64.243	67.179	70.466	72.617	76.798	13
13	23.202	27.383	29.534	32.821	35.757	39.245	45.254	52.055	58.804	64.683	68.058	70.876	74.006	76.038	79.952	12
14	26.533	30.961	33.218	36.643	39.678	43.258	49.357	56.162	62.815	68.525	71.764	74.447	77.401	79.304	82.931	11
15	30.044	34.698	37.049	40.594	43.711	47.359	53.507	60.271	66.779	72.279	75.361	77.891	80.650	82.410	85.729	10
16	33.744	38.601	41.035	44.678	47.858	51.551	57.708	64.379	70.695	75.942	78.843	81.200	83.746	85.354	88.345	9
17	37.646	42.683	45.185	48.905	52.128	55.840	61.961	68.487	74.557	79.507	82.204	84.370	86.680	88.122	90.766	8
18	41.767	46.958	49.516	53.289	56.532	60.236	66.270	72.594	78.361	82.966	85.432	87.385	89.439	90.704	92.979	7
19	46.137	51.453	54.047	57.848	61.086	64.753	70.644	76.702	82.099	86.306	88.509	90.228	92.001	93.075	94.962	6
20	50.797	56.205	58.819	62.615	65.820	69.411	75.095	80.808	85.760	89.503	91.412	92.869	94.338	95.206	96.688	5
21	55.814	61.274	63.883	67.638	70.773	74.245	79.639	84.913	89.324	92.527	94.099	95.264	96.403	97.053	98.110	4
22	61.300	66.761	69.337	73.003	76.020	79.314	84.310	89.013	92.759	95.318	96.505	97.344	98.121	98.540	99.174	3
23	67.484	72.875	75.375	78.880	81.711	84.738	89.170	93.105	95.991	97.762	98.499	98.974	99.370	99.560	99.807	2
24	74.989	80.191	82.540	85.753	88.265	90.852	94.387	97.153	98.808	99.562	99.787	99.895	99.958	99.979	99.996	1
	99.9	99.5	99.0	97.5	95.0	90.0	75.0	50.0	25.0	10.0	5.0	2.5	1.0	0.5	0.1	

SAMPLE SIZE = 25

ORDER NO.	0.1	0.5	1.0	2.5	5.0	10.0	25.0	50.0	75.0	90.0	95.0	97.5	99.0	99.5	99.9	ORDER NO.
1	0.004	0.020	0.040	0.101	0.205	0.421	1.144	2.735	5.394	8.799	11.293	13.719	16.824	19.098	24.142	25
2	0.185	0.422	0.605	0.984	1.440	2.148	3.849	6.623	10.412	14.687	17.612	20.352	23.749	26.176	31.422	24
3	0.791	1.398	1.801	2.546	3.352	4.491	6.950	10.554	15.087	19.913	23.104	26.031	29.594	32.101	37.431	23
4	1.809	2.824	3.448	4.537	5.656	7.166	10.244	14.492	19.581	24.802	28.172	31.219	34.878	37.426	42.771	22
5	3.170	4.589	5.422	6.831	8.230	10.062	13.662	18.435	23.957	29.467	32.961	36.083	39.793	42.352	47.663	21
6	4.816	6.625	7.655	9.356	11.005	13.123	17.171	22.379	28.242	33.965	37.540	40.704	44.427	46.976	52.216	20
7	6.709	8.889	10.102	12.072	13.947	16.317	20.754	26.324	32.456	38.331	41.952	45.129	48.838	51.357	56.493	19
8	8.817	11.350	12.733	14.950	17.031	19.625	24.398	30.270	36.610	42.582	46.221	49.387	53.056	55.532	60.535	18
9	11.121	13.987	15.530	17.971	20.238	23.031	28.097	34.216	40.710	46.734	50.364	53.500	57.107	59.524	64.371	17
10	13.607	16.787	18.476	21.125	23.559	26.530	31.844	38.161	44.761	50.796	54.394	57.479	61.002	63.349	68.018	16
11	16.264	19.738	21.563	24.403	26.986	30.111	35.638	42.107	48.766	54.772	58.317	61.335	64.756	67.020	71.490	15
12	19.083	22.836	24.786	27.796	30.512	33.773	39.474	46.053	52.728	58.668	62.138	65.072	68.374	70.545	74.793	14
13	22.065	26.075	28.141	31.306	34.139	37.514	43.353	50.001	56.647	62.486	65.861	68.694	71.859	73.925	77.935	13
14	25.207	29.455	31.626	34.928	37.862	41.332	47.272	53.947	60.526	66.227	69.488	72.204	75.214	77.164	80.917	12
15	28.510	32.980	35.244	38.665	41.683	45.228	51.234	57.893	64.362	69.889	73.014	75.597	78.437	80.262	83.736	11
16	31.982	36.651	38.998	42.521	45.606	49.204	55.239	61.839	68.156	73.470	76.441	78.875	81.524	83.213	86.393	10
17	35.629	40.476	42.893	46.500	49.636	53.266	59.290	65.784	71.903	76.969	79.762	82.029	84.470	86.013	88.879	9
18	39.465	44.468	46.944	50.613	53.779	57.418	63.390	69.730	75.602	80.375	82.969	85.050	87.267	88.650	91.183	8
19	43.507	48.643	51.162	54.871	58.048	61.669	67.544	73.676	79.246	83.683	86.053	87.928	89.898	91.111	93.291	7
20	47.784	53.024	55.573	59.296	62.460	66.035	71.758	77.621	82.829	86.877	88.995	90.644	92.345	93.375	95.184	6
21	52.337	57.648	60.207	63.917	67.039	70.533	76.043	81.565	86.338	89.938	91.770	93.169	94.578	95.411	96.830	5
22	57.229	62.574	65.122	68.781	71.828	75.198	80.419	85.508	89.756	92.834	94.344	95.463	96.552	97.176	98.191	4
23	62.569	67.899	70.406	73.969	76.896	80.087	84.913	89.446	93.050	95.509	96.648	97.454	98.199	98.602	99.209	3
24	68.578	73.824	76.251	79.648	82.388	85.313	89.588	93.377	96.151	97.852	98.560	99.016	99.395	99.578	99.815	2
25	75.858	80.902	83.176	86.281	88.707	91.201	94.606	97.265	98.856	99.579	99.795	99.899	99.960	99.980	99.996	1
	99.9	99.5	99.0	97.5	95.0	90.0	75.0	50.0	25.0	10.0	5.0	2.5	1.0	0.5	0.1	ORDER NO.

APPENDIX B

STANDARDIZED NORMAL DISTRIBUTION'S AREA TABLES

$$1 - F(z) = \int_z^\infty \frac{1}{\sqrt{2\pi}} e^{-\frac{t^2}{2}} \, dt.$$

USAGE

For $z \geq 3.0$, entries are in abbreviated notation x. x x x x $- p$, where

$$1 - F(z) = \text{x. x x x x} \times 10^{-p}.$$

EXAMPLE 1

X is normally distributed with mean $\mu = 27$ and standard deviation $\sigma = 4$. What is the probability X will exceed 41?

Answer: $z = (41 - 27)/4 = 3.50$,

$$\begin{aligned} P(X \geq 41) &= 1 - F(3.50), \\ &= 2.3263 \times 10^{-4} = 0.00023263. \end{aligned}$$

EXAMPLE 2

From Example 1, what is the probability that X will be less than 41?

$$\begin{aligned} P(X < 41) &= 1 - [1 - F(3.50)], \\ &= 1 - 0.00023263, \\ &= 0.99976737. \end{aligned}$$

z	$1 - F(z)$	z	$1 - F(z)$	z	$1 - F(z)$	z	$1 - F(z)$
0.00	0.50000	0.40	0.34458	0.80	0.21186	1.20	0.11507
0.01	0.49601	0.41	0.34090	0.81	0.20897	1.21	0.11314
0.02	0.49202	0.42	0.33724	0.82	0.20611	1.22	0.11123
0.03	0.48803	0.43	0.33360	0.83	0.20327	1.23	0.10935
0.04	0.48405	0.44	0.32997	0.84	0.20045	1.24	0.10749
0.05	0.48006	0.45	0.32636	0.85	0.19766	1.25	0.10565
0.06	0.47608	0.46	0.32276	0.86	0.19489	1.26	0.10383
0.07	0.47210	0.47	0.31918	0.87	0.19215	1.27	0.10204
0.08	0.46812	0.48	0.31561	0.88	0.18943	1.28	0.10027
0.09	0.46414	0.49	0.31207	0.89	0.18673	1.29	0.098525
0.10	0.46017	0.50	0.30854	0.90	0.18406	1.30	0.096800
0.11	0.45620	0.51	0.30503	0.91	0.18141	1.31	0.095093
0.12	0.45224	0.52	0.30153	0.92	0.17879	1.32	0.093418
0.13	0.44828	0.53	0.29806	0.93	0.17619	1.33	0.091759
0.14	0.44433	0.54	0.29460	0.94	0.17361	1.34	0.090123
0.15	0.44038	0.55	0.29116	0.95	0.17106	1.35	0.088508
0.16	0.43644	0.56	0.28774	0.96	0.16853	1.36	0.086915
0.17	0.43251	0.57	0.28434	0.97	0.16602	1.37	0.085343
0.18	0.42858	0.58	0.28096	0.98	0.16354	1.38	0.083793
0.19	0.42465	0.59	0.27760	0.99	0.16109	1.39	0.082264
0.20	0.42074	0.60	0.27425	1.00	0.15866	1.40	0.080757
0.21	0.41683	0.61	0.27093	1.01	0.15625	1.41	0.079270
0.22	0.41294	0.62	0.26763	1.02	0.15386	1.42	0.077804
0.23	0.40905	0.63	0.26435	1.03	0.15151	1.43	0.076359
0.24	0.40517	0.64	0.26109	1.04	0.14917	1.44	0.074934
0.25	0.40129	0.65	0.25785	1.05	0.14686	1.45	0.073529
0.26	0.39743	0.66	0.25463	1.06	0.14457	1.46	0.072145
0.27	0.39358	0.67	0.25143	1.07	0.14231	1.47	0.070781
0.28	0.38974	0.68	0.24825	1.08	0.14007	1.48	0.069437
0.29	0.38591	0.69	0.24510	1.09	0.13786	1.49	0.068112
0.30	0.38209	0.70	0.24196	1.10	0.13567	1.50	0.066807
0.31	0.37828	0.71	0.23885	1.11	0.13350	1.51	0.065522
0.32	0.37448	0.72	0.23576	1.12	0.13136	1.52	0.064255
0.33	0.37070	0.73	0.23270	1.13	0.12924	1.53	0.063008
0.34	0.36693	0.74	0.22965	1.14	0.12714	1.54	0.061780
0.35	0.36317	0.75	0.22663	1.15	0.12507	1.55	0.060571
0.36	0.35942	0.76	0.22363	1.16	0.12302	1.56	0.059380
0.37	0.35569	0.77	0.22065	1.17	0.12100	1.57	0.058208
0.38	0.35197	0.78	0.21770	1.18	0.11900	1.58	0.057053
0.39	0.34827	0.79	0.21476	1.19	0.11702	1.59	0.055917

z	$1 - F(z)$	z	$1 - F(z)$	z	$1 - F(z)$	z	$1 - F(z)$
1.60	0.054799	2.00	0.022750	2.40	0.0081975	2.80	0.0025551
1.61	0.053699	2.01	0.022216	2.41	0.0079763	2.81	0.0024771
1.62	0.052616	2.02	0.021692	2.42	0.0077603	2.82	0.0024012
1.63	0.051551	2.03	0.021178	2.43	0.0075494	2.83	0.0023274
1.64	0.050503	2.04	0.020675	2.44	0.0073436	2.84	0.0022557
1.65	0.049471	2.05	0.020182	2.45	0.0071428	2.85	0.0021860
1.66	0.048457	2.06	0.019699	2.46	0.0069469	2.86	0.0021182
1.67	0.047460	2.07	0.019226	2.47	0.0067557	2.87	0.0020524
1.68	0.046479	2.08	0.018763	2.48	0.0065691	2.88	0.0019884
1.69	0.045514	2.09	0.018309	2.49	0.0063872	2.89	0.0019262
1.70	0.044565	2.10	0.017864	2.50	0.0062097	2.90	0.0018658
1.71	0.043633	2.11	0.017429	2.51	0.0060366	2.91	0.0018071
1.72	0.042716	2.12	0.017003	2.52	0.0058677	2.92	0.0017502
1.73	0.041815	2.13	0.016586	2.53	0.0057031	2.93	0.0016948
1.74	0.040930	2.14	0.016177	2.54	0.0055426	2.94	0.0016411
1.75	0.040059	2.15	0.015778	2.55	0.0053861	2.95	0.0015889
1.76	0.039204	2.16	0.015386	2.56	0.0052336	2.96	0.0015382
1.77	0.038364	2.17	0.015003	2.57	0.0050849	2.97	0.0014890
1.78	0.037538	2.18	0.014629	2.58	0.0049400	2.98	0.0014412
1.79	0.036727	2.19	0.014262	2.59	0.0047988	2.99	0.0013949
1.80	0.035930	2.20	0.013903	2.60	0.0046612	3.00	1.3499 -3
1.81	0.035148	2.21	0.013553	2.61	0.0045271	3.01	1.3062
1.82	0.034380	2.22	0.013209	2.62	0.0043965	3.02	1.2639
1.83	0.033625	2.23	0.012874	2.63	0.0042692	3.03	1.2228
1.84	0.032884	2.24	0.012545	2.64	0.0041453	3.04	1.1829
1.85	0.032157	2.25	0.012224	2.65	0.0040246	3.05	1.1442 -3
1.86	0.031443	2.26	0.011911	2.66	0.0039070	3.06	1.1067
1.87	0.030742	2.27	0.011604	2.67	0.0037926	3.07	1.0703
1.88	0.030054	2.28	0.011304	2.68	0.0036811	3.08	1.0350
1.89	0.029379	2.29	0.011011	2.69	0.0035726	3.09	1.0008
1.90	0.028717	2.30	0.010724	2.70	0.0034670	3.10	9.6760 -4
1.91	0.028067	2.31	0.010444	2.71	0.0033642	3.11	9.3544
1.92	0.027429	2.32	0.010170	2.72	0.0032641	3.12	9.0426
1.93	0.026803	2.33	0.099031	2.73	0.0031667	3.13	8.7403
1.94	0.026190	2.34	0.096419	2.74	0.0030720	3.14	8.4474
1.95	0.025588	2.35	0.0093867	2.75	0.0029798	3.15	8.1635 -4
1.96	0.024998	2.36	0.0091375	2.76	0.0028901	3.16	7.8885
1.97	0.024419	2.37	0.0088940	2.77	0.0028028	3.17	7.6219
1.98	0.023852	2.38	0.0086563	2.78	0.0027179	3.18	7.3638
1.99	0.023295	2.39	0.0084242	2.79	0.0026354	3.19	7.1136

z	$1 - F(z)$	z	$1 - F(z)$	z	$1 - F(z)$	z	$1 - F(z)$
3.20	6.8714 -4	3.60	1.5911 -4	4.00	3.1671 -5	4.40	5.4125 -6
3.21	6.6367	3.61	1.5310	4.01	3.0359	4.41	5.1685
3.22	6.4095	3.62	1.4730	4.02	2.9099	4.42	4.9350
3.23	6.1895	3.63	1.4171	4.03	2.7888	4.43	4.7117
3.24	5.9765	3.64	1.3632	4.04	2.6726	4.44	4.4979
3.25	5.7703 -4	3.65	1.3112 -4	4.05	2.5609 -5	4.45	4.2935 -6
3.26	5.5706	3.66	1.2611	4.06	2.4536	4.46	4.0980
3.27	5.3774	3.67	1.2128	4.07	2.3507	4.47	3.9110
3.28	5.1904	3.68	1.1662	4.08	2.2518	4.48	3.7322
3.29	5.0094	3.69	1.1213	4.09	2.1569	4.49	3.5612
3.30	4.8342 -4	3.70	1.0780 -4	4.10	2.0658 -5	4.50	3.3977 -6
3.31	4.6648	3.71	1.0363	4.11	1.9783	4.51	3.2414
3.32	4.5009	3.72	9.9611 -5	4.12	1.8944	4.52	3.0920
3.33	4.3423	3.73	9.5740	4.13	1.8138	4.53	2.9492
3.34	4.1889	3.74	9.2010	4.14	1.7365	4.54	2.8127
3.35	4.0406 -4	3.75	8.8417 -5	4.15	1.6624 -5	4.55	2.6823 -6
3.36	3.8971	3.76	8.4957	4.16	1.5912	4.56	2.5577
3.37	3.7584	3.77	8.1624	4.17	1.5230	4.57	2.4386
3.38	3.6243	3.78	7.8414	4.18	1.4575	4.58	2.3249
3.39	3.4946	3.79	7.5324	4.19	1.3948	4.59	2.2162
3.40	3.3693 -4	3.80	7.2348 -5	4.20	1.3346 5	4.60	2.1125 -6
3.41	3.2481	3.81	6.9483	4.21	1.2769	4.61	2.0133
3.42	3.1311	3.82	6.6726	4.22	1.2215	4.62	1.9187
3.43	3.0179	3.83	6.4072	4.23	1.1685	4.63	1.8283
3.44	2.9086	3.84	6.1517	4.24	1.1176	4.64	1.7420
3.45	2.8029 -4	3.85	5.9059 -5	4.25	1.0689 -5	4.65	1.6597 -6
3.46	2.7009	3.86	5.6694	4.26	1.0221	4.66	1.5810
3.47	2.6023	3.87	5.4418	4.27	9.7736 -6	4.67	1.5060
3.48	2.5071	3.88	5.2228	4.28	9.3447	4.68	1.4344
3.49	2.4151	3.89	5.0122	4.29	8.9337	4.69	1.3660
3.50	2.3263 -4	3.90	4.8096 -5	4.30	8.5399 -6	4.70	1.3008 -6
3.51	2.2405	3.91	4.6148	4.31	8.1627	4.71	1.2386
3.52	2.1577	3.92	4.4274	4.32	7.8015	4.72	1.1792
3.53	2.0778	3.93	4.2473	4.33	7.4555	4.73	1.1226
3.54	2.0006	3.94	4.0741	4.34	7.1241	4.74	1.0686
3.55	1.9262 -4	3.95	3.9076 -5	4.35	6.8069 -6	4.75	1.0171 -6
3.56	1.8543	3.96	3.7475	4.36	6.5031	4.76	9.6796 -7
3.57	1.7849	3.97	3.5936	4.37	6.2123	4.77	9.2113
3.58	1.7180	3.98	3.4458	4.38	5.9340	4.78	8.7648
3.59	1.6534	3.99	3.3037	4.39	5.6675	4.79	8.3391

z	$1 - F(z)$	z	$1 - F(z)$	z	$1 - F(z)$	z	$1 - F(z)$
4.80	7.9333 -7	5.20	9.9644 -8	5.60	1.0718 -8	6.00	9.8659 -10
4.81	7.5465	5.21	9.4420	5.61	1.0116	6.01	9.2761
4.82	7.1779	5.22	8.9462	5.62	9.5479 -9	6.02	8.7208
4.83	6.8267	5.23	8.4755	5.63	9.0105	6.03	8.1980
4.84	6.4920	5.24	8.0288	5.64	8.5025	6.04	7.7057
4.85	6.1731 -7	5.25	7.6050 -8	5.65	8.0224 -9	6.05	7.2423 -10
4.86	5.8693	5.26	7.2028	5.66	7.5687	6.06	6.8061
4.87	5.5799	5.27	6.8212	5.67	7.1399	6.07	6.3955
4.88	5.3043	5.28	6.4592	5.68	6.7347	6.08	6.0091
4.89	5.0418	5.29	6.1158	5.69	6.3520	6.09	5.6455
4.90	4.7918 -7	5.30	5.7901 -8	5.70	5.9904 -9	6.10	5.3034 -10
4.91	4.5538	5.31	5.4813	5.71	5.6488	6.11	4.9815
4.92	4.3272	5.32	5.1884	5.72	5.3262	6.12	4.6788
4.93	4.1115	5.33	4.9106	5.73	5.0215	6.13	4.3939
4.94	3.9061	5.34	4.6473	5.74	4.7338	6.14	4.1261
4.95	3.7107 -7	5.35	4.3977 -8	5.75	4.4622 -9	6.15	3.8742 -10
4.96	3.5247	5.36	4.1611	5.76	4.2057	6.16	3.6372
4.97	3.3476	5.37	3.9368	5.77	3.9636	6.17	3.4145
4.98	3.1792	5.38	3.7243	5.78	3.7350	6.18	3.2050
4.99	3.0190	5.39	3.5229	5.79	3.5193	6.19	3.0082
5.00	2.8665 -7	5.40	3.3320 -8	5.80	3.3157 -9	6.20	2.8231 -10
5.01	2.7215	5.41	3.1512	5.81	3.1236	6.21	2.6492
5.02	2.5836	5.42	2.9800	5.82	2.9424	6.22	2.4858
5.03	2.4524	5.43	2.8177	5.83	2.7714	6.23	2.3321
5.04	2.3277	5.44	2.6640	5.84	2.6100	6.24	2.1878
5.05	2.2091 -7	5.45	2.5185 -8	5.85	2.4579 -9	6.25	2.0523 -10
5.06	2.0963	5.46	2.3807	5.86	2.3143	6.26	1.9249
5.07	1.9891	5.47	2.2502	5.87	2.1790	6.27	1.8052
5.08	1.8872	5.48	2.1266	5.88	2.0513	6.28	1.6929
5.09	1.7903	5.49	2.0097	5.89	1.9310	6.29	1.5873
5.10	1.6983 -7	5.50	1.8990 -8	5.90	1.8175 -9	6.30	1.4882 -10
5.11	1.6108	5.51	1.7942	5.91	1.7105	6.31	1.3952
5.12	1.5277	5.52	1.6950	5.92	1.6097	6.32	1.3078
5.13	1.4487	5.53	1.6012	5.93	1.5147	6.33	1.2258
5.14	1.3737	5.54	1.5124	5.94	1.4251	6.34	1.1488
5.15	1.3024 -7	5.55	1.4283 -8	5.95	1.3407 -9	6.35	1.0765 -10
5.16	1.2347	5.56	1.3489	5.96	1.2612	6.36	1.0088
5.17	1.1705	5.57	1.2737	5.97	1.1863	6.37	9.4514 -11
5.18	1.1094	5.58	1.2026	5.98	1.1157	6.38	8.8544
5.19	1.0515	5.59	1.1353	5.99	1.0492	6.39	8.2943

z	$1 - F(z)$	z	$1 - F(z)$	z	$1 - F(z)$	z	$1 - F(z)$
6.40	7.7689 -11	6.80	5.2310 -12	7.20	3.0106 -13	7.60	1.4807 -14
6.41	7.2760	6.81	4.8799	7.21	2.7976	7.61	1.3705
6.42	6.8137	6.82	4.5520	7.22	2.5994	7.62	1.2684
6.43	6.3802	6.83	4.2457	7.23	2.4150	7.63	1.1738
6.44	5.9737	6.84	3.9597	7.24	2.2434	7.64	1.0861
6.45	5.5925 -11	6.85	3.6925 -12	7.25	2.0839 2-13	7.65	1.0049 -14
6.46	5.2351	6.86	3.4430	7.26	1.9355	7.66	9.2967
6.47	4.9001	6.87	3.2101	7.27	1.7974	7.67	8.5998
6.48	4.5861	6.88	2.9926	7.28	1.6691	7.68	7.9544
6.49	4.2918	6.89	2.7896	7.29	1.5498	7.69	7.3568
6.50	4.0160 -11	6.90	2.6001 -12	7.30	1.4388 -13	7.70	6.8033 -15
6.51	3.7575	6.91	2.4233	7.31	1.3357	7.71	6.2909
6.52	3.5154	6.92	2.2582	7.32	1.2399	7.72	5.8165
6.53	3.2885	6.93	2.1042	7.33	1.1508	7.73	5.3773
6.54	3.0759	6.94	1.9605	7.34	1.0680	7.74	4.9708
6.55	2.8769 -11	6.95	1.8264 -12	7.35	9.9103 -14	7.75	4.5946 -15
6.56	2.6904	6.96	1.7014	7.36	9.1955	7.76	4.2465
6.57	2.5158	6.97	1.5847	7.37	8.5314	7.77	3.9243
6.58	2.3522	6.98	1.4759	7.38	7.9145	7.78	3.6262
6.59	2.1991	6.99	1.3744	7.39	7.3414	7.79	3.3505
6.60	2.0558 -11	7.00	1.2798 -12	7.40	6.8092 -14	7.80	3.0954 -15
6.61	1.9216	7.01	1.1916	7.41	6.3150	7.81	2.8594
6.62	1.7960	7.02	1.1093	7.42	5.8560	7.82	2.6412
6.63	1.6784	7.03	1.0327	7.43	5.4299	7.83	2.4394
6.64	1.5684	7.04	9.6120 -13	7.44	5.0343	7.84	2.2527
6.65	1.4655 -11	7.05	8.9459 -13	7.45	4.6670 -14	7.85	2.0802 -15
6.66	1.3691	7.06	8.3251	7.46	4.3261	7.86	1.9207
6.67	1.2790	7.07	7.7467	7.47	4.0097	7.87	1.7732
6.68	1.1947	7.08	7.2077	7.48	3.7161	7.88	1.6369
6.69	1.1159	7.09	6.7056	7.49	3.4437	7.89	1.5109
6.70	1.0421 -11	7.10	6.2378 -13	7.50	3.1909 -14	7.90	1.3945 -15
6.71	9.7312 -12	7.11	5.8022	7.51	2.9564	7.91	1.2869
6.72	9.0862	7.12	5.3964	7.52	2.7388	7.92	1.1876
6.73	8.4832	7.13	5.0184	7.53	2.5370	7.93	1.0957
6.74	7.9193	7.14	4.6665	7.54	2.3499	7.94	1.0109
6.75	7.3923 -12	7.15	4.3389 -13	7.55	2.1763 -14	7.95	9.3256 -16
6.76	6.8996	7.16	4.0339	7.56	2.0153	7.96	8.6020
6.77	6.4391	7.17	3.7499	7.57	1.8661	7.97	7.9337
6.78	6.0088	7.18	3.4856	7.58	1.7278	7.98	7.3167
6.79	5.6067	7.19	3.2396	7.59	1.5995	7.99	6.7469

z	$1 - F(z)$	z	$1 - F(z)$	z	$1 - F(z)$	z	$1 - F(z)$
8.00	6.2210 -16	8.40	2.2324 -17	8.80	6.8408 -19	9.20	1.7897 -20
8.01	5.7354	8.41	2.0501	8.81	6.2573	9.21	1.6306
8.02	5.2873	8.42	1.8824	8.82	5.7230	9.22	1.4855
8.03	4.8736	8.43	1.7283	8.83	5.2338	9.23	1.3532
8.04	4.4919	8.44	1.5867	8.84	4.7859	9.24	1.2325
8.05	4.1397 -16	8.45	1.4565 -17	8.85	4.3760 -19	9.25	1.1225 -20
8.06	3.8147	8.46	1.3369	8.86	4.0007	9.26	1.0222
8.07	3.5149	8.47	1.2270	8.87	3.6573	9.27	9.3073 -21
8.08	3.2383	8.48	1.1260	8.88	3.3430	9.28	8.4739
8.09	2.9832	8.49	1.0332	8.89	3.0554	9.29	7.7144
8.10	2.7480 -16	8.50	9.4795 -18	8.90	2.7923 -19	9.30	7.0223 -21
8.11	2.5310	8.51	8.6967	8.91	2.5516	9.31	6.3916
8.12	2.3309	8.52	7.9777	8.92	2.3314	9.32	5.8170
8.13	2.1465	8.53	7.3174	8.93	2.1300	9.33	5.2935
8.14	1.9764	8.54	6.7111	8.94	1.9459	9.34	4.8167
8.15	1.8196 -16	8.55	6.1544 -18	8.95	1.7774 -19	9.35	4.3824 -21
8.16	1.6751	8.56	5.6434	8.96	1.5234	9.36	3.9868
8.17	1.5419	8.57	5.1743	8.97	1.4826	9.37	3.6266
8.18	1.4192	8.58	4.7437	8.98	1.3538	9.38	3.2986
8.19	1.3061	8.59	4.3485	8.99	1.2362	9.39	3.0000
8.20	1.2019 -16	8.60	3.9858 -18	9.00	1.1286 -19	9.40	2.7282 -21
8.21	1.1059	8.61	3.6530	9.01	1.0303	9.41	2.4807
8.22	1.0175	8.62	3.3477	9.02	9.4045 -20	9.42	2.2554
8.23	9.3607 -17	8.63	3.0676	9.03	8.5836	9.43	2.0504
8.24	8.6105	8.64	2.8107	9.04	7.8336	9.44	1.8639
8.25	7.9197 -17	8.65	2.5750 -18	9.05	7.1484 -20	9.45	1.6942 -21
8.26	7.2836	8.66	2.3588	9.06	6.5225	9.46	1.5397
8.27	6.6980	8.67	2.1606	9.07	5.9509	9.47	1.3992
8.28	6.1588	8.68	1.9788	9.08	5.4287	9.48	1.2614
8.29	5.6624	8.69	1.8122	9.09	4.9520	9.49	1.1552
8.30	5.2056 -17	8.70	1.6594 -18	9.10	4.5166 -20	9.50	1.0495 -21
8.31	4.7851	8.71	1.5194	9.11	4.1191	9.51	9.5331 -22
8.32	4.3982	8.72	1.3910	9.12	3.7562	9.52	8.6590
8.33	4.0421	8.73	1.2734	9.13	3.4250	9.53	7.8642
8.34	3.7145	8.74	1.1656	9.14	3.1226	9.54	7.1416
8.35	3.4131 -17	8.75	1.0668 -18	9.15	2.8467 -20	9.55	6.4848 -22
8.36	3.1359	8.76	9.7625 -19	9.16	2.5949	9.56	5.8878
8.37	2.8809	8.77	8.9333	9.17	2.3651	9.57	5.3453
8.38	2.6464	8.78	8.1737	9.18	2.1555	9.58	4.8522
8.39	2.4307	8.79	7.4780	9.19	1.9642	9.59	4.4043

z	$1 - F(z)$	z	$1 - F(z)$	z	$1 - F(z)$	z	$1 - F(z)$
9.60	3.9972 -22	9.70	1.5075 -22	9.80	5.6293 -23	9.90	2.0814 -23
9.61	3.6274	9.71	1.3667	9.81	5.0984	9.91	1.8832
9.62	3.2916	9.72	1.2389	9.82	4.6172	9.92	1.7038
9.63	2.9865	9.73	1.1230	9.83	4.1809	9.93	1.5413
9.64	2.7094	9.74	1.0178	9.84	3.7855	9.94	1.3941
9.65	2.4578 -22	9.75	9.2234 -23	9.85	3.4272 -23	9.95	1.2609 -23
9.66	2.2293	9.76	8.3578	9.86	3.1025	9.96	1.1403
9.67	2.0219	9.77	7.5726	9.87	2.8082	9.97	1.0311
9.68	1.8336	9.78	6.8605	9.88	2.5416	9.98	9.3233 -24
9.69	1.6626	9.79	6.2148	9.89	2.3001	9.99	8.4291
						10.00	7.6199 -24

APPENDIX C
STANDARDIZED NORMAL DISTRIBUTION'S ORDINATE VALUES OR PROBABILITY DENSITIES

$$\phi(z) = \frac{1}{\sqrt{2\pi}} e^{-\frac{1}{2}z^2}, \text{ for } 0 \leq z \leq 4.99, \phi(-z) = \phi(z)$$

z	0.00	0.01	0.02	0.03	0.04	0.05	0.06	0.07	0.08	0.09
0.0	0.3989	0.3989	0.3989	0.3988	0.3986	0.3984	0.3982	0.3980	0.3977	0.3973
0.1	0.3970	0.3965	0.3961	0.3956	0.3951	0.3945	0.3939	0.3932	0.3925	0.3918
0.2	0.3910	0.3902	0.3894	0.3885	0.3876	0.3867	0.3857	0.3847	0.3836	0.3825
0.3	0.3814	0.3802	0.3790	0.3778	0.3765	0.3752	0.3739	0.3725	0.3712	0.3697
0.4	0.3683	0.3668	0.3653	0.3637	0.3621	0.3605	0.3589	0.3572	0.3555	0.3538
0.5	0.3521	0.3503	0.3485	0.3467	0.3448	0.3429	0.3410	0.3391	0.3372	0.3352
0.6	0.3332	0.3312	0.3292	0.3271	0.3251	0.3230	0.3209	0.3187	0.3166	0.3144
0.7	0.3123	0.3101	0.3079	0.3056	0.3034	0.3011	0.2989	0.2966	0.2943	0.2920
0.8	0.2897	0.2874	0.2850	0.2827	0.2803	0.2780	0.2756	0.2732	0.2709	0.2685
0.9	0.2661	0.2637	0.2613	0.2589	0.2565	0.2541	0.2516	0.2492	0.2468	0.2444
1.0	0.2420	0.2396	0.2371	0.2347	0.2323	0.2299	0.2275	0.2251	0.2227	0.2203
1.1	0.2179	0.2155	0.2131	0.2107	0.2083	0.2059	0.2036	0.2012	0.1989	0.1965
1.2	0.1942	0.1919	0.1895	0.1872	0.1849	0.1826	0.1804	0.1781	0.1758	0.1736
1.3	0.1714	0.1691	0.1669	0.1647	0.1626	0.1604	0.1582	0.1561	0.1539	0.1518
1.4	0.1497	0.1476	0.1456	0.1435	0.1415	0.1394	0.1374	0.1354	0.1334	0.1315
1.5	0.1295	0.1276	0.1257	0.1238	0.1219	0.1200	0.1182	0.1163	0.1145	0.1127
1.6	0.1109	0.1092	0.1074	0.1057	0.1040	0.1023	0.1006	0.09893	0.09728	0.09566
1.7	0.09405	0.09246	0.09089	0.08933	0.08780	0.08628	0.08478	0.08329	0.08183	0.08038
1.8	0.07895	0.07754	0.07614	0.07477	0.07341	0.07206	0.07074	0.06943	0.06814	0.06687
1.9	0.06562	0.06438	0.06316	0.06195	0.06077	0.05959	0.05844	0.05730	0.05618	0.05508
2.0	0.05399	0.05292	0.05186	0.05082	0.04980	0.04879	0.04780	0.04682	0.04586	0.04491
2.1	0.04398	0.04307	0.04217	0.04128	0.04041	0.03955	0.03871	0.03788	0.03706	0.03626
2.2	0.03547	0.03470	0.03394	0.03319	0.03246	0.03174	0.03103	0.03034	0.02965	0.02898
2.3	0.02833	0.02768	0.02705	0.02643	0.02582	0.02522	0.02463	0.02406	0.02349	0.02294
2.4	0.02239	0.02186	0.02134	0.02083	0.02033	0.01984	0.01936	0.01888	0.01842	0.01797
2.5	0.01753	0.01709	0.01667	0.01625	0.01585	0.01545	0.01506	0.01468	0.01431	0.01394
2.6	0.01358	0.01323	0.01289	0.01256	0.01223	0.01191	0.01160	0.01130	0.01100	0.01071
2.7	0.010421	0.010143	$0.0^2 9871$	$0.0^2 9606$	$0.0^2 9347$	$0.0^2 9094$	$0.0^2 8846$	$0.0^2 8605$	$0.0^2 8370$	$0.0^2 814$
2.8	$0.0^2 7915$	$0.0^2 7697$	$0.0^2 7483$	$0.0^2 7274$	$0.0^2 7071$	$0.0^2 6873$	$0.0^2 6679$	$0.0^2 6491$	$0.0^2 6307$	$0.0^2 612$
2.9	$0.0^2 5953$	$0.0^2 5782$	$0.0^2 5616$	$0.0^2 5454$	$0.0^2 5296$	$0.0^2 5143$	$0.0^2 4993$	$0.0^2 4847$	$0.0^2 4705$	$0.0^2 456$
3.0	$0.0^2 4432$	$0.0^2 4301$	$0.0^2 4173$	$0.0^2 4049$	$0.0^2 3928$	$0.0^2 3810$	$0.0^2 3695$	$0.0^2 3584$	$0.0^2 3475$	$0.0^2 337$
3.1	$0.0^2 3267$	$0.0^2 3167$	$0.0^2 3070$	$0.0^2 2975$	$0.0^2 2884$	$0.0^2 2794$	$0.0^2 2707$	$0.0^2 2623$	$0.0^2 2541$	$0.0^2 246$
3.2	$0.0^2 2384$	$0.0^2 2309$	$0.0^2 2236$	$0.0^2 2165$	$0.0^2 2096$	$0.0^2 2029$	$0.0^2 1964$	$0.0^2 1901$	$0.0^2 1840$	$0.0^2 178$
3.3	$0.0^2 1723$	$0.0^2 1667$	$0.0^2 1612$	$0.0^2 1560$	$0.0^2 1508$	$0.0^2 1459$	$0.0^2 1411$	$0.0^2 1364$	$0.0^2 1319$	$0.0^2 127$
3.4	$0.0^2 1232$	$0.0^2 1191$	$0.0^2 1151$	$0.0^2 1112$	$0.0^2 1075$	$0.0^2 1038$	$0.0^2 1003$	$0.0^3 9689$	$0.0^3 9358$	$0.0^3 903$
3.5	$0.0^3 8727$	$0.0^3 8426$	$0.0^3 8135$	$0.0^3 7853$	$0.0^3 7581$	$0.0^3 7317$	$0.0^3 7061$	$0.0^3 6814$	$0.0^3 6575$	$0.0^3 634$
3.6	$0.0^3 6119$	$0.0^3 5902$	$0.0^3 5693$	$0.0^3 5490$	$0.0^3 5294$	$0.0^3 5105$	$0.0^3 4921$	$0.0^3 4744$	$0.0^3 4573$	$0.0^3 440$
3.7	$0.0^3 4248$	$0.0^3 4093$	$0.0^3 3944$	$0.0^3 3800$	$0.0^3 3661$	$0.0^3 3526$	$0.0^3 3396$	$0.0^3 3271$	$0.0^3 3149$	$0.0^3 303$
3.8	$0.0^3 2919$	$0.0^3 2810$	$0.0^3 2705$	$0.0^3 2604$	$0.0^3 2506$	$0.0^3 2411$	$0.0^3 2320$	$0.0^3 2232$	$0.0^3 2147$	$0.0^3 206$
3.9	$0.0^3 1987$	$0.0^3 1910$	$0.0^3 1837$	$0.0^3 1766$	$0.0^3 1698$	$0.0^3 1633$	$0.0^3 1569$	$0.0^3 1508$	$0.0^3 1449$	$0.0^3 139$
4.0	$0.0^3 1338$	$0.0^3 1286$	$0.0^3 1235$	$0.0^3 1186$	$0.0^3 1140$	$0.0^3 1094$	$0.0^3 1051$	$0.0^3 1009$	$0.0^4 9687$	$0.0^4 929$
4.1	$0.0^4 8926$	$0.0^4 8567$	$0.0^4 8222$	$0.0^4 7890$	$0.0^4 7570$	$0.0^4 7263$	$0.0^4 6967$	$0.0^4 6683$	$0.0^4 6410$	$0.0^4 614$
4.2	$0.0^4 5894$	$0.0^4 5652$	$0.0^4 5418$	$0.0^4 5194$	$0.0^4 4979$	$0.0^4 4772$	$0.0^4 4573$	$0.0^4 4382$	$0.0^4 4199$	$0.0^4 402$
4.3	$0.0^4 3854$	$0.0^4 3691$	$0.0^4 3535$	$0.0^4 3386$	$0.0^4 3242$	$0.0^4 3104$	$0.0^4 2972$	$0.0^4 2845$	$0.0^4 2723$	$0.0^4 260$
4.4	$0.0^4 2494$	$0.0^4 2387$	$0.0^4 2284$	$0.0^4 2185$	$0.0^4 2090$	$0.0^4 1999$	$0.0^4 1912$	$0.0^4 1829$	$0.0^4 1749$	$0.0^4 167$
4.5	$0.0^4 1598$	$0.0^4 1528$	$0.0^4 1461$	$0.0^4 1396$	$0.0^4 1334$	$0.0^4 1275$	$0.0^4 1218$	$0.0^4 1164$	$0.0^4 1112$	$0.0^4 106$
4.6	$0.0^4 1014$	$0.0^5 9684$	$0.0^5 9248$	$0.0^5 8830$	$0.0^5 8430$	$0.0^5 8047$	$0.0^5 7681$	$0.0^5 7331$	$0.0^5 6996$	$0.0^5 667$
4.7	$0.0^5 6370$	$0.0^5 6077$	$0.0^5 5797$	$0.0^5 5530$	$0.0^5 5274$	$0.0^5 5029$	$0.0^5 4796$	$0.0^5 4573$	$0.0^5 4360$	$0.0^5 415$
4.8	$0.0^5 3961$	$0.0^5 3775$	$0.0^5 3598$	$0.0^5 3428$	$0.0^5 3267$	$0.0^5 3112$	$0.0^5 2965$	$0.0^5 2824$	$0.0^5 2690$	$0.0^5 256$
4.9	$0.0^5 2439$	$0.0^5 2322$	$0.0^5 2211$	$0.0^5 2105$	$0.0^5 2003$	$0.0^5 1907$	$0.0^5 1814$	$0.0^5 1727$	$0.0^5 1643$	$0.0^5 156$

APPENDIX D

PERCENTAGE POINTS, F DISTRIBUTION
FOR $F(F) = 0.50$

ν_1 / ν_2	1	2	3	4	5	6	7	8	9
1	1.0000	1.5000	1.7092	1.8227	1.8936	1.9420	1.9771	2.0038	2.0247
2	0.6667	1.0000	1.1349	1.2071	1.2519	1.2824	1.3046	1.3213	1.3344
3	0.5851	0.8811	1.0000	1.0632	1.1023	1.1289	1.1481	1.1628	1.1741
4	0.5486	0.8284	0.9405	1.0000	1.0367	1.0616	1.0797	1.0933	1.1040
5	0.5281	0.7988	0.9072	0.9646	1.0000	1.0240	1.0414	1.0545	1.0648
6	0.5149	0.7798	0.8858	0.9419	0.9765	1.0000	1.0170	1.0297	1.0398
7	0.5057	0.7665	0.8710	0.9262	0.9602	0.9833	1.0000	1.0126	1.0224
8	0.4990	0.7568	0.8600	0.9146	0.9483	0.9711	0.9876	1.0000	1.0097
9	0.4938	0.7494	0.8517	0.9058	0.9391	0.9618	0.9780	0.9904	1.0000
10	0.4897	0.7435	0.8451	0.8988	0.9320	0.9543	0.9705	0.9828	0.9923
11	0.4864	0.7387	0.8398	0.8931	0.9261	0.9484	0.9644	0.9766	0.9861
12	0.4837	0.7348	0.8353	0.8885	0.9213	0.9434	0.9594	0.9715	0.9809
13	0.4814	0.7315	0.8316	0.8845	0.9171	0.9392	0.9552	0.9672	0.9766
14	0.4794	0.7286	0.8284	0.8812	0.9137	0.9357	0.9516	0.9636	0.9730
15	0.4778	0.7262	0.8257	0.8783	0.9107	0.9327	0.9485	0.9605	0.9698
16	0.4763	0.7241	0.8233	0.8758	0.9082	0.9300	0.9458	0.9577	0.9671
17	0.4750	0.7222	0.8212	0.8735	0.9059	0.9276	0.9434	0.9553	0.9646
18	0.4738	0.7205	0.8194	0.8717	0.9038	0.9256	0.9413	0.9532	0.9624
19	0.4728	0.7191	0.8177	0.8698	0.9020	0.9238	0.9394	0.9513	0.9606
20	0.4719	0.7177	0.8162	0.8682	0.9004	0.9221	0.9378	0.9496	0.9588
21	0.4711	0.7165	0.8149	0.8669	0.8989	0.9206	0.9362	0.9480	0.9573
22	0.4703	0.7155	0.8137	0.8656	0.8975	0.9192	0.9348	0.9466	0.9559
23	0.4696	0.7145	0.8126	0.8644	0.8963	0.9180	0.9336	0.9454	0.9546
24	0.4690	0.7136	0.8115	0.8633	0.8952	0.9168	0.9324	0.9442	0.9534
25	0.4684	0.7127	0.8106	0.8624	0.8942	0.9159	0.9314	0.9431	0.9524
26	0.4679	0.7120	0.8097	0.8615	0.8933	0.9149	0.9305	0.9421	0.9514
27	0.4674	0.7113	0.8089	0.8606	0.8924	0.9140	0.9295	0.9412	0.9505
28	0.4670	0.7106	0.8082	0.8599	0.8916	0.9132	0.9287	0.9404	0.9496
29	0.4665	0.7100	0.8075	0.8592	0.8909	0.9124	0.9279	0.9396	0.9488
30	0.4662	0.7094	0.8069	0.8584	0.8902	0.9117	0.9272	0.9389	0.9480
40	0.4633	0.7053	0.8023	0.8536	0.8851	0.9065	0.9219	0.9336	0.9427
60	0.4605	0.7012	0.7977	0.8487	0.8802	0.9014	0.9168	0.9283	0.9375
120	0.4577	0.6972	0.7932	0.8439	0.8752	0.8964	0.9116	0.9232	0.9322
∞	0.45494	0.69315	0.78866	0.83917	0.87029	0.89135	0.90654	0.91802	0.92698

ν_2 \\ ν_1	10	12	15	20	24	30	40	60	120	∞
1	2.0416	2.0676	2.0922	2.1191	2.1337	2.1450	2.1575	2.1717	2.1848	2.1981
2	1.3450	1.3610	1.3771	1.3933	1.4014	1.4096	1.4178	1.4261	1.4344	1.427
3	1.1833	1.1972	1.2111	1.2252	1.2323	1.2393	1.2465	1.2536	1.2608	1.2680
4	1.1126	1.1255	1.1386	1.1517	1.1583	1.1650	1.1716	1.1782	1.1849	1.1916
5	1.0730	1.0855	1.0981	1.1106	1.1170	1.1233	1.1298	1.1361	1.1426	1.1490
6	1.0478	1.0600	1.0722	1.0845	1.0907	1.0969	1.1032	1.1093	1.1156	1.1219
7	1.0304	1.0423	1.0543	1.0664	1.0724	1.0785	1.0847	1.0907	1.0969	1.1031
8	1.0175	1.0293	1.0411	1.0531	1.0591	1.0651	1.0711	1.0772	1.0832	1.0893
9	1.0077	1.0194	1.0312	1.0430	1.0488	1.0548	1.0607	1.0667	1.0727	1.0788
10	1.0000	1.0116	1.0232	1.0349	1.0408	1.0467	1.0526	1.0585	1.0645	1.0705
11	0.9937	1.0053	1.0168	1.0285	1.0342	1.0401	1.0460	1.0519	1.0578	1.0637
12	0.9886	1.0000	1.0115	1.0231	1.0289	1.0347	1.0405	1.0463	1.0523	1.0582
13	0.9842	0.9956	1.0071	1.0186	1.0243	1.0301	1.0359	1.0418	1.0477	1.0535
14	0.9805	0.9919	1.0033	1.0147	1.0205	1.0263	1.0321	1.0379	1.0436	1.0495
15	0.9773	0.9886	1.0000	1.0114	1.0172	1.0229	1.0287	1.0345	1.0402	1.0461
16	0.9746	0.9858	0.9972	1.0085	1.0143	1.0200	1.0257	1.0316	1.0373	1.0431
17	0.9721	0.9833	0.9947	1.0060	1.0117	1.0174	1.0232	1.0290	1.0347	1.0405
18	0.9699	0.9811	0.9924	1.0038	1.0095	1.0152	1.0209	1.0267	1.0325	1.0382
19	0.9680	0.9792	0.9905	1.0018	1.0075	1.0132	1.0189	1.0246	1.0303	1.0361
20	0.9663	0.9774	0.9887	1.0000	1.0057	1.0114	1.0171	1.0228	1.0286	1.0343
21	0.9647	0.9759	0.9871	0.9984	1.0041	1.0098	1.0154	1.0211	1.0269	1.0326
22	0.9633	0.9745	0.9856	0.9969	1.0026	1.0082	1.0139	1.0197	1.0254	1.0311
23	0.9620	0.9732	0.9843	0.9956	1.0012	1.0069	1.0126	1.0182	1.0240	1.0297
24	0.9608	0.9719	0.9831	0.9944	1.0000	1.0057	1.0113	1.0170	1.0227	1.0284
25	0.9597	0.9709	0.9820	0.9932	0.9989	1.0045	1.0102	1.0159	1.0215	1.0273
26	0.9587	0.9698	0.9810	0.9922	0.9978	1.0035	1.0091	1.0148	1.0205	1.0262
27	0.9578	0.9689	0.9801	0.9912	0.9969	1.0025	1.0081	1.0138	1.0195	1.0252
28	0.9569	0.9680	0.9791	0.9904	0.9960	1.0016	1.0073	1.0129	1.0186	1.0243
29	0.9561	0.9672	0.9783	0.9896	0.9951	1.0008	1.0064	1.0120	1.0178	1.0234
30	0.9554	0.9664	0.9776	0.9887	0.9944	1.0000	1.0056	1.0113	1.0170	1.0226
40	0.9500	0.9611	0.9721	0.9832	0.9888	0.9944	1.0000	1.0056	1.0113	1.0169
60	0.9448	0.9557	0.9667	0.9777	0.9833	0.9889	0.9944	1.0000	1.0056	1.0112
120	0.9394	0.9503	0.9613	0.9722	0.9778	0.9833	0.9888	0.9945	1.0000	1.0056
∞	0.93418	0.94503	0.95592	0.96687	0.97236	0.97787	0.98338	0.98891	0.99445	1.0000

APPENDIX E

CRITICAL VALUES FOR THE KOLMOGOROV–SMIRNOV GOODNESS-OF-FIT TEST

Sample size N	Level of significance				
	0.20	0.15	0.10	0.05	0.01
1	0.900	0.925	0.950	0.975	0.995
2	0.684	0.726	0.776	0.842	0.929
3	0.565	0.597	0.642	0.708	0.828
4	0.494	0.575	0.564	0.624	0.733
5	0.446	0.424	0.510	0.454	0.669
6	0.410	0.436	0.470	0.521	0.618
7	0.381	0.405	0.438	0.486	0.577
8	0.358	0.381	0.411	0.457	0.543
9	0.339	0.360	0.388	0.432	0.514
10	0.322	0.342	0.368	0.410	0.490
11	0.307	0.326	0.452	0.391	0.468
12	0.295	0.313	0.338	0.375	0.405
13	0.284	0.302	0.325	0.361	0.433
14	0.274	0.292	0.314	0.349	0.418
15	0.266	0.293	0.304	0.338	0.404
16	0.258	0.274	0.295	0.328	0.392
17	0.250	0.266	0.286	0.318	0.381
18	0.244	0.259	0.278	0.309	0.371
19	0.237	0.252	0.272	0.301	0.363
20	0.231	0.246	0.264	0.294	0.356
25	0.21	0.22	0.24	0.27	0.32
30	0.19	0.20	0.22	0.24	0.29
35	0.18	0.19	0.21	0.23	0.27
> 35	$\frac{1.07}{\sqrt{N}}$	$\frac{1.14}{\sqrt{N}}$	$\frac{1.22}{\sqrt{N}}$	$\frac{1.36}{\sqrt{N}}$	$\frac{1.63}{\sqrt{N}}$

Index

ABOUT THE AUTHOR

This book is based on the following extensive experience of the author in Reliability Engineering:

1. He initiated and was the Director of the Corporate Reliability Engineering Program at the Allis-Chalmers Manufacturing Co., Milwaukee, Wisconsin, from 1960 to 1963.

2. He started the Reliability Engineering Instructional Program at The University of Arizona in 1963, which now has more than ten courses in it. A Master's Degree with a Reliability Engineering Option is currently being offered in the Aerospace and Mechanical Engineering Department at The University of Arizona. This option started in 1969. A Master's Degree in Reliability Engineering is also being offered now in the Systems and Industrial Engineering Department at The University of Arizona. This degree started in January 1987.

3. He conceived and directed the first two Summer Institutes for College Teachers in reliability engineering ever to be supported by the National Science Foundation. The first was in 1965 and the second in 1966, for 30 college and university faculty each summer. These faculty started teaching reliability engineering courses at their respective universities and/or incorporating reliability engineering concepts into their courses.

4. In 1963 he conceived, initiated, and has directed since then the now internationally famous *Annual Reliability Engineering and Management Institute* at The University of Arizona.

5. In 1975 he conceived, initiated, and has directed since then the now internationally famous *Annual Reliability Testing Institute* at The University of Arizona.

6. He has lectured extensively and conducted over 280 training courses, short courses and seminars worldwide and has exposed over 10,000 reliability, maintainability, test, design, and product assurance engineers to the concepts in this book.

7. He has been the principal investigator of mechanical reliability research for the NASA-Lewis Research Center, the Office of Naval

Research, and the Naval Weapons Engineering Support Activity for ten years.

8. He has been consulted extensively by over 80 industries and government agencies worldwide on reliability engineering, reliability and life testing, maintainability engineering, and mechanical reliability matters.

9. He has been active in the Reliability and Maintainability Symposia and Conferences dealing with reliability engineering since 1963.

10. He founded the Tucson Chapter of the Society of Reliability Engineers in 1974 and was its first President. He also founded the first and very active Student Chapter of the Society of Reliability Engineers at The University of Arizona.

11. He has authored or co-authored over 120 papers and articles, of which over 103 are in all areas of reliability engineering.

12. In addition to this book, he authored or contributed to the following books:

    (1) *Bibliography on Plasticity - Theory and Applications,* Dimitri Kececioglu, published by the American Society of Mechanical Engineers, New York, 191 pp., 1950.

    (2) *Manufacturing, Planning and Estimating Handbook,* Dimitri Kececioglu and Lawrence Karvonen contributed part of Chapter 19, pp. 19-1 to 19-12, published by McGraw-Hill Book Co., Inc., New York, 864 pp., 1963.

    (3) *Introduction to Probabilistic Design for Reliability,* Dimitri Kececioglu, published by the United States Army Management Engineering Training Agency, Rock Island, Illinois, contributed Chapter 7 of 109 pp., and Chapter 8 of 137 pp., May 1974.

    (4) *Manual of Product Assurance Films on Reliability Engineering and Management, Reliability Testing, Maintainability, and Quality Control,* Dimitri Kececioglu, printed by Dr. Dimitri Kececioglu, 7340 N. La Oesta Avenue, Tucson, Arizona 85704, 178 pp., 1976.

(5) *Manual of Product Assurance Films and Videotapes*, Dimitri Kececioglu, printed by Dimitri Kececioglu, 7340 N. La Oesta Avenue, Tucson, Arizona 85704, 327 pp., 1980.

(6) *Reliability and Life Testing*, Dimitri Kececioglu, will be published in 1991 by Prentice Hall.

13. He has received over thirty prestigious awards and has been recognized for his valuable contributions to the field of reliability engineering. Among these are the following: (1) Fulbright Scholar in 1971. (2) Ralph Teetor Award of the Society of Automotive Engineers as "Outstanding Engineering Educator" in 1977. (3) Certificate of Excellence by the Society of Reliability Engineers for his "personal contributions made toward the advancement of the philosophy and principles of reliability engineering" in 1978. (4) ASQC-Reliability Division, Reliability Education Advancement Award for his "outstanding contributions to the development and presentation of meritorious reliability educational programs" in 1980. (5) ASQC Allen Chop Award for his "outstanding contributions to Reliability Science and Technology" in 1981. (6) The University of Arizona College of Engineering Anderson Prize for "engineering the Master's Degree program in the Reliability Engineering Option" in 1983. (7) Designation of "Senior Extension Teacher" by Dr. Leonard Freeman, Dean, Continuing Education and University Extension, the University of California, Los Angeles in 1983. (8) Honorary Member, Golden Key National Honor Society in 1984. (9) Honorary Professor, Shanghai University of Technology in 1984. (10) Honorary Professor, Phi Kappa Phi Honor Society in 1988.

14. He conceived and established *The Dr. Dimitri Basil Kececioglu Reliability Engineering Research Fellowships Endowment Fund* in 1987. The cosponsors of his institutes, mentioned in Items 4 and 5, have contributed generously to this fund which has now crossed the $230,000 mark.

15. He received his BSME from Robert College, Istanbul, Turkey in 1942, and his M.S. in Industrial Engineering in 1948 and his Ph.D. in Engineering Mechanics in 1953, both from Purdue University, Lafayette, Indiana.

16. He has also been granted five patents.